Handbook of Statistical Procedures and Their Computer Applications to Education and the Behavioral Sciences

Handbook of Statistical Procedures and Their Computer Applications to Education and the Behavioral Sciences

Melvyn N. Freed, Ph.D.
Editor

Joseph M. Ryan, Ph.D.
Robert K. Hess, Ph.D.
Authors

American Council on Education   Macmillan Publishing Company
NEW YORK
Maxwell Macmillan Canada
TORONTO
Maxwell Macmillan International
NEW YORK   OXFORD   SINGAPORE   SYDNEY

Copyright © 1991 by American Council on Education and
　　　　　Macmillan Publishing Company,
　　　　　A Division of Macmillan, Inc.

All rights reserved. No part of this book may be reproduced or transmitted in any form or by any means, electronic or mechanical, including photocopying, recording, or by any information storage and retrieval system, without permission in writing from the Publisher.

Macmillan Publishing Company

866 Third Avenue, New York, N.Y. 10022

Maxwell Macmillan Canada, Inc.

1200 Eglinton Avenue East, Suite 200
Don Mills, Ontario M3C 3N1

Macmillan, Inc. is part of the Maxwell Communication Group of Companies

Library of Congress Catalog Card Number: 90-20436

Printed in the United States of America

printing number
1　2　3　4　5　6　7　8　9　10

LIBRARY OF CONGRESS CATALOGING-IN-PUBLICATION DATA

Freed, Melvyn N.
　　Handbook of statistical procedures and their computer applications to education and the behavioral sciences / Melvyn N. Freed, editor; Joseph M. Ryan, Robert K. Hess, authors.
　　　　p.　　cm. — (American Council on Education/Macmillan series in higher education)
　　Includes index.
　　ISBN 0-02-897147-7
　　1. Education—Research—Statistical methods.　2. Educational statistics—Data processing.　3. Statistics—Data processing.
　　I. Ryan, Joseph M.　II. Hess, Robert K.　III. Title.　IV. Series.
　　LB1028.F72　1991
　　370′.7′8—dc20　　　　　　　　　　　　　　　　　　　90-20436
　　　　　　　　　　　　　　　　　　　　　　　　　　　　　CIP

The paper used in this publication meets the minimum requirements of American National Standard for Information Sciences—Permanence of Paper for Printed Library Materials. ANSI Z39.48-1984.

Contents

Acknowledgments   *xi*
Introduction   *xiii*

SECTION A: Research Processes   1

CHAPTER 1   Introduction   3

Purpose   3
Section Overview   3

CHAPTER 2   Linking Research Questions to Research Designs and Statistical Procedures   6

Introduction   6
How To Use This Chapter   6
Part 1: Research Questions Dealing with Description   8
Part 2: Research Questions Dealing with Relationships   8
Part 3: Research Questions Dealing with Differences   9

CHAPTER 3   A Summary of Research Designs   15

Introduction: Purpose of Research Designs   15
External and Internal Validity   15
Descriptive and Experimental Research   16
Research Designs and Analyses   16
The Design Perspective   17
The Statistical Perspective   18
Research Design Profiles   19
References and Resources   38

CHAPTER 4   A Summary of Statistical Procedures   39

Introduction   39
Computer Applications   39
Content Overview   40
Profiles of Statistical Procedures   44
References and Resources   120

CHAPTER 5   A Summary of Sampling Techniques   122

Introduction   122
Selection and Assignment   123
Sampling Terms and Procedures   123
References and Resources   133

CHAPTER 6   A Research Process Checklist   134

Introduction   134
The Checklist   135

SECTION B: Microcomputer Software Profiles   139

CHAPTER 1   Nonstatistical Packages for Microcomputers   141

Introduction   141
Spreadsheets   142
Profiles of Spreadsheets   142
Additional Spreadsheet Programs   146
Database Managers   147
Profiles of Database Managers   148
Integrated Packages   155
Profiles of Integrated Packages   156

CHAPTER 2   Statistical Packages for Microcomputers   163

Introduction   163
Profiles of Statistical Packages   164
Table of Critical Features for Statistical Programs   176

SECTION C: Software Applications in Statistics 179

CHAPTER 1 Introduction 181

Purpose   181
Intended Audience   181
Section Overview   182
Delimitations   184
Using Section C and the Index   184

CHAPTER 2 Statistical Analysis System (SAS) 185

Introduction   185
Text Format   185
Some *SAS* Basics   186
Some Helpful *SAS* Features   187
Data Entry (Input)   187
Data Manipulation (Transformation)   190
Descriptive Statistics   192
Measures of Relationships   194
Regression Analysis   196
Comparisons: One Sample   200
Comparisons: Two Samples (Parametric)   202
Significance Levels for z-, Chi-square, t-, and F-statistics   205
Comparisons: Two Samples (Nonparametric)   206
Comparisons: Two or More Samples (Nonparametric)   207
Comparisons: Two or More Samples—Analysis of Variance (ANOVA)   207
Multiple-Comparison Procedures   220

CHAPTER 3 SYSTAT 223

Introduction   223
Text Format   224
Some *SYSTAT* Basics   224
Descriptive Statistics   234
Measures of Relationships   238
Regression Analysis   244
Using Command Programs in *SYSTAT*   249
Comparisons: One Sample   250
Comparisons: Two Samples (Parametric)   253
Significance Levels for z-, Chi-square, t-, and F-statistics   256
Comparisons: Two Samples (Nonparametric)   256
Comparisons: Two or More Samples (Nonparametric)   258

Comparisons: Two or More Samples—Analysis of Variance (ANOVA) 259
Multiple-Comparison Procedures 274

CHAPTER 4 SPSS-X 277

Introduction 277
Text Format 277
Some *SPSS-X* Basics 277
Some Helpful *SPSS-X* Features 279
Data Entry (Input) 281
Data Manipulation (Transformation) 285
Descriptive Statistics 288
Measures of Relationships 292
Regression Analysis 295
Comparison: One Sample 300
Comparisons: Two Samples (Parametric) 302
Comparisons: Two Samples (Nonparametric) 305
Comparisons: Two or More Samples (Nonparametric) 307
Comparisons: Two or More Samples—Analysis of Variance (ANOVA) 308
Multiple-Comparison Procedures 312

CHAPTER 5 Minitab 313

Introduction 313
Text Format 313
Some *Minitab* Basics 314
Data Entry (Input) 315
Descriptive Statistics 327
Measures of Relationships 332
Regression Analysis 333
Using Command Programs in Minitab 337
Comparison: One Sample 339
Comparison: Two Samples (Parametric) 342
Significance Levels for z-, Chi-square, t-, and F-statistics 345
Comparisons: Two Samples (Nonparametric) 346
Comparisons: Two or More Samples (Nonparametric) 347
Comparisons: Two or More Samples—Analysis of Variance (ANOVA) 349
Multiple-Comparison Procedures 363

APPENDIX A Normal Distribution Curve: Ordinates and Areas 367

APPENDIX B	t-Distribution: Critical Values	369
APPENDIX C	Percentage Points for F-distribution	371
APPENDIX D	Probability Values for the Distribution of Chi-square	378

Index *381*

Acknowledgments

Important to the preparation of this book has been the cooperation that was received from software publishers. This assistance took different forms. Some provided complimentary software packages for review, while others sent printed descriptive materials. These and the numerous telephone-assistance calls made this book possible. To all those who responded to the authors' call for cooperation, we say thank you.

We hereby acknowledge receipt, for review purposes, of complimentary full sets of software packages from SYSTAT, Inc. (1800 Sherman Ave., Evanston, IL 60201, [708] 864-5670; Fax: [708] 492-3567) and Minitab, Inc. (3081 Enterprise Dr., State College, PA 16801, [814] 238-3280; Fax: [814] 238-4383).

The authors also express their appreciation to the Biometrika Trustees (England) and the Houghton Mifflin Publishing Company (USA) for permission to reprint their copyrighted materials. (See tables in appendixes for full credit citation.) Also, the authors are grateful to the Literary Executor of the late Sir Ronald A. Fisher, F.R.S., to Dr. Frank Yates, F.R.S., and the Longman Group Ltd., London, for permission to reprint tables III and IV from the book *Statistical Tables for Biological, Agricultural, and Medical Research* (6th ed., 1974).

Special thanks are due to Ms. Lily Chang, University of South Carolina, for her invaluable assistance with the preparation of the chapters on *SAS* and *SPSS-X*.

Introduction

The successful pursuit of educational research requires skills with research designs, statistical procedures, sampling techniques, and computers. The investigation of a large number of research questions necessitates having the competence to use computers in the research process. Inherent in this is the need to be familiar with the software packages that are available to the researcher. It is this interplay of research techniques, statistical procedures, and computerization that makes possible meaningful advances into understanding learning and other areas of human behavior.

Building on the need to synthesize design, statistics, and computer analysis, there has been assembled in the pages of this book an array of information previously not integrated under a single title. This book serves the student and the professional in education and the behavioral sciences by providing a ready-reference handbook that presents in a succinct and direct style material on research designs, summaries of statistical procedures and their computer applications, sampling techniques, profiles of statistical and nonstatistical software packages, and guidelines for conducting educational research.

This handbook will assist with the planning, implementation, and evaluation of research. It provides summary profiles that will trigger the memories of those who have been trained in such procedures, and it offers clarification for the student. This book synthesizes large quantities of technical information and brings together statistical procedures and their computer applications to meet the practical needs of researchers.

Among the initial responsibilities of the educational researcher is the formulation of an operational statement of the research question and the selection of the proper design and statistical procedure. To facilitate this process, a table is presented that links the research question to the research design and statistical procedure. The researcher need only place the research question in the appropriate category for the design and statistical procedure to be readily identified.

Research is an orderly process. To assist with the implementation of this sequence of steps, a research checklist has been included. Its guidance addresses each of the five major phases of the research process. This instrument will help the educational researcher stay on the right path.

Section A presents brief and informative profiles of 22 research designs, 97 statistical procedures, and an assortment of sampling techniques. Descriptions are given, applications discussed, and strengths and limitations disclosed.

Section B, using succinct profiles, introduces the reader to computer software spreadsheets, database managers, integrated packages, and statistical packages. It presents the functions performed, hardware requirements, cost, and offers helpful comments that will assist the user.

Section C is designed to help the educational researcher execute selected statistical procedures using computers. It explains how to perform this task with *SAS, SYSTAT, SPSS-X,* and *Minitab*. Statistical procedures are explained along with the operational commands of these computer programs.

In conclusion, this handbook brings together an explanation of the research process, its statistical tools, and the computer adaptation to this process. Its purpose is to assist the researcher pursue an understanding of the educational process by using the tools of statistics and computers.

Handbook of Statistical Procedures and Their Computer Applications to Education and the Behavioral Sciences

SECTION A

Research Processes

CHAPTER 1

Introduction

Purpose

A vast array of research procedures can be used to gather needed information about important educational questions. Becoming facile with the wide variety of useful educational research procedures and remaining so, however, is a difficult, time-consuming, and, perhaps, impossible task. Researchers often face the necessity of reviewing various textbooks in research design, statistics, and sampling.

The purpose of Section A, "Research Processes," is to summarize the most commonly used educational research procedures and compile this material into a concise reference compendium. Students who are novices at educational research can use Section A as a simplified supplement to the traditional comprehensive and detailed textbook treatment of the various methodologies. Educators can use these chapters to refresh their understanding of research procedures, to clarify vaguely remembered details of various methodologies, to verify the appropriateness of using a procedure in a particular situation, and to review limitations or assumptions that need to be considered.

Section Overview

Section A contains six chapters. To help the reader, a brief overview of these chapters is provided.

CHAPTER 1: INTRODUCTION

This introductory chapter describes the purpose, content, and delimitations of this section, the content and organization of the different chapters, and some suggestions about how the information might be most effectively used.

CHAPTER 2: LINKING RESEARCH QUESTIONS TO RESEARCH DESIGNS AND STATISTICAL PROCEDURES

This chapter is designed to function as a "directory" of educational research questions or research problems. A classification system for categorizing research questions is developed and explained. An educator wanting to investigate a research problem uses information in Chapter 2 to classify the research question. Based on this classification, the researcher is directed to a specific portion of Chapter 3, where the appropriate research design for addressing the question is described and, in addition, the researcher is directed to specific portions of Chapter 4 where the appropriate statistical procedure is described. Thus Chapter 2 facilitates the investigation of a research problem by identifying the appropriate design and statistical tool.

CHAPTER 3: A SUMMARY OF RESEARCH DESIGNS

This chapter begins with a brief review of major concepts of research designs. The review is followed by detailed descriptions of 22 common research designs and their variations. Each description includes a (1) visual representation of the design, (2) discussion of situations in which the design is commonly used, and (3) commentary on the strengths and weaknesses of the design. The descriptions conclude with references to specific statistical procedures in Chapter 4 that can be used to analyze data related to the design. This chapter functions as a summary of research designs for both the student and those who have been trained and need only a refresher.

CHAPTER 4: A SUMMARY OF STATISTICAL PROCEDURES

This chapter contains descriptions of 97 commonly used statistical procedures. These procedures are presented in the following nine categories:

1. Descriptive Statistics
2. Measures of Relationships
3. Regression Analysis
4. Comparisons: One Sample
5. Comparisons: Two Samples (Parametric)
6. Comparisons: Two Samples (Nonparametric)
7. Comparisons: Two or More Samples (Nonparametric)
8. Comparisons: Two or More Samples—Analysis of Variance (ANOVA)
9. Multiple-Comparison Procedures

The information for each statistical procedure includes a (1) description, (2) common applications, (3) the type of data to which it is applied, and (4) limitations of the procedure, including assumptions.

CHAPTER 5: A SUMMARY OF SAMPLING TECHNIQUES

This chapter discusses basic sampling concepts and sampling techniques. The information about each concept or technique includes (1) a description, (2) common applications, and (3) strengths and weaknesses.

CHAPTER 6: A RESEARCH PROCESS CHECKLIST

The Research Process Checklist is a general guide for conducting educational research. The checklist is a series of statements and questions that calls attention to major decisions and issues that must be addressed in the research process. The checklist is organized under the headings of five major steps in the research process:

1. Defining the research question and the nature of the research
2. Defining variables, subjects, and the research design
3. Verifying the objectivity, reliability, and validity of observation instruments and procedures
4. Analyzing data
5. Interpreting research results

CHAPTER "ENTRY" CODES

Each research design in Chapter 3, statistical procedure in Chapter 4, and sampling concept and technique in Chapter 5 is listed as an "entry" in its respective chapter. Each entry has a unique code for ready reference. An example of a typical entry code is "A4-17." Each entry code provides three pieces of information: the letter "A" indicates Section A; the number "3," "4," or "5," indicates which chapter (3, 4, or 5, respectively); and the one- or two-digit number after the dash is the item identification number. For example, the entry code, "A4-17," refers to Section A, Chapter 4, entry 17.

DELIMITATIONS

Although qualitative/ethnographic research methods and multivariate statistical procedures are useful and appropriate for educators in many research settings, the major focus in this section is on traditional, hypothesis-testing research methods that use univariate statistical procedures. When space limitations demanded some sort of focus, the procedures summarized here were chosen for their widespread applicability in educational research situations and their appropriateness to the background of many educators.

USING SECTION A

A considerable amount of information is contained in the six chapters of this section, each of which begins with an introduction. Regardless of the reader's interests, the usefulness of the information would be increased by reading all the introductions before reading any of the chapters in complete detail.

CHAPTER

2

Linking Research Questions to Research Designs and Statistical Procedures

Introduction

Research questions in education are numerous and are spread across widely diverse areas of interest. Most of them can be classified, however, as belonging in one of three general categories. Research questions are concerned with (1) description, (2) relationships, and (3) differences. Questions of description generally ask about some characteristic of a group of subjects, such as their average performance or the variability in their performance. Questions about relationships are generally concerned with whether two different characteristics of a group of subjects are systematically related (correlated) to each other. Questions about differences generally focus on whether two or more groups are different on some characteristic of interest.

The purpose of this chapter is to provide a link among (1) research questions that fall into these three broad categories, (2) the research designs in Chapter 3, and (3) the statistical tools in Chapter 4. Following is a step-by-step explanation of how to use this chapter. This is followed by parts 1, 2, and 3, one for each of the three categories of research questions. At the beginning of each part the nature of the research question is defined, the appropriate research design(s) for studying the research question is identified (entries in Chapter 3, "A Summary of Research Designs"), and the appropriate statistical procedure(s) is identified (entries in Chapter 4, "A Summary of Statistical Procedures").

How To Use This Chapter

The following is an eight-step procedure for using this chapter to identify research questions and link them to research designs (Chapter 3) and statistical procedures (Chapter 4). The eight steps are:

Step 1: Read the introductions to Chapters 3 and 4 to become familiar with the format and organization of the entries.
Step 2: Explicitly state the research question.
Step 3: Classify the research question into one of the three question categories: (1) description, (2) relationships, or (3) differences. The classification of research questions can be facilitated by examining the definitions found at the beginning of the sections on description, relationships, and differences.
Step 4: Turn to the part (Part 1, 2, or 3) in this chapter which corresponds to the category identified in Step 3.
Step 5: Read the description of the research question for the section. Notice that subsections are identified and described.
Step 6: Locate the appropriate subsection as required.
Step 7: In the appropriate part (Step 5) or subsection (Step 6), the design(s) that should be used for the research question is identified by reference to an entry number in Chapter 3. Turn to the entry in Chapter 3 to read about the design.
Step 8: In the appropriate part (Step 5) or subsection (Step 6), the statistical procedure(s) that should be used for the research question is identified by reference to an entry number in Chapter 4. Turn to the entry in Chapter 4 to read about the procedure.

The following example uses the eight-step procedure.

Step 1: Read the introductions to Chapter 3 and 4.
Step 2: Explicitly state the research question.
"Is there a significant difference in the achievement of three groups of sixth-grade students who study using calculators (group 1), microcomputers (group 2), and no electronic devices (group 3), on a test of arithmetic computation?"
Step 3: Classify the research question into one of three question categories: (1) description, (2) relationships, or (3) differences. The research question belongs in category 3, research questions about differences.
Step 4: Go to Part 3 of this chapter.
Step 5: Read the description of the research question for Part 3. Note that subsection 3-A of Part 3 refers to research questions about differences among multiple groups, observed once.
Step 6: Locate the appropriate subsection as required. Since the research question concerns three groups measured once, the appropriate subsection is 3-A of Part 3.
Step 7: Identify the appropriate research design in Chapter 3. The appropriate research design for research questions in subsection 3-A is described in **A3-7,** Multiple Groups, One Observation. Turn to **A3-7** to study this design.
Step 8: Identify the appropriate statistical procedures in Chapter 4. The appropriate statistical procedures for research questions in subsection 3-A are **A4-71,** One-Way Analysis of Variance (ANOVA), and **A4-87,** Randomized-Blocks Designs. Turn to **A4-71** and **A4-87** to study these procedures.

Part 1: Research Questions Dealing with Description

Research dealing with description is the most basic type of research activity. The accurate and comprehensive description of a sample of subjects with respect to some characteristic of interest is critical in the early stages of research on any topic. Careful description is always important, even in areas that have been thoroughly researched and in areas where experimental research designs are used to test specific hypotheses. Description provides a background and context within which to understand other research findings.

Research questions that deal with description generally pertain to one sample of subjects and one characteristic of that sample. A typical research question of this type might be, "What are the characteristics of basic skills performance at Woodlawn Junior High School?" In answering this question, the researcher might divide the sample into three groups (seventh, eighth, and ninth graders) and then obtain reading and mathematics basic skills test scores. Even with three groups and two variables, the description would be provided for one group and one variable at a time, for instance, the reading scores of seventh graders. In this example, the one-group, one-variable approach could be applied eight times to provide a complete description. These would include the three classes on reading, the three classes on mathematics, all students on reading, and all students on mathematics.

Design
A3-1: One Group, One Observation
Statistical Procedures
A4-1 to **A4-4:** Central Tendency
A4-5 to **A4-10:** Variability or Dispersion
A4-11 to **A4-13:** Distribution Shapes

Part 2: Research Questions Dealing with Relationships

Questions about relationships between and among characteristics or variables can be categorized into three subsections. These are (1) relationships between two variables, (2) relationships among sets of three or more variables, and (3) relationships involving the prediction of one variable from one or more variables.

SUBSECTION 1: RELATIONSHIPS BETWEEN TWO VARIABLES

Relationships between two variables are called correlations and can be examined (1) when one group is observed once and measured on several variables, or (2) when one group is observed twice and variables from the two observations are correlated.

Designs
A3-1: One Group, One Observation (Multiple Variables)
A3-2: One Group, Two Observations

Statistical Procedures
A4-14 to **A4-29:** Correlation coefficients and other measures of relationship selected based on the nature of the variables involved.

SUBSECTION 2: RELATIONSHIPS AMONG SETS OF THREE OR MORE VARIABLES

Relationships among groups of variables are generally studied when one sample of subjects is observed on one occasion and measured on several variables. The research question is concerned with whether three or more variables are so highly related to each other that they are all basically measuring one single characteristic of a sample.

Design
A3-1: One Group, One Observation (Multiple Measurements)
Statistical Procedure
A4-30: Factor Analysis

SUBSECTION 3: RELATIONSHIPS INVOLVING THE PREDICTION OF ONE VARIABLE FROM ONE OR MORE VARIABLES

Research questions about relationships involving predictions ask whether and how well some characteristic of a sample of subjects can be predicted from some other characteristic(s) of the subjects. The prediction of one variable from one or more other variables is performed (1) when one group is observed once and measured on multiple variables, or (2) when one group is observed more than once and prediction is made over time, for example, from a pretest to a posttest.

Designs
A3-1: One Group, One Observation (Multiple Measurements)
A3-2: One Group, Two Observations
Statistical Procedures
A4-31 to **A4-39:** Regression Analysis

Part 3: Research Questions Dealing with Differences

There is a wide variety of research questions that deal with differences. For this reason, these research questions are organized into seven major subsections which are further subdivided.

The seven major subsections address differences related to:

1. One Group
2. Two Groups
3. Multiple Groups
4. Groups in Factorial Arrangements
5. One or More Groups Measured on Two or More Occasions

6. Groups in Nested or Hierarchical Arrangements
7. Group Differences Involving a Control Variable

These subsections are not mutually exclusive, and the reader is urged to refer to all subsections that might be relevant. For example, a researcher who wishes to know if four groups of students differ on a posttest might examine subsections 3 and 4. In most cases, the subsection information overlaps and will direct the researcher to the same design entries in Chapter 3 and statistical entries in Chapter 4.

SUBSECTION 1: DIFFERENCES RELATED TO ONE GROUP

A. One Group, One Observation   Research questions about one group of subjects generally ask whether some statistical characteristic of the sample, such as the group's average performance (i.e., sample mean), differs from some particular value of interest, such as a population average (i.e., population mean). The sample statistical characteristic of interest might be the mean, variance, proportion, correlation, or the overall sample distribution.

Design
A3-1: One Group, One Observation

Statistical Procedures
A4-40 to **A4-47:** Comparisons for One Group, including the mean, variance, proportion, correlation coefficient, and overall distribution.

B. One Group, Two Observations   This situation is common in pretest–posttest designs with one group. The research question asks, "Is there a significant difference in some sample characteristic when it is measured during two different observations, normally on two different occasions?" Statistical tests used in this situation must be designed for dependent or correlated samples.

Design
A3-2: One Group, Two Observations

Statistical Procedures
A4-50, 52, 54, 56: Parametric tests for means, variances, proportions, and correlations, using dependent or correlated samples.
A4-61, 63, 64, 65: Nonparametric tests for dependent or correlated samples.

C. One Group, Multiple Observations   This situation is concerned with research questions about differences in some sample characteristic when it is measured during several different observations, usually on different occasions. Such questions are common in longitudinal research in which one group is followed across many occasions and hypotheses about changes in the characteristics of the group over time are tested.

Design
A3-3: One Group, Multiple Observations over Time

Statistical Procedures
A4-79: One Group, Repeated-Measures Analysis of Variance
A4-90: Time-Series Analysis

SUBSECTION 2: DIFFERENCES RELATED TO TWO GROUPS

A. Two Groups, One Observation   This situation is very common in educational research and is generally encountered when a researcher wants to know if two groups differ with respect to some statistical characteristic of a variable. The performance of two groups on some variable can be compared in terms of means, variability in performance, some proportion, correlation, or the overall distribution of performance.

Design
A3-4: Two Groups, One Observation

Statistical Procedures
A4-48 to **A4-56:** Comparisons for Two Groups, including the means, variances, proportions, correlation coefficients, and overall distributions.
A4-57 to **A4-65:** Nonparametric tests for two groups.

B. Two Groups, Two Observations   Research questions about differences between two groups observed twice generally involve a pretest and a posttest administered to the two groups. Often the two groups are an experimental group and a control group.

Designs
A3-5: Two Groups, Two Observations

Statistical Procedures
A4-49: Independent Sample t-test on Gain Scores
A4-71: One-Way Analysis of Variances on Gain Scores
A4-80: Two-Way Split-Plot Analysis of Variance
A4-89: Analysis of Covariance

C. Two Groups, Multiple Observations   Research questions that ask about differences between two groups observed on multiple occasions are common in longitudinal research. Such research questions are also appropriate in experimental research when an experimental group and a control group are compared before, during, immediately after, and sometimes considerably after the application of the experimental treatment.

Design
A3-6: Two Groups, Multiple Observations

Statistical Procedures
A4-80: Split-Plot Design
A4-90: Time-Series Analysis

SUBSECTION 3: DIFFERENCES RELATED TO MULTIPLE GROUPS

Research questions about three or more groups generally ask about differences among group means. Either naturally occurring groups or experimentally created groups might be compared. An example using naturally occurring groups might involve comparing students from rural, suburban, and urban settings. An example using experimentally created groups might involve comparing students in two treatment groups and a control group.

A. *Multiple Groups, One Observation* Research questions about differences among multiple groups observed once are often concerned with group differences on a posttest after some experimental treatment has been applied. Comparisons among naturally occurring groups are also common.

Design
A3-7: Multiple Groups, One Observation

Statistical Procedures
A4-71: One-Way Analysis of Variance
A4-87: Randomized-Blocks Designs

B. *Multiple Groups, Two Observations* Research questions about differences among multiple groups (three or more) observed twice are generally asked when the first observation refers to a pretest and the second observation refers to a posttest. The groups often include experimental groups and a control group, observed before and after some experimental treatment has been applied.

Design
A3-8: Multiple Groups, Two Observations

Statistical Procedures
A4-80: Two-Way, Split-Plot Analysis of Variance
A4-89: Analysis of Covariance

SUBSECTION 4: DIFFERENCES RELATED TO GROUPS IN FACTORIAL ARRANGEMENTS

Factorial arrangements of groups refers to situations in which groups are viewed as differing along a common dimension or common factor. For example, a group of males and a group of females differ on the common dimension or factor, gender. Gender is a two-level factor because there are two groups. Three groups of students, each taught by a different instructional method, can be represented on one factor, method of instruction. Method of instruction, in this example, has three levels. Groups can be organized or arranged on two, three, or more factors simultaneously. A two-factor arrangement or design, for example, might include "gender" and "method of instruction." In such a case, each of the three instructional treatment groups would have males and females, the two levels of the "gender" factor. The critical attribute of factorial arrangements is that all levels of each factor are represented in all levels of all other factors. For the example males and females, both levels of "gender" appear in all three instructional groups.

Designs
A3-10: One-Way Factorial Analysis of Variance
A3-11: Two-Way Factorial Analysis of Variance
A3-12: Three-Way Factorial Analysis of Variance

Statistical Procedures
A4-71 to **A4-78:** One-, Two-, and Three-Way Analysis of Variance

SUBSECTION 5: DIFFERENCES RELATED TO ONE OR MORE GROUPS MEASURED ON TWO OR MORE OCCASIONS (REPEATED-MEASURES/SPLIT-PLOT ARRANGEMENTS)

A "repeated-measures" arrangement or design refers to a situation in which a sample of subjects is measured on the same variable on two or more occasions. For example, subjects are measured repeatedly. This situation is very common in longitudinal research. Research questions related to repeated measures arrangements are concerned with changes or growth in the performance of subjects over time.

A "split-plot" arrangement or design refers to a situation in which subjects in two or more groups are measured on the same variable on two or more occasions. Research questions related to split-plot arrangements are concerned with (1) differences between the groups regardless of when they are measured, (2) differences between the measurements on different occasions regardless of the groups measured, and (3) differences between the occasions in the differences between the groups.

Designs
A3-13 to **A3-16**: Repeated-Measures and Two-, Three-, and Four-Way Split-Plot Designs

Statistical Procedures
A4-79 to **A4-83**: Repeated-Measures and Two-, Three-, and Four-Way Split-Plot Analyses

SUBSECTION 6: DIFFERENCES RELATED TO GROUPS IN NESTED OR HIERARCHICAL ARRANGEMENTS

Nested or hierarchal arrangements of groups describes situations in which groups of subjects at one level—for example classrooms—are organized within groups at a second level—for example, schools. A research question in this situation might ask about differences in the performance of students in five classrooms in schools 1, 2, and 3. "Classrooms" and "schools" constitute two factors in this situation. This is not a factorial arrangement because only some (5 out of 15) of the groups in the classroom factor appear under each group of the school factor.

Designs
A3-17 to **A3-19**: Two- and Three-Factor Hierarchical Designs

Statistical Procedures
A4-84 to **A4-86**: Two- and Three-Factor Hierarchical Analysis of Variance

SUBSECTION 7: GROUP DIFFERENCES INVOLVING CONTROL VARIABLES

Many research questions about differences between groups are asked in a research situation in which a control variable(s) is used to minimize the effect of an extraneous variable(s) that might obscure the group comparisons. For example, a research question might ask, "Are there differences in the performance of three groups of students studying under different methods of instruction when controlling for initial differences among the students?" Control variables can be incorporated into the research design as independent variables, matching variables, or blocking variables. Control variables can be incorporated into statistical analyses as covariates.

Designs
A3-20: Randomized-Blocks Designs
A3-21: Latin Square Designs
A3-22: Analysis of Covariance Designs

Statistical Procedures
A4-87: Randomized-Blocks Analyses
A4-88: Latin Square Analyses
A4-89: Analysis of Covariance

The following table contains a synopsis of the information needed to link research questions to research designs and statistical procedures without narrative explanations. It is designed to facilitate the use of the information in this chapter and in Chapters 3 and 4.

Linking Research Questions, Designs, and Statistical Procedures

Chapter 2 *Research Questions* *Dealing with:*	*Chapter 3* *Research Designs*	*Chapter 4* *Statistical Procedures*
Part 1. Description	A3-1	A4-1 to A4-13
Part 2. Relationships		
S_1 Two Variables	A3-1, A3-2	A4-14 to A4-29
S_2 Three or More Variables	A3-1	A4-30
S_3 Prediction	A3-1, A3-2	A4-31 to A4-39
Part 3. Differences		
S_1 One Group		
A. One Observation	A3-1	A4-40 to A4-47
B. Two Observations	A3-2	A4-50, 52, 54, 56, 61, 63, 64, 65
C. Multiple Observations	A3-3	A4-79, A4-90
S_2 Two Groups		
A. One Observation	A3-4	A4-48 to A4-56; A4-57 to A4-65
B. Two Observations	A3-5	A4-49, 71, 80, 89
C. Multiple Observations	A3-6	A4-80, A4-90
S_3 Multiple Groups		
A. One Observation	A3-7	A4-71, 87
B. Two Observations	A3-8	A4-80, 89
S_4 Factorial Arrangements	A3-10 to A3-12	A4-71 to A4-78
S_5 One or More Groups, Two or More Occasions	A3-13 to A3-16	A4-79 to A4-83
S_6 Nested or Hierarchical Arrangements	A3-17 to A3-19.	A4-84 to A4-86
S_7 Involving Control Variables	A3-20 to A3-22	A4-87 to A4-89

CHAPTER

3

A Summary of Research Designs

Introduction: Purpose of Research Designs

The purpose of a research design is to isolate and study the influence of an independent variable on some dependent measure. The research design is constructed to increase the validity of the claim that variation in the dependent measure can be attributed exclusively to variation in the independent variable. The influence of all other extraneous variables on the dependent measure must, accordingly, be eliminated, held constant, or controlled in some other fashion. Research designs are used to achieve the desired control over extraneous variables that could confound the independent-dependent variable relationship. The efficiency of designs varies according to the degree to which they control the influence of extraneous variables.

External and Internal Validity

For a variety of reasons, the control of extraneous variables is difficult to achieve in many educational research settings. The general difficulty is that educational researchers rarely have control over all the factors that need to be manipulated in order to create a true experimental situation. It is often impossible for a researcher to randomly select subjects from a population or randomly assign subjects to experimental treatment conditions. The lack of both random selection and random assignment of subjects constitutes the major roadblock to conducting true experimental research in education. The lack of random selection greatly limits external validity, while the lack of random assignment greatly limits internal validity of research studies.

"External validity" refers to the extent to which the results of a study can be generalized to a larger population. The major tool for increasing external validity is random selection from the population of the sample being used in the study. Random sampling **(A5-5)** increases the probability that the sample is truly representative of the population with respect to the relevant variables so that the research results for the sample can be generalized to the population.

"Internal validity" refers to the extent to which variation in the dependent measure can be attributed exclusively to the independent variable. A major tool for increasing internal validity is random assignment, generally called randomization, which controls the influence of extraneous variables by causing them to be equally present in all treatment conditions. In this way the influence of the extraneous variables on the dependent measure is constant across all treatment conditions. Thus, any difference between treatment groups must be due to treatment conditions since the influence of the extraneous variables is a constant.

There are many specific threats to external validity and internal validity. These are examined in great detail in the classic *Experimental and Quasi-Experimental Designs for Research,* by D. T. Campbell and J. C. Stanley (1966). The interested reader is strongly urged to study this informative work. (See "References and Resources" section at the end of this chapter for full citation.)

Descriptive and Experimental Research

In studying research designs, it is useful to imagine that designs run along a continuum from purely descriptive at one end to purely experimental on the other. Descriptive studies simply present the characteristics of the sample being studied. There is no attempt to show causal relationships among variables, even if two variables are highly correlated. The description of a sample may be generalizable to a population if the sample was randomly selected. Experimental studies, however, are designed to study a causal relationship between independent and dependent variables, in which variation in the dependent variable may be attributed to variation in the independent variable. Experimental studies provide generalizable results when conducted using random samples.

As mentioned before, it is useful to think of descriptive and experimental research designs as varying in degree along a continuum from purely descriptive to purely experimental, rather than as being mutually exclusive dichotomous categories. Between the two ends of the descriptive-experimental continuum are a wide variety of designs referred to as "quasi-" or "pseudo-" experimental designs. Exactly where a research design is located on the continuum depends on how well the design controls for the various threats to internal and external validity. The intricacies of these issues, while beyond the scope of this book, are discussed in detail by Campbell and Stanley (1966) and Cook and Campbell (1979).

Research Designs and Analyses

Research designs are often discussed from two slightly different perspectives. When examined from a strictly design point of view, research designs are often discussed in terms of arrangements of subjects or groups of subjects, the application of independent variable treatment conditions, the measurement of control variables, and dependent variable outcome measures. Using this approach, a study might be described as being an experimental group-control group design or one-group pretest-posttest design. When examined from a strictly statistical point of view, however, research designs are often discussed in terms of the statistical procedures used to analyze data obtained in a study. For example, a study which has two independent variables, namely gender and treatment, and is designed so that all levels of each variable appear along with all levels of the other independent variable, might be described simply as a two-way factorial analysis of variance, gender by treatment. Both the "design" and the "statistical" approaches will be used in this chapter.

The Design Perspective

From the design perspective, research designs are organized in the following general categories:

Designs involving

1. One Group, One Observation
2. One Group, Two Observations (Pretest-Posttest)
3. One Group, Multiple Observations over Time
4. Two Groups, One Observation (Posttest)
5. Two Groups, Two Observations (Pretest-Posttest)
6. Two Groups, Multiple Observations over Time
7. Multiple Groups, One Observation (Posttest)
8. Multiple Groups, Two Observations (Pretest-Posttest)
9. Multiple Groups, One and Two Observations over Time

A number of abbreviation and graphic formats will be used in presenting these designs. (These are the symbols used by Campbell and Stanley, with modifications in some cases.) The letter "O" indicates that an observation or measurement has been made on some group. An "X" indicates that some experimental treatment condition has been applied to a group. A "C" indicates a control group and generally appears opposite an "X." The control group does not receive the experimental treatment. The sequence of events for a single group is represented on one horizontal line. The temporal order in which events occurred is reflected in the sequence, from left to right, on each line. Events that occurred for different groups are shown on different horizontal lines. Events that occurred at the same time, but for different groups, are on different horizontal lines but appear one over the other. Parallel rows separated by a dashed line indicate comparison groups of subjects who were not randomly assigned to groups. For one-sample designs, the distinction between randomly selected and nonrandomly selected samples is made explicit. In all other cases, the reader must understand that the designs could be presented twice, once for randomly and once from nonrandomly selected samples.

Designs are described in terms of the number of groups involved and the number of observations made of each group. In the context of research designs, an "observation" refers to a single period of observation. Within this single observational period, however, multiple measurements might be made. Thus, a two-observations design includes two periods of observation, within each of which multiple measurements might be made.

The concept of matching is used in two different ways when describing research designs. First, matching used with random assignment involves grouping subjects together into "blocks," so that all the subjects in a "block" are as similar to each other as possible, with respect to the matching variable. Subjects are then randomly assigned from these blocks to different groups so that each group has a subject or subjects from each block. These procedures constitute randomized-blocks designs **(A3-20)**. Second, matching is also used in a less rigorous and less useful way in designs that employ a "matched" comparison group. In these designs, the subjects receiving the experimental treatment condition are compared to some nonrandomly selected and nonrandomly assigned group. This comparison group is used because it "matches" the treatment group with respect to variables that might influence the dependent measure. This so-called matching is thought to "control" these confounding variables because the two groups are equated on these measures. This type of matching is vastly overrated as a mechanism for improving research designs.

The Statistical Perspective

From the statistical analysis perspective, research designs are organized into the following general categories, listed by entry number:

Factorial Analysis of Variance (ANOVA) Designs

10. One-Way ANOVA
11. Two-Way ANOVA
12. Three-Way ANOVA

Repeated-Measures/Split-Plot Analysis of Variance (ANOVA) Designs

13. One-Way Repeated-Measures Designs
14. Two-Way Split-Plot Designs
15. Three-Way Split-Plot Designs (A and B)
16. Four-Way Split-Plot Design

Nested/Hierarchical and Other Designs

17. Two Nested Factors
18. Three Nested Factors
19. Cross and Nested Designs
20. Randomized-Blocks Designs
21. Latin Square Designs
22. Analysis of Covariance Designs

Research Design Profiles

A3-1

Category: One Group, One Observation

Designs: A. Nonrandom Selection   O
         B. Random Selection   O
         C. Nonrandom Selection   X   O
         D. Random Selection   X   O

Description: The single-group, one-observation designs involve a single measurement or set of measurements for one sample. The sample may be a nonrandom convenience sample (A and C) or a true random sample (B and D). The observations may be preceded by an experimental treatment (C and D).

Applications: Single-group, one-observation studies are commonly used in the earliest stages of research in an area for obtaining basic descriptive information. Such designs are generally employed when the researcher has no control over the research setting but would like to begin collecting baseline information. Often the subjects in a one-group, one-observation design are a convenience sample. One-group one-observation designs are often used for correlational studies when multiple measurements are made during the observational period.

Comments: Clearly, this is the weakest of all designs. Nonrandom single observation studies are strictly descriptive in nature and should always be viewed as exploratory. Very useful information can be obtained from such designs but caution should be exercised in making strong claims from such studies. Findings from single-group, one-observation designs can be very helpful in generating hypotheses to be explored in subsequent research and for identifying possible extraneous variables that should be controlled in future research.

Analysis: One-sample designs are analyzed using a variety of procedures depending on the purpose of the research. Purely descriptive statistics are routinely calculated, using **A4-1** to **A4-13**. Inferences related to one sample, generally comparing a sample statistic to some value of interest, like a population parameter, are tested with **A4-40** to **A4-47**. When several different variables are measured for one sample, the relationships or correlations between pairs of variables are examined with **A4-14** to **A4-30**. Regression analysis might also be conducted with multiple measurements taken on one sample using **A4-31** to **A4-39**.

A3-2

Category: One Group, Two Observations (Pretest-Posttest Design)

Designs: A. Nonrandom Selection O_1 O_2
         B. Random Selection O_1 O_2
         C. Nonrandom Selection O_1 X O_2
         D. Random Selection O_1 X O_2

Description: The pretest-posttest design involves two observations of the same sample of subjects. The samples may be random (B and D) or nonrandom (A and C). In some cases, no specific experimental manipulation occurs between the two observations (A and B). In many cases, an observation or set of observations is made prior to some experimental treatment condition and then again following the application of the treatment.

Applications: The pretest-posttest designs are commonly used in education to examine changes over time (A and B) and the effects due to some treatment (C and D). The treatments of interest are often curriculum programs or instructional strategies. Such approaches are very common in evaluation designs that are devised to demonstrate the impact of new programs.

Comments: The pretest-posttest design, frequently used in educational research, is a very weak design. The major problem is substantiating the claim that the experimental treatment X actually accounts for observed pretest-posttest differences. In most cases, many plausible explanations can be advanced to explain pretest-posttest differences. These are at least as likely as the explanation that the experimental treatment caused the difference. In most cases, findings based on pretest-posttest designs should be considered descriptive and exploratory. However, such designs can be useful in generating or refining hypotheses for more rigorous future study and for identifying extraneous variables that need to be controlled in subsequent research.

Analysis: Pretest-posttest designs are analyzed using procedures for dependent or correlated samples. Dependent sample tests for means, variances, proportions and correlations are described, respectively, in **A4-50, A4-52, A4-54,** and **A4-56.** Nonparametric tests for dependent samples are found in **A4-63** to **A4-65.** A one-way repeated measures analysis, **A4-79,** might be applied to test pretest-posttest differences.

A3-3

Category: One Group, Multiple Observations over Time

Designs:
A. Nonrandom Selection O_1 O_2 O_i .... O_L
B. Random Selection O_1 O_2 O_i .... O_L
C. Nonrandom Selection O_1 O_2 O_i .... X .... O_L
D. Random Selection O_1 O_2 O_i .... X .... O_L

Description: The one-group, multiple-observations designs involve repeated measurements of the same subjects. The sample can be random (B and D) or nonrandom (A and C). The repeated observation may occur over time without any intervening experimental treatment condition (A and B), or some experimental treatment condition may occur within the series of measurements. When there are many observations, these designs are also called "time-series designs."

Applications: Designs involving repeated measurements of a single sample are commonly used in longitudinal research, especially in longitudinal developmental research. Designs with multiple observations and an intervening experimental treatment are quasi-experimental designs generally referred to as "interrupted time-series designs." Such designs are applicable when periodic baseline measurements have been made prior to some treatment condition, and measurements continue to be made after the treatment has been implemented. These designs can be used to assess the impact of social policy changes or changes in regulations, such as changes in the highway speed limit. These types of changes can be considered an experimental treatment, and data associated with these changes, for example, rate of highway accidents by month or year before and after the change, are regarded as the periodic measures.

Comments: Repeated measures and time-series designs are more powerful than the simple pretest-posttest designs. These multiple-observations designs allow for a careful analysis of systematic trends before, during, and after the introduction of the experimental treatment condition. The impact of certain threats to internal validity can be examined using these designs and thus their influence can be ruled out in some cases. It is important to remember, however, that these are quasi-experimental designs.

Analysis: One-group, multiple-observations designs are analyzed using one-way repeated-measures analysis of variance **(A4-79)** or time-series analysis procedures **(A4-90)**.

A3-4

Category: Two Groups, One Observation

Designs: A. Randomized   X   O
            Randomized   C   O

        B. Matched   X   O
           Matched   C   O

        C. Matched and Randomized   X   O
           Matched and Randomized   C   O

        D. Treatment   X   O
           Convenience   C   O

Description: The two-groups, one-observation designs involve a single measurement or set of measurements on two groups. The single observation or measurement is generally a posttest as depicted in designs A, B, C, and D. The two groups are referred to as the experimental group, which receives treatment X, and the control group, C, which does not receive the treatment. The classification of groups as experimental and control groups can be based on (A) randomization, (B) matching on what are thought to be salient background variables, (C) sequential matching and randomization together, and (D) convenience or mere availability.

Applications: The two-groups, one-observation designs are commonly used to assess the effects of some treatment condition, X, by comparing two groups of subjects in a situation where one group receives the treatment and one group does not. All other things being equal, the difference between the two groups after the treatment reflects the impact of the treatment. These designs are often used to examine the impact of different instructional strategies, curriculum materials, or training methods.

Design C, involving both matching and randomization, is a type of randomized-blocks design **(A3-20).** The variable (or variables) on which the subjects are matched is a blocking variable—the influence of which is controlled by matching. Subjects are matched to be as similar as possible on the blocking variable and are formed into blocks.

With a two-group design, the blocks contain two subjects who are as similar as possible on the blocking variable. The two subjects in each block are then randomly assigned to one of the two groups, experimental or control. The subjects from each block are treated as if they were the same subject measured twice, once under each condition. Thus, statistical comparisons appropriate for dependent samples are used with design C.

Comments: The two-groups, one-observation design with randomization, as in designs A and C, is a true experimental design despite the absence of a pretest. The random assignment of subjects to groups is used to create comparability between groups and equivalent distributions of confounding variables within the two groups. The randomization equalizes the influence of potential extraneous variables so that groups differ on the independent variable, but all other characteristics of the groups are equivalent.

An approximation to randomization involves matching (design B) two groups based on demographic variables or some other variables that might influence the dependent measure. The matching is thought to minimize the differences between the groups prior to the application of the experimental treatment. The validity of the inference that posttreatment differences are due to the independent variable is improved by documenting pretreatment similarities through matching. Despite some apparent advantages, matching is not an adequate substitute for randomization and cannot be used to support independent-dependent variable causal inferences.

The use of a convenience sample (design D) as a comparison group, or pseudo-control group, is common in educational research. Often, such designs include the suggestion that the treatment group and comparison group were similar prior to the treatment, and thus are "matched" samples. These designs are very weak and are most usefully thought of as single-group, one-observation designs. Information about the control group can be used to describe the general context within which the treatment group is studied. This information cannot be used to strengthen independent-dependent variable causal inferences.

Analysis: Two-groups, posttest-only designs are generally used to test differences between groups after some treatment has been applied. Parametric procedures for comparing means, variances, proportions, and correlations for two groups are described in **A4-48** to **A4-56**. Nonparametric tests for two groups are described in **A4-57** to **A4-65**. The difference between two groups might be examined using a one-way analysis of variance **(A4-71)** or a randomized-blocks design **(A4-87).**

A3-5

Category: Two Groups, Two Observations
(Pretest-Posttest Designs)

Designs: A. Randomized O_1 X O_2
Randomized O_1 C O_2

B. Matched O_1 X O_2
Matched O_1 C O_2

C. Matched and Random O_1 X O_2
Matched and Random O_1 C O_2

D. Treatment O_1 X O_2
Convenience O_1 C O_2

E. Randomized   X O_2 Treatment Group
Randomized O_1 C   Control Group

Description: Designs involving two groups and two observations generally include a pretest and posttest for an experimental group and a control group. Subjects can be assigned to experimental or control groups at random, based on matching, or based on convenience sampling. The pretest and posttest are frequently the same test or alternate forms of the same test used to assess subjects' knowledge before and after the treatment.

Design C involves matching and randomization. In a two-groups design, subjects most similar to each other on the matching variable(s) are paired together into blocks. From each block each subject is randomly assigned to either the experimental group or the control group. In this randomized-blocks design, subjects from the same block are treated as the same subject measured twice. Statistical procedures for dependent samples are applied.

Campbell and Stanley refer to design E as a "separate-sample pretest-posttest" design. In this design, a randomly assigned experimental group is measured after the treatment has been administered and a randomly assigned control group is measured before the treatment. The posttest for the treatment group is compared to the pretest for the control group. The effect of the "treatment" on the control group is not assessed.

Applications: The pretest-posttest experimental group-control group design is very commonly used to test the impact of an experimental treatment. In educational research, the experimental treatment of interest is often an instructional strategy, a set of curriculum materials, or some teacher-training procedure.

Comments: The value of the design for testing hypotheses about treatment effects and for making independent-dependent variable causal inferences depends directly on how the groups are assigned. Designs in which subjects are randomly assigned to treatment and control group (designs A and C) are true experiments and allow for independent-dependent variable causal inferences. Randomization, however, is not practical in many educational settings.

Matching designs (design B) can be informative but are clearly limited by the nature of the matching variables and the care with which matching is performed. Even under the best of conditions, matching designs should be considered more descriptive than experimental. Matching designs are very common when a teacher or group of teachers is implementing some new program (e.g., teaching strategy or curriculum), and a "matched" group of teachers who do not use the new program is identified.

Design C is a randomized-blocks design that is very powerful. The test statistic for such designs basically eliminates between-subject variation in the estimate of random error that is used as the denominator in hypothesis testing.

Design D represents a pretest-posttest design with an experimental group and a comparison group that is used simply because it is conveniently available. Often, such designs include the suggestion that the treatment group and comparison group were similar prior to the treatment, and thus are "matched" samples. Such comparison groups provide little, if any, useful information. Indeed, the use of a convenience sample can easily be misleading since it gives the design the appearance of an experiment without providing the controls over confounding variables that true experimental designs provide. Moreover, the convenience sample, itself, may actually introduce additional confounding variables.

Analysis: Two-groups, pretest-posttest designs are used to assess the impact of some experimental treatment applied to one group compared to a second group that does not receive the treatment. Data from these designs can be analyzed using a variety of procedures. Gain scores defined as the difference between posttest and pretest can be examined using a t-test, **A4-49,** or **A4-50** (for matched samples in design C above). A one-way analysis of variance, **A4-71,** might also be used to examine the gain scores. Gain scores are not very reliable, however. Many researchers would examine these designs using analysis of covariance, **A4-89,** or a split-plot design, **A4-80.**

A3-6

Category: Two Groups, Multiple Observations over Time

Designs: A. Randomized $O_1\ O_2\ O_3\quad X\quad O_4\ O_5\ O_6$
           Randomized $O_1\ O_2\ O_3\quad X\quad O_4\ O_5\ O_6$

B. Matched $O_1\ O_2\ O_3\quad X\quad O_4\ O_5\ O_6$
   Matched $O_1\ O_2\ O_3\quad X\quad O_4\ O_5\ O_6$

C. Treatment $O_1\ O_2\ O_3\quad X\quad O_4\ O_5\ O_6$
   Convenience $O_1\ O_2\ O_3\quad X\quad O_4\ O_5\ O_6$

Description: The two-groups, multiple-observations designs involve a treatment group that receives some experimental treatment condition, and a control group that does not receive the treatment. Observations of both groups are made at different times before ($O_1\ O_2\ O_3$) and after ($O_4\ O_5\ O_6$) the treatment is applied. As with the preceding two-groups, two-observations designs, the samples may be randomly assigned, matched, or convenience samples.

Applications: Two-groups, multiple-observations designs are commonly used to assess the impact of some experimental treatment condition applied to one group but not to the other. Designs of this sort are very useful in determining long-term treatment effects in contrast to pretest-posttest designs which assess the impact of an experimental treatment only once, immediately following the application of the treatment. In addition, the multiple observations preceding the treatment are useful in detecting emerging pretreatment trends or shifts.

Comments: The two-groups, multiple-observations designs are also called "time-series designs" if there are a large number of observations. In general, these designs are very useful when the experimental treatment can be randomly assigned to one of the two groups.

Analysis: Two-groups, multiple-observations designs are analyzed using a split-plot analysis, **A4-80,** or time-series analysis procedures, **A4-90.**

A3-7

Category: Multiple Groups, One Observation (Posttest)

Designs: A. Randomized   X   O
            Randomized   X   O
            Randomized   C   O

         B. Matched   X   O
            Matched   X   O
            Matched   C   O

         C. Matched and Randomized X   O
            Matched and Randomized X   O
            Matched and Randomized C   O

         D. Nonrandomized   X   O
            Nonrandomized   X   O
            Nonrandomized   C   O

Description: The multiple-groups, one-observation designs generally involve a posttest administered to three or more groups. The specific designs illustrated above show two experimental groups and one control group (group 3). There can be any number of groups in varying combinations of experimental and control groups. Subjects can be randomly assigned to groups (as in A and C). Groups can be matched on some confounding variable the influence of which the researcher wishes to control (designs B and C). Design C is a randomized-blocks design (see entries **A3-20** and **A4-87**). Design D involves the comparison of convenience samples or nonequivalent groups.

Applications: Multiple-groups, one-observation (posttest) designs are extremely common in educational research. Such designs are used to compare different experimental treatments to each other and to a control group. Often, the different experimental groups represent subjects who are receiving different variations of the same experimental treatment. Multiple control groups are also frequently incorporated into such designs.

Comments: The value of multiple-groups, one-observation (posttest) designs depends directly on how subjects are assigned to groups. Design A involves randomization and is a true experimental design. Design C is a randomized-blocks design (**A3-20** and **A4-87**). Such designs are not always practical since they require considerable control over the experimental setting. The randomized-blocks design is a powerful experimental design that should be considered whenever possible. Multiple-groups designs often involve "matching" based on the fact that groups are similar with respect to background variables or other measures. Comparing these matched groups is not equivalent to the experimental control of extraneous variables achieved through randomization. Studies using this type of matching or convenience samples may be informative; however, they should be interpreted with great caution. Often it is useful to consider such studies as descriptive studies even if they are analyzed using inferential statistical procedures.

Analysis: Multiple-groups, posttest-only designs are generally examined using one-way analysis of variance procedures, **A4-71,** or a randomized-blocks analysis, **A4-87.**

A3-8

Category: Multiple Groups, Two Observations (Pretest-Posttest Designs)

Designs: A. Randomized O_1 X O_2
           Randomized O_1 X O_2
           Randomized O_1 C O_2

Designs: B. Matched O_1 X O_2
           Matched O_1 X O_2
           Matched O_1 C O_2

Designs: C. Matched and Randomized O_1 X O_2
           Matched and Randomized O_1 X O_2
           Matched and Randomized O_1 C O_2

Designs: D. Nonrandomized O_1 X O_2
           Nonrandomized O_1 X O_2
           Nonrandomized O_1 C O_2

Description: Multiple-groups, two-observations designs are generally pretest-posttest designs with three or more groups. The designs above are shown as having two experimental groups (groups 1 and 2) and one control group (group 3). There can be any number of groups in varying combinations of experimental and control groups. Subjects can be assigned to experimental or control groups in different ways. In designs A and C, subjects are randomly assigned to groups. In design B, groups are matched on a variable the influence of which the researcher wishes to control. Design C is a randomized-blocks design, the details of which are described elsewhere **(A3-20, A4-87),** and D involves the comparison of convenience samples or nonequivalent groups.

Applications: Multiple-groups, two-observations (pretest-posttest) designs are often used to test the impact of experimental treatments. In educational research, the experimental treatments of interest are often instructional strategies, curriculum materials, or teacher-training procedures.

These designs can be used in many different ways. An analysis of students' gains can be performed by testing hypotheses about the differences between posttest and pretest performance. Pretest and posttest scores can be used in a repeated-measures and split-plot analyses (**A3-13** to **A3-16**). Finally, the pretest can be used as a control variable in an analysis of covariance procedure (**A3-22**).

Multiple-groups, two-observations (pretest-posttest) designs are used to compare different experimental treatments to each other and to a control group. Multiple control groups are occasionally incorporated into such designs.

Comments: The value of the multiple-groups, pretest-posttest designs is directly related to how subjects are assigned to the groups. Randomization, as always, is the key. Designs A and C, which include randomization, are more useful and are true experimental designs. These designs can support independent-dependent variable causal inferences. Design B involves comparisons among groups that are "matched," based on the fact that they are similar with respect to some relevant variables that the researcher thinks should be controlled. Such matching is not very useful in controlling the effects of extraneous variables unless it is accompanied by random assignment, as in design C. Design D involves the comparison of convenience samples or nonequivalent groups. Such designs are very weak and should be considered descriptive in nature even if they are analyzed using inferential techniques.

Analysis: Multiple-groups, two-observations (pretest-posttest) designs can be analyzed in several ways. Gain scores can be examined using a one-way analysis of variance, but the lack of reliability typical of gain scores makes this approach undesirable. Analysis of covariance, **A4-89**, is commonly used for these designs. A split-plot analysis, **A4-80**, with one between-subjects factor and one within-subjects factor would also be appropriate.

A3-9

Category: Multiple Groups, One and Two Observations over Time

Design: Solomon Four-Group Design

Randomized	O_1	X	O_2	Group 1
Randomized	O_1	C	O_2	Group 2
Randomized		X	O_2	Group 3
Randomized		C	O_2	Group 4

Description: The Solomon Four-Group Design involves four randomly assigned groups. This is a true experimental design since all subjects are randomly assigned to groups. There are two experimental groups (groups 1 and 3), one of which takes a pretest and posttest (group 1), the other of which takes the posttest only (group 3). There are two control groups (groups 2 and 4), one of which takes a pretest and posttest (group 2), the other of which takes the posttest only (group 4).

Applications: The Solomon Four-Group Design is used to examine the effects of experimental treatment condition, X, while controlling for a variety of extraneous variables. The design requires considerable control over the experimental situation.

Comments: The Solomon Four-Group Design is useful and powerful in detecting the effects of experimental treatment conditions. A comparison of the posttest for groups 1 and 2 examines treatment effects in the presence of a pretest. A comparison of groups 3 and 4 examines the treatment effects without the influence of a pretest. The effects of the experimental treatment are tested when groups 1 and 3 are compared to groups 2 and 4. The effects of the pretest are examined with a comparison of the posttests of groups 1 and 2 to the posttests of groups 3 and 4.

This is an informative and useful design which requires a large number of subjects as well as considerable control over the experimental situation.

Analysis: The Solomon Four-Group Design is generally analyzed using a one-way analysis of variance, **A4-71,** with four groups, using the posttest as the dependent measure. Scheffé contrasts, **A4-94,** are used to compare groups 1 and 2 against groups 3 and 4. This examines the effect of the pretest. The contrast of groups 1 and 3 against 2 and 4 examines the treatment effect.

A3-10

Category: Factorial Analysis of Variance (ANOVA) Designs

Design: One-Way Analysis of Variance

Description: The one-way analysis of variance design contains a single independent variable, Factor A. The independent variable can have two or more levels forming different groups. The groups are compared when the analysis of variance is performed **(A4-71).**

The basic arrangement for a one-way analysis of variance is shown below.

```
      Factor A, A Levels
       1  2  3  .  .  A
      ┌──┬──┬──┬──┬──┐
      │  │  │  │  │  │
      └──┴──┴──┴──┴──┘
```

Applications: This one-way design and accompanying analysis has many applications in educational research and evaluation. The one-way analysis of variance is appropriate anytime two or more groups are being compared on some measure. The procedure is commonly used with multiple-group, posttest-only designs.

Comments: The one-way analysis of variance is one of the most commonly used research designs. For details, see **A4-71.**

Analysis: One-Way Analysis of Variance, **A4-71**

A3-11

Category: Factorial Analysis of Variance (ANOVA) Designs

Design: Two-Way Analysis of Variance

Description: The two-way analysis of variance contains two independent variables, or factors, on the basis of which subjects are classified. Each factor can have two or more levels. If Factor A has A levels, and Factor B has B levels, then subjects are classified into A × B individual cells.

The basic arrangement for a two-way analysis of variance is shown below. Factor A is shown with A levels and Factor B is shown with 3 levels.

```
                    Factor A, A Levels
                    1   2   3  . .  A
                 1 ┌───┬───┬───┬───┬───┐
                   │   │   │   │   │   │
                   ├───┼───┼───┼───┼───┤
       Factor B  2 │   │   │   │   │   │
                   ├───┼───┼───┼───┼───┤
                 3 │   │   │   │   │   │
                   └───┴───┴───┴───┴───┘
```

Applications: Two-way analysis of variance designs with accompanying analyses are used whenever there are two independent variables, and all levels of each variable appear under all levels of the other variable. Such designs are very common when subjects in different treatment groups, represented by the levels of Factor A, can be classified on some second variable, represented by Factor B. For example, a two-way analysis of variance, treatment × gender, might have 3 levels of treatment and 2 levels of gender. Such a design would be a 3 × 2 two-way analysis of variance. Two-way analysis of variance is commonly used in aptitude-treatment research in which aptitude levels are represented by Factor A and treatment groups by Factor B.

Comments: Two-way analysis of variance designs are used to test hypotheses about the two independent variables and their interaction. The levels of the factors can be fixed, random, or mixed. For details, see **A4-72, A4-73,** and **A4-74,** respectively.

Analysis: Two-Way Analysis of Variance
Fixed Effects Model, **A4-72**
Random Effects Model, **A4-73**
Mixed Effects Model, **A4-74**

A3-12

Category: Factorial Analysis of Variance (ANOVA) Designs

Design: Three-Way Analysis of Variance

Description: The three-way analysis of variance contains three independent variables or factors on the basis of which subjects are classified. Each factor can have two or more levels. If Factor A has A levels, Factor B has B levels, and Factor C has C levels, then subjects are classified into A × B × C individual cells.

The basic arrangement for a three-way analysis of variance is shown below. Factor A is shown with A levels, Factor B is shown with 3 levels, and Factor C is shown with 3 levels.

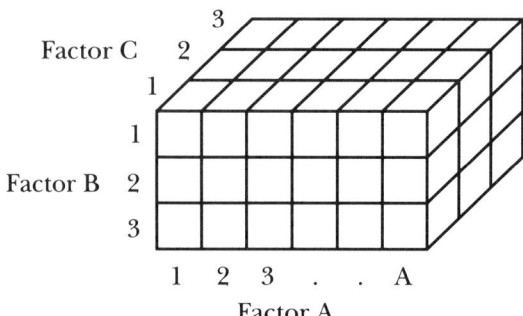

Applications: Three-way analysis of variance designs with accompanying analyses are used whenever there are three independent variables and all levels of each variable appear under all levels of the other variable. Such designs are used when subjects can be classified on three different variables simultaneously. For example, a three-way ANOVA, treatment × gender × SES (socioeconomic status), might have 3 levels of treatment, 2 levels of gender, and 3 levels of SES. Such a design would be a 3 × 2 × 3 three-way ANOVA.

Comments: Three-way analysis of variance designs are used to test hypotheses about three independent variables, their two-way interactions, and their three-way interaction. The levels of the factors can be fixed, random, or mixed. For details, see **A4-75** to **A4-78,** respectively.

Analysis: Three-Way Analysis of Variance
Fixed Effects Model, **A4-75**
Mixed Effects Model
  One Fixed, Two Random Factors, **A4-76**
  Two Fixed, One Random Factor, **A4-77**
Random Effects Model, **A4-78**

A3-13

Category: Repeated-Measures and Split-Plot Analysis of Variance (ANOVA) Designs

Design: One-Way Repeated-Measures Design

Description: One-way repeated-measures designs involve the repeated measurement of the same group of subjects over several occasions. From a design perspective, there is one factor, occasions, with A levels (see **A4-79**). From an analysis perspective, these designs are treated as a two-way mixed model (see **A4-74**). The two factors are subjects (the random factor) and occasions (the fixed factor).

The basic arrangement for a one-way repeated measures analysis of variance is illustrated below with Factor A, Occasions, consisting of 4 levels.

SUMMARY OF RESEARCH DESIGNS   31

Factor A, Occasions (4 Levels)

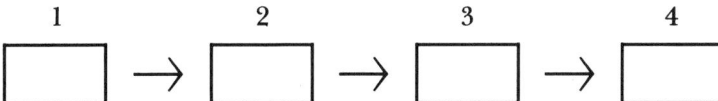

Applications: Repeated-measures designs are used anywhere a group of subjects is measured on more than one occasion. Such designs are common in developmental and longitudinal research.

Comments: Repeated-measures designs are statistically very powerful. They are the extension of the dependent sample t-test **(A4-50)**. For details, see entry **A4-79**.

Analysis: One-Way Repeated-Measures Analysis of Variance, **A4-79**

A3-14

Category: Repeated-Measures and Split-Plot Analysis of Variance (ANOVA) Designs

Design: Two-Way Split-Plot Design, One Between-Subjects Factor, One Within-Subjects Factor

Description: The two-way, split-plot design with one within-subjects and one between-subjects factor involves repeated measurements (Factor B) for subjects arranged into different groups (Factor A). The repeated measurements across occasions is the within-subjects factor (B), and the classification of subjects into different groups is the between-subjects factor (A).

The basic arrangement for this design is illustrated below with 4 levels of Factor B (Occasions) and 2 levels of Factor A (Groups).

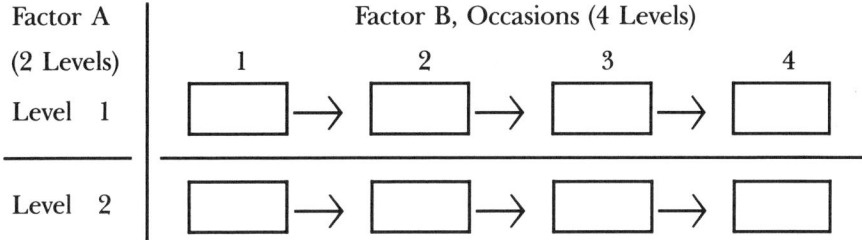

Applications: This design is used when groups of subjects defined on some independent variable (Factor A) are measured on more than two occasions (Factor B). By way of illustration, a very common application might have three levels of Factor B; namely, measurements taken before, during, and after an experimental intervention, and two levels of Factor A, a treatment group and a control group.

Longitudinal research on groups that develop at different rates with respect to the dependent variable would employ this design.

Comments: The one-between one-within, two-factor, split-plot design is an informative, useful, and powerful design. For details, see entry **A4-80**.

Analysis: Two-Way Split-Plot Analysis of Variance, One Between-Subjects Factor, One Within-Subjects Factor, **A4-80**

A3-15A

Category: Repeated-Measures and Split-Plot Analysis of Variance (ANOVA) Designs

Design: Three-Way Split-Plot Design, One Between-Subjects Factor, Two Within-Subjects Factors

Description: This design is used when subjects in different groups (between-subjects, Factor A), are measured two or more times (within-subjects, Factor B), on two or more occasions (within-subjects, Factor C). Such situations arise when subjects in different groups (Factor A) attempt a task several times under different conditions (Factor B), before, during, and after some experimental intervention (Factor C).

The basic arrangement for this design is illustrated below. Factor A, between-subjects, has 2 levels; Factor B, within-subjects, has 3 levels; and Factor C, within-subjects, has 4 levels.

Applications: A three-way, split-plot design with one between-subjects factor and two within-subjects factors is used when subjects in different groups are measured repeatedly (e.g., trials), on different occasions.

Comments: For details, see entry **A4-81**.

Analysis: Three-Way Split-Plot Analysis of Variance, One Between-Subjects Factor, Two Within-Subjects Factors, **A4-81**

A3-15B

Category: Repeated-Measures and Split-Plot Analysis of Variance (ANOVA) Designs

Design: Three-Way Split-Plot Design, Two Between-Subjects Factors, One Within-Subjects Factor

Description: This design is used when subjects in different groups (between-subjects, Factor A), are at the same time assigned to different levels of a second between-groups factor (Factor B), and all subjects in the A × B between-groups design are measured two or more times (within-subjects, Factor C). Such situations arise when subjects in a two-way factorial ANOVA are measured on repeated occasions. The basic arrangement for this design is illustrated below. Factor A, between-subjects, has 2 levels; Factor B, between-subjects, has 3 levels; and Factor C, within-subjects, has 3 levels.

Factor C, Within Subjects, 3 Levels

C-1 → C-2 → C-3

Factor A, Between S's, 2 Levels (1, 2) | Factor A, Between S's, 2 Levels (1, 2) | Factor A, Between S's, 2 Levels (1, 2)

Factor B Between Subjects 3 Levels

Applications: A three-way split-plot design with two between-subjects factors and one within-subjects factor is used when subjects in a two-way factorial ANOVA design are measured on different occasions. Such applications occur when subjects in a two-way design (e.g., gender × treatment), are measured before, during, and after some experimental treatment is applied.

Comments: For details, see entry **A4-82**.

Analysis: Three-Way Split-Plot Analysis of Variance, Two Between-Subjects Factors, One Within-Subjects Factor, **A4-82**.

A3-16

Category: Repeated-Measures and Split-Plot Analysis of Variance (ANOVA) Designs

Design: Four-Way Split-Plot Design, Two Between-Subjects Factors, Two Within-Subjects Factors

Description: This four-way split-plot design involves two between-subjects factors and two within-subjects factors. The between-subjects factors have the form of a two-way factorial analysis of variance. The two within-subjects factors indicate that subjects, in the two-way factorial design, are measured two or more times, on two or more occasions.

Applications: By way of illustration, this four-way design would be used in an experiment in which gender is crossed with treatment in the form of a two-way factorial ANOVA. Both gender and treatment are between-subjects factors. All subjects are measured over repeated trials of some tasks, and the measurements over trials occur before, during, and after the application of the experimental treatment conditions. Trials and occasions are within-subjects factors.

Comments: For details, see **A4-83**.

Analysis: Four-Way Split-Plot Analysis of Variance, Two Between-Subjects Factors, Two Within-Subjects Factors, **A4-83**

A3-17

Category: Nested (Hierarchical) Analysis of Variance (ANOVA) Designs

Design: Two-Factor Nested Design, Factor B Nested in Factor A

Description: Two-factor nested designs involve two independent variables. Only some of the levels of one variable appear under the levels of the other variable. For example, if Factor A has 2 levels, and Factor B nested in Factor A has 10 levels, then levels 1, 2, 3, 4, and 5 of Factor B might appear under level 1 of Factor A. Levels 6, 7, 8, 9, and 10 of Factor B might appear under level 2 of Factor A. This differs from a two-way factorial design with factors A and B. In a factorial design, all the levels of Factor B would appear under all the levels of Factor A.

The general arrangement of a two-factor nested design is illustrated below for Factor A with 2 levels and Factor B with 10 levels.

Factor A	
Level 1	Level 2
Factor B	Factor B
Level 1 □	Level 6 □
Level 2 □	Level 7 □
Level 3 □	Level 8 □
Level 4 □	Level 9 □
Level 5 □	Level 10 □

Applications: Two-factor nested designs are used when only some of the levels of one factor appear under the levels of the other factor. This situation commonly occurs when experimental treatments are administered to different sets of subjects. The example illustrated above could be such a situation.

The illustration applies if Factor A represents levels of treatment, with level 1 being an experimental treatment and level 2 being a control. The 10 levels of Factor B are classrooms, with classrooms 1 through 5 receiving the treatment, and classrooms 6 through 10 acting as the control. Two-factor nested designs are common in educational research when new instructional methods or curriculum materials are assigned to experimental classes and other classes serve as controls.

Comments: Two-factor nested designs, with classrooms nested within treatment conditions, are common in education. These designs are frequently analyzed incorrectly as a one-way analysis of variance, with treatment conditions as the one independent variable. For details, see **A4-84**.

Analysis: Two-Factor Nested Analysis of Variance, Factor B Nested in Factor A, **A4-84**

A3-18

Category: Nested (Hierarchical) Analysis of Variance (ANOVA) Designs

Design: Three-Factor Nested Design, Factor C Nested in Factor B, Factor B Nested in Factor A

Description: Three-factor nested designs involve a double nesting. One factor is nested within a second, and the second is nested in the third. The basic arrangement for such a design follows. Factor A has 2 levels, Factor B has 4 levels, and Factor C has 12 levels.

Factor C, levels 1, 2, and 3, are nested under level 1 of Factor B. Factor C, levels 4, 5, and 6, are nested under level 2 of Factor B. Factor C, levels 7, 8, and 9, are nested under Factor B, level 3. Factor C, levels 10, 11, and 12, are nested under Factor B, level 4. Factor B, levels 1 and 2, are nested under Factor A, level 1. Factor B, levels 3 and 4, are nested under Factor A, level 2.

```
                              Factor A
             ┌──────────────────────┬──────────────────────┐
                Factor A, Level 1       Factor A, Level 2
             ┌──────────────────────┐┌──────────────────────┐
                              Factor B
             ─────────────┬─────────────┬─────────────┬─────────────
               B, Level 1    B, Level 2    B, Level 3    B, Level 4
             ┌──────────┐ ┌──────────┐ ┌──────────┐ ┌──────────┐
  Factor C    □  □  □     □  □  □     □  □  □     □  □   □
               1  2  3     4  5  6     7  8  9    10 11  12
             └─────────────── Factor C, 12 Levels ────────────────┘
```

Applications: The illustration above will be used to describe an application of this design in educational research. In this sample application, Factor A is treatment, with 2 levels (a new experimental curriculum and the current curriculum). Factor B is schools, with 4 levels. Two schools, levels 1 and 2 of factor B, are assigned to use the new curriculum, and two schools, levels 3 and 4 of Factor B, are assigned the current curriculum. There are 12 different teachers in the study. Teachers are Factor C. Three different teachers are nested under each of the 4 levels of Factor B.

Comments: For details, see **A4-85**

Analysis: Three-Factor Nested Analysis of Variance, Factor C Nested in Factor B, Factor B Nested in Factor A, **A4-85**

A3-19

Category: Nested (Hierarchical) Analysis of Variance (ANOVA) Designs

Design: Three-Factor Design, Two Factors Nested, One Factor Crossed

Description: The three-factor design with two factors nested and one factor crossed is a combination of a two-factor nested design and a factorial design. In this design, one factor is crossed with the second, and the second is nested under the third. The basic arrangement for this type of design is illustrated below. In this illustration, Factor C, with 2 levels, is crossed with Factor B, with 6 levels. These factors are crossed because all levels of Factor C appear under all levels of Factor B. Factor B is nested in Factor A, with 2 levels. These factors are nested since only some of the levels of Factor B appear under some of the levels of Factor A. Levels 1, 2, and 3 of Factor B are nested under level 1 of Factor A. Levels 4, 5, and 6 of Factor B are nested under level 2 of Factor A.

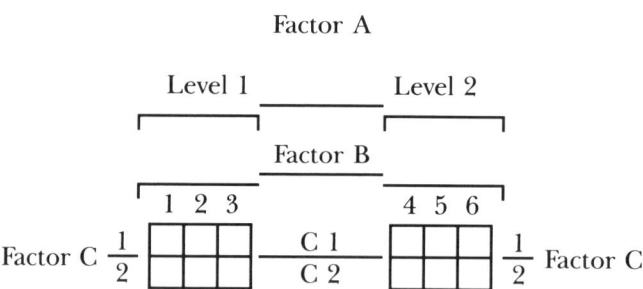

Applications: The illustration will be used to describe an application of this design in educational research. Factor A is Treatment, with 2 levels (a new experimental curriculum and the current curriculum). Factor B represents teachers, each of whom teaches a group of students using one of the two curriculum approaches. Teachers 1, 2, and 3 use the new curriculum, while teachers 4, 5, and 6 use the current curriculum. Factor C is students' gender, which is crossed with "teacher," since the teachers all have female and male students in their classes. This design allows the researcher to examine differences in the impact of the treatments, gender differences, and the interaction of gender and treatment.

Comments: For details, see **A4-86.**

Analysis: Three-Factor Analysis of Variance, Two Factors Nested, One Factor Crossed, **A4-86**

A3-20

Category: Randomized-Blocks Designs

Designs: Randomized-Blocks Designs in General

Description: Randomized-blocks designs represent a large class of research designs in which an independent variable is used as a blocking variable to control the influence of a possible extraneous variable and to increase the statistical power of the analysis. A blocking variable can be added to almost any type of analysis of variance design. A blocking variable is selected based on the researcher's belief that it is related, possibly highly related, to the dependent measure. A general procedure for a simple randomized-blocks design begins by ordering the subjects on the blocking variable. Subjects who are adjacent to each other in this ordering are then grouped into blocks. In some cases, the number of subjects in each block is equal to the number of treatment conditions. In other cases, the number of subjects in each block is a multiple of the number of treatments. Subjects are then randomly assigned from each block to each treatment group. As a result of this procedure, each group has an equal number of subjects from each block and each group has subjects from all blocks. In this way, the effects of the blocking variable are spread evenly across all treatment groups. Thus, differences between groups cannot be due to the blocking variable.

It is imperative that the blocking variable be incorporated as an independent variable (factor) in the statistical analysis of a randomized-blocks design. Variation on the dependent measure attributable to the blocking variable is removed from the estimate of random error in the analysis. This increases the statistical power of the analysis since the estimate of random error (reduced) is the denominator in testing hypotheses about treatment effects. The use of a randomized-blocks design without incorporating the blocking variable into the statistical analysis inflates the estimate of random error because it increases within-cell variance. This reduces the power of the statistical analysis.

Applications: Randomized-blocks designs can be incorporated into virtually any type of analysis of variance design by adding the blocking factor to the design and analysis. Randomized-blocks designs are used when the researcher wants to control for the effects of a confounding variable and increase the statistical power of the analysis.

Comments: For a description of additional applications and details, see entry **A4-87**.

Analysis: Randomized-Blocks Designs, **A4-87**

A3-21

Category: Latin Square Designs

Designs: Latin Square Designs in General

Description: Latin square designs represent a large class of analysis of variance procedures used to control for potential sources of extraneous variation. Latin square designs incorporate as independent variables factors which otherwise might confound the analysis of the effects due to some experimental treatment. The effects due to the order in which treatment conditions are presented, and the effects due to the practice students receive under different treatment conditions, are often controlled using Latin square designs. These designs are analyzed as fixed-effects, factorial Analysis of Variance models with no interactions among the independent variables.

Applications: Latin square designs are commonly used to control variation due to order, sequence, practice effects, and other possible confounding effects in experimental studies.

Comments: For details, see **A4-88**.

Analysis: Latin Square Designs, **A4-88**

A3-22

Category: Analysis of Covariance (ANCOVA)

Designs: Analysis of Covariance Designs in General

A3-17

Description: Analysis of covariance represents a large class of designs that are structurally identical to analysis of variance designs but include an additional variable. The additional variable is a continuous independent variable. This continuous independent variable is called a covariate. The covariate is systematically related to the dependent measure. The variation in the dependent measure that can be attributed to the covariate is estimated and thus removed from the estimate of random error. This reduces the size of the estimate of random error, which is used as the denominator in testing various hypotheses. The statistical power of the design is thereby increased.

Applications: Analysis of covariance is used in a wide range of educational research situations. It is commonly used in pretest-posttest design with the pretest as a covariate. Scores on alternate forms of the dependent measure can be used as covariates.

Analysis of covariance is sometimes used to adjust for sampling bias or sampling error. This application can be misleading because it statistically creates groups that are equivalent on the covariate. However, the groups in question might rarely, if ever, be equivalent on the covariate in actual practice. Thus, inferences based on an analysis of covariance generalize to the statistical condition created by the procedure but may have no real-life applicability.

Comments: For details, see **A4-89**.

Analysis: Analysis of Covariance, **A4-89**

References and Resources

Campbell, D. T., and Stanley, J. C. *Experimental and Quasi-Experimental Designs for Research.* Chicago: Rand McNally, 1966.

Cook, T. D., and Campbell, D. T. *Quasi-Experimentation: Design and Analysis Issues for Field Settings.* Boston: Houghton Mifflin, 1979.

Keppel, G. *Design and Analysis: A Researcher's Handbook.* 2d edition. Englewood Cliffs, NJ: Prentice-Hall, 1982.

Kirk, R. E. *Experimental Designs: Procedures for the Behavioral Sciences.* 2d edition. Monterey, CA: Brooks/Cole Publishing Co., 1982.

Leedy, P. D. *Practical Research: Planning and Design.* 3d edition. New York: Macmillan Publishing Company, 1985.

Spector, P. E. *Research Designs.* Sage University Paper series on Quantitative Applications in the Social Sciences, 07-023. Beverly Hills and London: Sage Publications, 1981.

CHAPTER 4

A Summary of Statistical Procedures

Introduction

The purpose of this chapter is to provide the reader with a concise summary of the statistical procedures commonly used in educational, psychological, and sociological research. It is designed from an operational point of view for easily familiarizing the reader with the purpose and application of each statistical procedure, its limitations, and other essential characteristics. This chapter's scope and arrangement offer a wide range of statistical topics to serve the student both during his or her period of training in statistical procedures and as a reference and review source after the student has become a practicing researcher.

In this chapter, a balance is struck between breadth and depth of content coverage. A broad range of statistical procedures is covered at a depth designed to be useful and informative. Each statistical procedure is presented as a separate "entry," consistent with the format used throughout the book. Each entry contains eight sections:

1. CATEGORY—basic classification for the procedure
2. STATISTIC or STATISTICAL PROCEDURE—the name of the procedure
3. DESCRIPTION—basic information about the procedure
4. APPLICATION(S)—research situations in which the procedure is used and how it is used
5. DATA—the nature of the data to which the procedure is applied
6. LIMITATIONS—assumptions and weaknesses of the procedure
7. COMMENTS—additional information
8. FORMULA—the statistical expression for the procedure

Computer Applications

The effective use of this chapter assumes that a researcher has access to a mainframe or microcomputer. This assumption has influenced the way certain procedures are described. For example, in presenting a formula for many procedures there was often a choice between an expression that reflected the statistical concept of the procedure and an expression that would be used if the procedure were being applied by hand. The expression of the basic statistical concept is presented for two reasons. First, for most procedures it is adequate for actual applications. Second, in most research situations actual calculations are performed with a computer or a powerful calculator with basic statistical functions.

The description of many procedures, especially different analysis of variance procedures, assumes that the researcher will use one of the available computer packages for actual applications. *BMDP* (Dixon, 1981), *SAS* (SAS Institute Inc., 1985), and *SPSS* (Nie, Hull, Jenkins, Steinbrenner, and Brent, 1975) are among the many available packages that might be used. Thus, the formula sections of the analysis of variance procedures emphasize showing what term goes in the numerator and what term goes in the denominator to test hypotheses about different factors. The actual computations are left to the computer.

Content Overview

Ninety-seven statistical procedures, organized into nine general categories, are described in this chapter. A complete listing of the procedures is presented here to serve as a chapter overview and directory.

These nine general categories of statistical procedures are assigned entry numbers and appear in the text with a listing of the statistical procedures for each category. These category listings serve as subdirectories to facilitate cross-referencing and using the book's index.

(A4-1) DESCRIPTIVE STATISTICS

Central Tendency

(A4-2) Mean
(A4-3) Median
(A4-4) Mode

Variability or Dispersion

(A4-5) Range
(A4-6) Mean Deviation
(A4-7) Variance
(A4-8) Standard Deviation

Disarray (Nonparametric)

(A4-9) Sum of D-squared
(A4-10) S

Distribution Shapes

- **(A4-11)** Frequency Distributions—Raw and Proportional Frequencies and Cumulative Frequencies
- **(A4-12)** Skewness
- **(A4-13)** Kurtosis

(A4-14) MEASURES OF RELATIONSHIPS

- **(A4-15)** Covariance
- **(A4-16)** Pearson Product-Moment Correlation
- **(A4-17)** Spearman's Rank Order Correlation—Rho
- **(A4-18)** Kendall's Tau
- **(A4-19)** Point Biserial Correlation
- **(A4-20)** Biserial Correlation
- **(A4-21)** Phi-Coefficient
- **(A4-22)** Tetrachoric Correlation
- **(A4-23)** Contingency Coefficient
- **(A4-24)** Kendall's Coefficient of Concordance
- **(A4-25)** Intraclass Correlation
- **(A4-26)** Part Correlation
- **(A4-27)** Partial Correlation
- **(A4-28)** Multiple Correlation
- **(A4-29)** Eta-squared
- **(A4-30)** Factor Analysis

(A4-31) REGRESSION ANALYSIS

- **(A4-32)** Simple Linear Regression
- **(A4-33)** Regression Coefficient
- **(A4-34)** Y-Intercept
- **(A4-35)** Standard Error of Estimate
- **(A4-36)** Multiple Regression
- **(A4-37)** R-squared
- **(A4-38)** Curvilinear Regression
- **(A4-39)** Stepwise Regression

(A4-40) COMPARISONS: ONE SAMPLE

- **(A4-41)** Sample Mean Compared to Some Value
- **(A4-42)** Sample Variance Compared to Some Value
- **(A4-43)** Sample Proportion Compared to Some Value
- **(A4-44)** Sample Correlation Compared to 0
- **(A4-45)** Sample Correlation Compared to Some Value

(A4-46) Chi-square Test for Goodness of Fit
(A4-47) Kolmogorov-Smirnov One-Sample Test

(A4-48) COMPARISONS: TWO SAMPLES (PARAMETRIC)

(A4-49) Comparing Two Sample Means—Independent Samples
(A4-50) Comparing Two Sample Means—Dependent or Correlated Samples
(A4-51) Comparing Two Sample Variances—Independent Samples
(A4-52) Comparing Two Sample Variances—Dependent or Correlated Samples
(A4-53) Comparing Two Sample Proportions—Independent Samples
(A4-54) Comparing Two Sample Proportions—Dependent or Correlated Samples
(A4-55) Comparing Two Sample Correlations—Independent Samples
(A4-56) Comparing Two Sample Correlations—Dependent or Correlated Samples

(A4-57) COMPARISONS: TWO SAMPLES (NONPARAMETRIC)

(A4-58) Median Test—Sign Test for Independent Samples
(A4-59) Wilcoxon Rank Sum Test—Rank Test for Independent Samples
(A4-60) Mann-Whitney U-Test—Signed Rank Test for Independent Samples
(A4-61) Chi-Square Test of Independence (Homogeneity)
(A4-62) Kolmogorov-Smirnov Two-Sample Test Comparing Distributions
(A4-63) McNemar's Test for Significance of Change—Dependent or Correlated Samples
(A4-64) Sign Test for Dependent Samples
(A4-65) Wilcoxon Matched-Pairs Signed-Rank Test for Dependent Samples

(A4-66) COMPARISONS: TWO OR MORE SAMPLES (NONPARAMETRIC)

(A4-67) Sign Test for K-Independent Samples
(A4-68) Kruskal-Wallis Rank Test for K-Independent Samples
(A4-69) Friedman Two-Way ANOVA by Ranks for Dependent or Correlated Samples

(A4-70) COMPARISONS: TWO OR MORE SAMPLES —ANALYSIS OF VARIANCE (ANOVA)

Factorial Analysis of Variance

(A4-71) One-Way Analysis of Variance
(A4-72) Two-Way Analysis of Variance—Fixed Effects Model
(A4-73) Two-Way Analysis of Variance—Random Effects Model

(A4-74) Two-Way Analysis of Variance—Mixed Effects Model
(A4-75) Three-Way Analysis of Variance—Fixed Effects Model
(A4-76) Three-Way Analysis of Variance—Mixed Model, One Factor Fixed, Two Factors Random
(A4-77) Three-Way Analysis of Variance—Mixed Model, Two Factors Fixed, One Factor Random
(A4-78) Three-Way Analysis of Variance—Random Effects Model

Repeated-Measures Analysis of Variance/Split-Plot Analyses

(A4-79) One-Way Repeated Measures—Subjects by Occasions
(A4-80) Two-Way Split Plot, One Between-Subjects Factor and One Within-Subjects Factor
(A4-81) Three-Way Split Plot, One Between-Subjects Factor and Two Within-Subjects Factors
(A4-82) Three-Way Split Plot, Two Between-Subjects Factors and One Within-Subjects Factor
(A4-83) Four-Way Split Plot, Two Between-Subjects Factors and Two Within-Subjects Factors

Nested/Hierarchical Analysis of Variance

(A4-84) Two-Factor Designs, B Nested in A
(A4-85) Three-Factor Designs, C Nested in B, B Nested in A
(A4-86) Three-Factor Designs, B Nested in A, C Crossed with B Nested in A

General Analysis of Variance Designs

(A4-87) Randomized-Blocks Designs
(A4-88) Latin Square Designs

Related Designs

(A4-89) Analysis of Covariance
(A4-90) Time-Series Analysis

(A4-91) MULTIPLE-COMPARISON PROCEDURES

A Priori Planned Comparisons

(A4-92) Orthogonal Contrasts
(A4-93) Orthogonal Polynomial Contrasts

Post Hoc Comparisons

(A4-94) Scheffé Contrasts
(A4-95) Tukey's Honestly Significant Difference

Profiles of Statistical Procedures

A4-1

DESCRIPTIVE STATISTICS

Central Tendency
- (A4-2) Mean
- (A4-3) Median
- (A4-4) Mode

Variability or Dispersion
- (A4-5) Range
- (A4-6) Mean Deviation
- (A4-7) Variance
- (A4-8) Standard Deviation

Disarray (Nonparametric)
- (A4-9) Sum of D-squared
- (A4-10) S

Distribution Shapes
- (A4-11) Frequency Distributions—Raw and Proportional Frequencies and Cumulative Frequencies
- (A4-12) Skewness
- (A4-13) Kurtosis

A4-2

Category: Descriptive, Measure of Central Tendency

Statistic: Mean (or Arithmetic Mean)

Description: The arithmetic average value of a variable in a sample or population.

Application: Used when the performance of a group is to be represented by a single value.

Data: Used with data that are interval or ratio level and normally or otherwise symmetrically distributed around the mean.

Limitations: Not an appropriate measure of central tendency if the data are skewed (A4-12). Sensitive to extreme cases.

Comments: Generally the statistic referred to when the term "average" is used.

Formula:

$$\overline{X} = \frac{\sum_{i=1}^{n} X_i}{n}$$

(A4-96) Newman-Keuls Test
(A4-97) Duncan Multiple Range Test

A4-3

Category: Descriptive, Measure of Central Tendency

Statistic: Median

Description: The value of a variable such that half the values in the distribution are above and half the values in the distribution are below. The value of a variable obtained by the subject in the 50th percentile.

Application: Used when the performance of a group is to be represented by a single value and the data are not normally or symmetrically distributed.

Data: Used when data are ordinal, interval, or ratio level. Used with raw distributions or frequency distributions.

Limitations: Can be difficult to calculate. The median does not reflect performance on the extreme ends of the distribution.

Comments: The median is not sensitive to the influence of extreme cases on either end of the distribution. Commonly used measure of central tendency when data are known to be skewed.

Formula: Median for Grouped Data

$$\text{Median} = L + \frac{(n/2) - F}{f_m} I$$

where
L = exact lower limit of interval containing median
n = number of cases
F = sum of all frequencies below L
I = width of the class intervals, and
f_m = frequency of the interval containing the median

A4-4

Category: Descriptive, Measure of Central Tendency

Statistic: Mode

Description: The most frequently occurring value of a variable in a distribution.

Application: The mode is used as a rough estimate of central tendency. Used to identify "typical" performance.

Data: Used with data that are nominal, ordinal, interval, or ratio.

Limitations: With small samples, the mode is very unstable.

Comments: Distributions can be unimodal, bimodal, or have no mode if all values of the variable occur equally often. The only measure of central tendency appropriate for nominal data.

Formula: Often determined by inspection without calculation.

A4-5

Category: Descriptive, Variability or Dispersion

Statistic: Range

Description: The difference between the highest and lowest values of a variable.

Application: Used as a rough estimate of variability with small samples.

Data: Used with ordinal, interval, or ratio-level data.

Limitations: The range is sensitive to extreme values in the distribution, generally unstable, and influenced by sample size.

Comments: Useful as a rough, first estimate of variability; more useful with small samples than large samples.

Formula:

$$\text{Range} = (\text{Largest Value} - \text{Smallest Value}) + 1$$

A4-6

Category: Descriptive, Variability or Dispersion

Statistic: Mean Deviation

Description: The average of the absolute values of the deviations from the mean.

Application: An estimate of the average deviation around the mean.

Data: Can be used with data that are interval or ratio level.

Limitations: Not easily manipulated algebraically; effected by extreme scores.

Comments: Used frequently when teaching the concept of variability, but rarely used in statistical practice.

Formula:

$$\text{Mean Deviation} = \frac{\sum_{i=1}^{n} |X_i - \overline{X}|}{n}$$

A4-7

Category: Descriptive, Variability or Dispersion

Statistic: Variance

Description: The average of the squared deviations from the mean.

Application: Used as a common index of variability in a wide range of situations, it is the basic element in a general set of procedures known as Analysis of Variance (ANOVA) **(A4-70)**.

Data: Used with interval or ratio data.

Limitations: Not in the metric or scale of the original data since all deviations from the mean have been squared. Sensitive to extreme values.

Comments: Dividing the sum of squared deviations by n yields a biased estimate of the population variance (systematically smaller than the true population variance); dividing the sum of squared deviations by $(n - 1)$ yields an unbiased estimate of the population variance.

Formula:

$$S^2 = \frac{\sum_{i=1}^{n} (X_i - \overline{X})^2}{n - 1}$$

A4-8

Category: Descriptive, Variability or Dispersion

Statistic: Standard Deviation

Description: The square root of the variance.

Application: Used as a standard index of variability.

Data: Used with interval or ratio level data.

Limitations: Sensitive to extreme cases, but otherwise broadly useful in describing the typical level of dispersion.

Comments: One of the most common measures of variability; preserves the original scale by taking the square root of the sum of squared deviations. Provides a standard unit or scale for interpreting departures from the mean.

Formula:

$$S = \sqrt{S^2} = \sqrt{\frac{\sum_{i=1}^{n} (X_i - \overline{X})^2}{n - 1}}$$

A4-9

Category: Descriptive, Measure of Disarray

Statistic: Sum of D-squared

Description: The sum of the squared differences in the rankings of subjects on two variables.

Application: Used as an index of differences in rank orderings; roughly analogous to the variance for interval or ratio data.

Data: Used when both variables are ordinal level.

Limitations: Is difficult to interpret.

Comments: Has a minimum value of 0; a maximum value of $(N(N^2 - 1))/3$; a random value of $(N(N^2 - 1))/6$.

Formula:

$$\sum d^2 = \sum_{i=1}^{n} (Rank_i \text{ (variable 1)} - Rank_i \text{ (variable 2)})^2$$

A4-10

Category: Descriptive, Measure of Disarray

Statistic: S

Description: The sum of the weights $+1$ and -1 assigned for each of $(N(N-1))/2$

ordered pairs for rankings on some variable, assigning +1 for a pair in natural order and −1 for a pair in reverse order.

Application: Used as an index of disarray in rank orderings; roughly analogous to the variance for interval or ratio level data.

Data: Ordinal

Limitations: Difficult to interpret.

Comments: Has a minimum value of $-(N(N-1))/2$; a maximum value of $+(N(N-1))/2$; and a random value of 0.

Formula:

$$S = \sum_{i=1}^{(N(N-1))/2} \text{Weight}_i, \text{ (for all pairs of values on variable Y)},$$

weight = +1, if pair of values is in natural order
weight = −1, if pair of values is in reverse order

A4-11

Category: Descriptive, Shapes of Distributions

Statistic: Raw and Proportional Frequencies and Cumulative Frequencies

Description: A basic descriptive device for displaying the distribution of some variable in table form. The table lists the frequency with which each value of the variable occurs in terms of raw and proportional frequency, and also commonly lists the raw and proportional cumulative frequency. Values of the variable are often grouped into class intervals.

Application: The frequency distribution is the most basic form of descriptive information. The mode and median generally can be read directly or easily estimated from the frequency table. The basic shape of the distribution also can be seen or approximated from the frequency table.

Data: The categories or class intervals used for the frequency distribution are discrete and may represent single values or intervals containing a range of values for the variable in question.

Limitations: With certain samples, the choice of class intervals can influence the general "picture" presented by the frequency distribution. Other descriptive measures should be used along with the examination of the frequency distribution.

Comments: The researcher interested in getting a basic description of a set of data should always examine the frequency distribution along with corresponding visual representations, such as the histogram, frequency polygon, and cumulative frequency polygon. The frequency distribution will provide a general picture of the distribution's skewness **(A4-12)** and kurtosis **(A4-13).**

Formula: The table below shows a standard presentation of a frequency distribution for 200 subjects on a variable that ranges from 1 to 70, grouped into class intervals of 5.

Sample Frequency Distribution

Class Interval	Raw Frequency	Proportional Frequency	Raw Cumulative Frequency	Proportional Cumulative Frequency
66–70	11	.055	200	1.000
61–65	16	.080	189	.945
56–60	19	.095	173	.865
51–55	29	.145	154	.770
46–50	26	.130	125	.625
41–45	27	.135	99	.495
36–40	22	.110	72	.360
31–35	18	.090	50	.250
26–30	11	.055	32	.160
21–25	9	.045	21	.105
16–20	7	.035	12	.060
11–15	4	.020	5	.025
6–10	1	.005	1	.005
1–5	0	.000	0	.000

A4-12

Category: Descriptive, Shape of Distribution

Statistic: Skewness

Description: An index of a distribution's asymmetry about the mean. The skewness is reflected in the average value of the cubed deviations from the mean, also known as the third moment about the mean. This index of skewness is generally standardized to be independent of the variable's scale.

Application: Commonly used to indicate whether or not the distribution of some variable is normal. In a normal distribution, the skew is 0.

Data: Used when the data are interval or ratio.

Limitations: Can be an artifact of the scale being used, especially if the scale has a limited range.

Comments: Very important in the application of inferential procedures that assume the dependent variable is normally distributed.

Formula:

$$\text{Skewness} = g_1 = \frac{m_3}{m_2 \sqrt{m_2}},$$

where

$$m_3 = \frac{\sum_{i=1}^{n} (X_i - \overline{X})^3}{n}, \text{ and}$$

$$m_2 = \frac{\sum_{i=1}^{n} (X_i - \overline{X})^2}{n}$$

A4-13

Category: Descriptive, Shape of Distribution

Statistic: Kurtosis

Description: An index of nonnormality in a distribution. Kurtosis indicates whether the distribution is nonnormally flat or peaked. The kurtosis is the average of the deviations from the mean raised to the fourth power and standardized.

Application: Used as an indication of normality in a distribution.

Data: Used with interval or ratio data.

Limitations: Can be an artifact of the scale being used, especially if the scale has a limited range.

Comments: A useful descriptive statistic especially in the application of inferential procedures assume the dependent variable is normally distributed.

Formula:

$$\text{Kurtosis} = g_2 = \frac{m_4}{m_2^2} - 3, \text{ where}$$

$$m_4 = \frac{\sum_{i=1}^{n}(X_i - \overline{X})^4}{n} \text{ and } m_2 = \frac{\sum_{i=1}^{n}(X_i - \overline{X})^2}{n}$$

A4-14

MEASURES OF RELATIONSHIPS

- **(A4-15)** Covariance
- **(A4-16)** Pearson Product-Moment Correlation
- **(A4-17)** Spearman's Rank Order Correlation–Rho
- **(A4-18)** Kendall's Tau
- **(A4-19)** Point Biserial Correlation
- **(A4-20)** Biserial Correlation
- **(A4-21)** Phi-Coefficient
- **(A4-22)** Tetrachoric Correlation
- **(A4-23)** Contingency Coefficient
- **(A4-24)** Kendall's Coefficient of Concordance
- **(A4-25)** Intraclass Correlation
- **(A4-26)** Part Correlation
- **(A4-27)** Partial Correlation
- **(A4-28)** Multiple Correlation
- **(A4-29)** Eta-squared
- **(A4-30)** Factor Analysis

A4-15

Category: Measures of Relationship

Statistic: Covariance

Description: A statistic that reflects the degree to which two variables are related. The covariance is the average product of deviation scores for two variables.

Application: Used as a general index of relationship between two variables and in the calculation of many other statistics.

Data: Used with interval or ratio data.

Limitations: The covariance is calculated and reported in the unstandardized scale of the original variables. Hence, comparisons among the covariances for different pairs of variables are difficult.

Comments: Not a commonly reported statistic. The covariance is standardized by dividing by the product of the standard deviations of the two variables involved. The standardized covariance is the Pearson product-moment correlation coefficient.

The covariance (**A4-15**), the statistical measure of the relationship between two variables, is often confused with the analysis of covariance (**A4-89**), a statistical procedure for testing hypotheses about differences between group means.

Formula:

$$S_{xy} = \frac{\sum_{i=1}^{n} (X_i - \overline{X})(Y_i - \overline{Y})}{n - 1}$$

A4-16

Category: Measures of Relationship

Statistic: Pearson Product-Moment Correlation Coefficient

Description: The average product of the deviation scores for two variables, divided by the product of their standard deviations. This is the same as the average product of the standardized (Z) scores for two variables.

Application: An informative and very commonly used measure of the relationship between two variables. Perhaps the most well-known measure of relationship. Generally, the statistic referred to when the phrase "the correlation" is used.

Data: Used with interval or ratio data.

Limitations: Reflects only linear relationships. Assumes both variables are normally distributed.

Comments: Generally described as ranging from -1 to $+1$. Can approximate these minimum and maximum values only when the assumption is true that both variables are normally distributed.

Formula:

$$r_{xy} = \frac{\sum_{i=1}^{n} (X_i - \overline{X})(Y_i - \overline{Y})}{(n-1) S_x S_y}$$

A4-17

Category: Measures of Relationship

Statistic: Spearman's Rank Order Correlation—Rho

Description: A special case of the Pearson product-moment correlation in which both variables are rankings on an ordinal scale.

Application: Commonly used as a measure of relationship when subjects are ranked on two variables.

Data: Both variables ordinal rankings.

Limitations: Some loss of information because ordinal level data, as opposed to interval or ratio data, are used.

Comments: Equivalent to the Pearson product-moment correlation except that the data are ranks. Tied ranks are assigned the average rank for the rank order position in which they fall.

Formula:

$$\rho = 1 - \frac{6 \sum_{i=1}^{n} d_i^2}{n(n^2 - 1)}$$

where $\sum d_i^2$ is the sum of the squared differences in rank **(A4-9)**

A4-18

Category: Measures of Relationship

Statistic: Kendall's Tau

Description: A measure of relationship for two variables ranked on ordinal level scales; Tau is the measure of disarray, S, divided by the maximum value S can take on **(A4-10).**

Application: An alternative to Spearman's Rho.

Data: Both variables ordinal rankings.

Limitations: Some information is lost because ordinal scales are used.

Comments: Tau is an ordering statistic and does not have the same meaning as product-moment correlations. Tied ranks are assigned the average rank order position into which they fall.

Formula:

$$\tau = \frac{S}{\{n(n-1)\}/2}$$

where S, a measure of disarray,
is defined as in **A4-10,**
n = number of S's

A4-19

Category: Measures of Relationship

Statistic: Point Biserial Correlation

Description: A special case of the Pearson product-moment correlation in which one variable is measured on an interval or ratio scale and the other variable is nominal and dichotomous with no underlying continuous normal distribution.

Application: Used to examine relationships between discrete dichotomous variables and continuous variables.

Data: One variable interval or ratio, the other variable nominal, discrete dichotomous.

Limitations: The direction of the correlation (+ or −) is arbitrary depending on how the values of the dichotomous variable are assigned.

Comments: Commonly used in measurement applications to correlate scored item responses (0 or 1) with total test score as an index of item discrimination.

Formula:

$$r_{pb} = \frac{\overline{X}_{\cdot 1} - \overline{X}_{\cdot 0}}{S_x} \sqrt{\frac{n_1 n_0}{n(n-1)}}, \text{ where}$$

$\overline{X}_{\cdot 1}$ = mean score on X of S's scoring 1 on variable Y
$\overline{X}_{\cdot 0}$ = mean score on X of S's scoring 0 on variable Y
S_x = Standard Deviation of X for all S's
n_1 = number of S's scoring 1 on Y
n_0 = number of S's scoring 0 on Y
n = total number of S's

A4-20

Category: Measures of Relationship

Statistic: Biserial Correlation

Description: A special case of the Pearson product-moment correlation in which one variable is measured on a continuous interval or ratio scale and the other variable is nominal and dichotomous. The dichotomous variable has an underlying continuous normal distribution, unlike the somewhat similar point biserial **(A4-19).**

Application: Used as a measure of relationship between a continuous variable and a dichotomous variable, when the dichotomous variable has an underlying continuous normal distribution. For example, students' SAT scores, a continuous variable, can be dichotomized into "above average" and "below average." The biserial correlation could be used to correlate SAT scores, in this dichotomous form, with students' grade-point average.

Data: One variable interval or ratio and the other variable is dichotomous with an underlying continuous normal distribution.

Limitations: Information is lost by representing a continuous normal variable as a dichotomous variable.

Comments: The biserial correlation is informative and useful when data from a dichotomous variable, with an underlying normal distribution, are readily available, or easy and inexpensive to collect.

54 RESEARCH PROCESSES

Formula:

$$r_{bis} = \left(\frac{\overline{X}_{\cdot 1} - \overline{X}_{\cdot 0}}{S_x}\right) \cdot \frac{n_1 n_0}{U n \sqrt{n^2 - n}} \text{ where}$$

$\overline{X}_{\cdot 1}$ = mean score on X for S's with a score of 1 on Y
$\overline{X}_{\cdot 0}$ = mean score on X for S's with a score of 0 on Y
S_x = standard deviation on X for all S's
n_1 = number of S's with a score of 1 on Y
n_0 = number of S's with a score of 0 on Y
n = total number of S's
U = Ordinate of the Z distribution, unit normal distribution, at the point above which is located $(n_n/n) \times 100$ percent of the area under the curve

A4-21

Category: Measures of Relationship

Statistic: Phi-Coefficient

Description: A measure of relationship between two variables used when both variables are discrete dichotomous variables. A special case of the Pearson product-moment correlation. The Phi-coefficient is calculated from a 2 × 2 contingency table with the following form:

		Variable 2		
		0	1	
Variable 1	1	A	B	A+B
	0	C	D	C+D
		A+C	B+D	N=A+B+C+D

Application: Used to measure the relationship between two nominal dichotomous variables.

Data: Both variables are measured on discrete dichotomous scales.

Limitations: Ranges between +1 and −1 only when the marginal proportions all equal .5. That is,

$$\frac{A+B}{N} = \frac{C+D}{N} = \frac{A+C}{N} = \frac{B+D}{N} = .5$$

Comments: The direction of the correlation, either + or −, is arbitrary and must be interpreted carefully inspecting the arrangement of the contingency table. The Phi-coefficient can be converted into a 1-degree of freedom Chi-square by multiplying the squared Phi-coefficient by N.

Formula:

$$\phi = \frac{BC - AD}{\sqrt{(A + C)(B + D)(C + D)(A + B)}}$$

A4-22

Category: Measures of Relationship

Statistic: Tetrachoric Correlation Coefficient

Description: A measure of relationship between two variables used when both variables are dichotomous but have underlying continuous normal distributions.

Application: Used as a measure of relationship between two variables that are thought to have continuous normal distributions but are measured on dichotomous scales. This is a commonly used statistic in studying the relationship between two test items. In such a situation, the data are configured as follows:

	Item 2		
	0	1	
Item 1 1	A	B	A+B
Item 1 0	C	D	C+D
	A+C	B+D	N=A+B+C+D

Data: Both variables have underlying continuous normal distributions but are operationalized on dichotomous scales.

Limitations: Information is lost when continuous normal variables are dichotomized.

Comments: The tetrachoric correlation coefficient was often used to facilitate computation of inter-item correlations before the widespread availability of computing devices. It is used when continuous data are difficult or expensive to collect on the variables of interest.

Formula:

$$r_{tet} = \text{cosine} \frac{180°}{1 + \sqrt{BC/AD}}$$

Where A, B, C, and D are defined as in the Application section above. Many statistics books have tables for the value of the tetrachoric correlation as a function of BC/AD.

A4-23

Category: Measures of Relationship

Statistic: Contingency Coefficient, C

Description: A measure of relationship when both variables are measured on scales that are discrete and nominal but not dichotomous.

Application: Used as an index of correlation for nominal variables that have two or more categories.

Data: Both variables are discrete, nominal, and have two or more categories.

Limitations: Does not range between −1 and +1. The direction of the correlation coefficient is arbitrary and must be determined from the contingency table. Contingency coefficients are comparable only for tables with the same numbers of rows and columns.

Comments: The maximum value of C is estimated by the square root of $((k - 1)/k)$, where k is the number of categories in the variable with the fewest categories.

Formula:

$$C = \sqrt{\frac{\chi^2}{N + \chi^2}}$$

where χ^2 is the Chi-square for the contingency table for the two discrete variables

A4-24

Category: Measures of Relationship

Statistic: Kendall's Coefficient of Concordance, W

Description: A measure of relationship or agreement among multiple rankings of a set of subjects on some criterion. Generally involves three or more rankings.

Application: Used as an index of similarity for multiple rankings of the same elements. Commonly used as an index of agreement among raters who are ranking several subjects or objects.

Data: Elements are arranged on an ordinal scale.

Limitations: Not directly comparable to correlation coefficients.

Comments: W = 0 when maximum disagreement among raters occurs, and W = 1 when maximum agreement occurs.

Formula:

$$W = \frac{12S}{m^2(n^3 - n)}$$

where

$$S = \sum_{j=1}^{n} \left(R_j - \frac{\sum R_j}{n} \right)^2$$

R_j = sum of the ranks assigned to the jth person,
m = number of judges
n = number of S's being ranked

A4-25

Category: Measures of Relationship

Statistic: Intraclass Correlation

Description: The correlation between two interval or ratio-level variables for subjects in different groups. The intraclass correlation is generally viewed from an analysis-of-variance perspective **(A4-71).** It is the correlation between two variables for subjects within groups, for all groups of subjects.

Application: Often used in studies of twins. Used for correlating variables within classrooms with different teachers and with nested designs **(A3-17).**

Data: Used with interval or ratio scales.

Limitations: Can be difficult to interpret and explain. Computationally more complex than typical correlation coefficients.

Comments: Generally conceptualized and calculated within an analysis-of-variance framework or component-analysis framework. The intraclass correlation is basically the within-group correlation removing between-group variation.

Formula:

$$r_I = \frac{MSB - MSW}{MSB + (n - 1) MSW}$$

where
MSB = Mean Square Between Groups
MSW = Mean Square Within Groups
n = number of S's in Groups

A4-26

Category: Measures of Relationship

Statistic: Part Correlation

Description: The correlation between two variables when the effect of some third variable has been removed from one of the two variables being correlated. This differs from the partial correlation **(A4-27),** which is the correlation between two variables with the effect of some third variable removed from both variables being correlated.

Application: Used to examine the influence of some variable on the correlation between two other variables, and to examine correlations removing the influence of a confounding variable.

Data: Used with interval or ratio scales.

Limitations: May result in a correlation based on a statistically created condition that does not have practical meaning.

Comments: Usually conceptualized in the context of regression analysis as the correlation between two variables, where one is the residual from the regression in which the control variable is the independent variable.

Formula:

$$r_{x(y \cdot z)} = \frac{r_{xy} - r_{xz} r_{yz}}{\sqrt{1 - r_{yz}^2}}$$

A4-27

Category: Measures of Relationship

Statistic: Partial Correlation

Description: The correlation between two variables removing from both the influence of a third variable. This differs from the part correlation (**A4-26**), which is the correlation between two variables when the effect of some third variable has been removed from one of the two variables being correlated.

Application: Used to examine the influence of some variable on the correlation between two other variables, and to examine correlations removing the influence of a confounding variable.

Data: Used with interval or ratio scales.

Limitations: May result in a correlation based on a statistically created condition that does not have any practical or real-life meaning.

Comments: Usually conceptualized in the context of regression analysis as the correlation between two variables both of which are residuals from separate regressions in which the control variable is the independent variable.

Formula:

$$r_{(xy) \cdot z} = \frac{r_{xy} - r_{xz} r_{yz}}{\sqrt{(1 - r_{xz}^2)(1 - r_{yz}^2)}}$$

A4-28

Category: Measures of Relationships

Statistic: Multiple Correlation, R

Description: The Pearson product-moment correlation between a criterion variable and its predicted value from a regression equation.

Application: Used as an index of how well the criterion variable is predicted.

Data: Used with interval or ratio scales.

Limitations: The multiple correlation is limited by the extent to which the variables in the regression equation meet the assumptions for regression analysis.

Comments: The multiple correlation is particularly useful when it is squared. R-squared is the ratio of the variance in the predicted values over the total variance. (See R-squared, **A4-37**.)

Formula:

$$R = r_{y \cdot \hat{y}} \quad \text{or}$$

$$R = \sqrt{\frac{\text{Sum of Squares Regression}}{\text{Sum of Squares Total}}} = \sqrt{\frac{\text{SS Reg}}{\text{SS Tot}}}$$

A4-29

Category: Measures of Relationship

Statistic: Eta-squared (Correlation Ratio)

Description: An index of relationship that does not assume a linear relationship. Eta-squared varies between 0 and 1 and reflects relationships that may be curvilinear. Eta-squared is also known as the correlation ratio. Eta-squared is comparable to R-squared **(A4-37)**, not simply to the correlation between two variables.

Application: Used when an examination of a scatterplot, previous research, or theoretical considerations suggest a nonlinear relationship.

Data: The Y variable is interval or ratio. The X variable can be any level, but Eta-squared is more interpretable if X is at least ordinal.

Limitations: Is not symmetric like other correlations. For example, Eta-squared (x,y) may not equal Eta-squared (y,x).

Comments: Very useful in describing many situations in which linear correlations indicate no relationship. It is essential to examine the scatterplot when interpreting Eta-squared.

Formula:

$$\eta_{yx}^2 = 1 - \frac{\text{Sum of Squares Within}}{\text{Sum of Squares Total}}$$

where S's are grouped on X values

A4-30

Category: Measures of Relationship

Statistical Procedure: Factor Analysis

Description: A large category of analytic procedures used to identify relationships among subsets of variables. In general, factor analysis is used to identify underlying traits or constructs called "factors" to which observed measures are related. Observed measures that are highly correlated to each other tend to be related to the same underlying factor. The correlations of the observed measures to the underlying factors are called factor loadings. The factors form a simplified structure within which the original measures can be plotted using a variety of procedures.

Applications: Factor analysis is used in a wide range of applications. In general, factor analysis is used to reduce the data from a large set of measures to a smaller set of factors that retain all the basic information from the measures, but does not reflect redundancies found in the original measures. Factor analysis can be used as an exploratory procedure to provide initial identification of factors and to identify commonalities among a set of measures. Factor analysis can also be used as a confirmatory procedure to verify theoretically or empirically derived constructs.

Data: Factor analysis is applied to the intercorrelations of the observed measures.

60 RESEARCH PROCESSES

Limitations: The details of factor analysis are difficult to explain and computationally complex. The results of factor-analysis procedures must be interpreted with special care and expertise. Technical expertise, alone, is not sufficient for providing appropriate interpretation of factor analyses.

Comments: The technical details of factor analysis are beyond the scope of this volume. The researcher interested in more details should examine Mulaik (1972) and Kim and Mueller (1978).

As suggested, exceptional care should be used in interpreting the results of factor analyses. Expertise with the variables, the constructs they measure, and the conceptual context within which the measures are made is necessary.

A4-31

REGRESSION ANALYSIS
- **(A4-32)** Simple Linear Regression
- **(A4-33)** Regression Coefficient
- **(A4-34)** Y-Intercept
- **(A4-35)** Standard Error of Estimate
- **(A4-36)** Multiple Regression
- **(A4-37)** R-squared
- **(A4-38)** Curvilinear Regression
- **(A4-39)** Stepwise Regression

A4-32

Category: Regression Analysis

Statistic: Simple Linear Regression

Description: A procedure for predicting some variable Y from variable X by using a linear weighting of X added to a constant. The weighting is the regression coefficient and the constant is the intercept.

Application: Commonly used to make predictions; to identify the proportion of one variable accounted for by knowledge of another; to estimate residual values on Y having removed the effects of X.

Data: Used with interval or ratio scales.

Limitations: Addresses only linear relationships. Uses only one predictor; therefore, it is not sensitive to more complex relationships. Assumes Y and X are normally distributed.

Comments: Despite some limitations, a very useful procedure for understanding the strength of the relationship.

Formula:

$$\hat{Y} = b_1 X + b_0, \text{ where}$$
$\hat{Y}$ = the predicted value of Y
b_1 = the regression coefficient **(A4-33)**
X = the independent variable
b_0 = the Y-intercept **(A4-34)**

A4-33

Category: Regression Analysis

Statistic: Regression Coefficient

Description: The regression coefficient is the amount of change in the dependent variable Y that occurs when the independent variable X changes by 1 scale unit. In simple linear regression, the regression coefficient is the slope of the "best fitting" line defined by the criterion of least squared error.

Application: The regression coefficient is the rate of change in Y as a function of X and is central to all regression analyses.

Data: Regression analysis assumes that both the Y and X variables are interval or ratio. The regression coefficient, a multiplier of X, is ratio in nature, (e.g., a regression coefficient of 4 is twice the size of a regression coefficient of 2).

Limitations: Regression coefficients from different analyses cannot be compared directly because they are scale specific.

Comments: The regression coefficient is the product of the ratio of the standard deviations of Y to X multiplied by the correlation between Y and X. When the standard deviations are equal, the regression coefficient and correlation coefficient are equal. The magnitude of the regression coefficient is directly proportional to the correlation coefficient.

Formula:

b_{yx} = the regression coefficient for predicting Y

$$b_{yx} = \frac{\sum_{i=1}^{n} (X_i - \overline{X})(Y_i - \overline{Y})}{\sum (X_i - \overline{X})^2}$$

also,

$$b_{yx} = \frac{S_y}{S_x} \cdot r_{xy}$$

A4-34

Category: Regression Analysis

Statistic: Y-Intercept

Description: In a regression analysis, the Y-intercept is the value of the predicted dependent variable, Y, when the independent predictor variable, X, is zero. Geometrically, the intercept is the place on the Y-axis that is crossed by the regression line.

Application: The Y-intercept is an integral part of any regression analysis since it defines the location or origin (X = 0) needed as the starting point for predicting Y.

Data: The Y-intercept is calculated and reported on the same scale as the Y variable.

Limitations: The Y-intercept is not very useful without the regression coefficient.

Formula:

$$b_0 = \overline{Y} - b_{yx} \overline{X}$$

A4-35

Category: Regression Analysis

Statistic: Standard Error of Estimate (S.E.)

Description: The standard error of estimate is the standard deviation of the residual difference between the observed and predicted values of the dependent variable in a regression analysis. This standard deviation of the residuals is the mean square due to error in the analysis of variance for the regression.

Application: The standard error of estimate is used to place confidence intervals of some desired range around predicted scores. For example, a predicted value plus and minus 1.96 × (S.E.) yields a 95% confidence interval.

Data: The standard error of estimate is reported in the scale of the predicted variable.

Limitations: Common practice generally involves reporting a single standard error for all possible predicted values. This practice ignores the fact that predicted values in the center of a distribution are known with more precision than values at the tails of a distribution.

Comments: The standard error of the estimate is very helpful in making decisions based on predicted values of some variable. Confidence intervals of any desired range can be established to control the likelihood of making false positive or false negative decisions based on the predicted values.

Formula:

$$S_e = \sqrt{\frac{\sum_{i=1}^{n}(e_i - \overline{e})^2}{n-1}} = \sqrt{\frac{\sum_{i=1}^{n} e_i^2}{n-1}}$$

where e_i is the residual for person i.

or

$$S_e = S_y \sqrt{1 - r_{xy}^2}$$

A4-36

Category: Regression Analysis

Statistic: Multiple Regression

Description: A procedure for predicting some dependent variable Y from a set of n independent variables, $\{X_1, X_2, X_3 \ldots X_n\}$, by using a weight on each X_i and a constant. The weights are regression coefficients and the constant is the intercept. Multiple regression is sometimes used to test theoretical constructs about relationships among variables.

Application: A powerful technique commonly used to (1) make predictions from several independent variables, (2) to identify the proportion of variance in one variable accounted for by knowledge of a set of other variables, and (3) to control the influence of confounding variables. Multiple regression is sometimes used to test theoretical constructs about relationships among variables.

Data: The dependent and independent variables are interval or ratio.

Limitations: The dependent and independent variables involved are assumed to be normally distributed. Multiple regression lends itself to "blind" empiricism in which some measures that happen to be available are used as independent variables to predict a dependent measure without any conceptual framework to suggest a rationale for such prediction. Multiple regression used in this fashion should be considered exploratory and descriptive.

Comments: The independent variables in a multiple-regression analysis can be included in the regression equation (1) one at a time in a series of steps **(A4-39)**, (2) grouped into subsets of variables and included in a series of steps, or (3) altogether at one time. Polynomial terms, as well as the liner term, for the independent variables can be used to perform curvilinear regression.

Formula: Multiple regression equations are of the form:

$\hat{Y} = b_1X_1 + b_2X_2 + ... + b_iX_i + ... + b_nX_n + b_0$, where

$\hat{Y}$ is the predicted dependent variable,

b_1 b_2 ... b_i ... b_n are the regression coefficients for the respective independent variables,

X_1 X_2 ... X_i ... X_n are the independent variables, and

b_0 is the Y-intercept

A4-37

Category: Regression Analysis

Statistic: R-squared (R^2)

Description: R-squared is a statistic used in regression analysis that provides an index of how well the independent variable(s) is predicting the dependent measure. R-squared is the proportion of the variation in the dependent measure that is accounted for by the prediction made from the independent variable(s). In regression analysis, R-squared is the sum of squares due to regression divided by the sum of squares total. R-squared is equal to the square of the correlation or multiple correlation between the dependent and independent variable(s).

Application: R-squared is a commonly used index of how well the dependent measure is predicted by the independent measure(s). Changes in R-squared are studied to determine if a variable added to a regression equation actually improves the predictive power of the regression equation.

Limitations: R-squared is a useful descriptive index that must not be interpretped as suggesting a causal explanation between the dependent and independent variable(s), even though R-squared is described as the proportion of variation in the dependent variable accounted for by the independent variables.

Comments: R-squared is a very useful index of the power of a regression equation and of each independent variable's contribution to the equation in terms of increased R-squared.

Formula: R-squared can be calculated in two ways:

1. $R^2 = r_{xy}^2$
2. $R^2 = \dfrac{\text{Sum of Squares Regression}}{\text{Total Sum of Squares}}$

A4-38

Category: Regression Analysis

Statistic: Curvilinear Regression

Description: A particular type of multiple regression in which the independent variables include the linear and polynomial representations of the predictor variables. The dependent variable Y is predicted from a set of n independent variables, { X_1 X_2 X_i ... X_n }, where these can be the original independent variables and the independent variables squared, cubed, raised to the fourth power, and so on. Other nonlinear transformations of the independent variables, such as logs, or trigonometric functions, may also be employed.

Application: This is a special application of multiple regression when the relationship between the dependent and independent variables is not linear. Curvilinear regression is commonly used when the relationship between the dependent variable and independent variable(s) is one of "diminishing returns." In such situations, increases in Y are associated with increases in X, up to a certain value of X. After the particular value of X has been reached, further increases in X are associated with a decrease in the values of Y. For example, agricultural yields, in terms of bushels per acre, have a simple curvilinear relationship with rainfall. Curvilinear regression is sometimes used to test theoretical constructs about relationships among variables.

Data: The dependent and independent variables are interval or ratio.

Limitations: The variables involved are assumed to be normally distributed in their linear form.

Comments: Curvilinear regression is helpful in exploring a wide range of dependent-independent variable relationships in which the values of the dependent variable, Y, increase or decrease as the values of the independent variable, X, increase up to a point on the X scale; after that point has been reached, the direction of the relationship reverses.

Formula: Curvilinear regression equations are of the form:

$$\hat{Y} = b_1 X_1 + b_2 X_1^2 + b_3 X_1^3 + \ldots + b_i X_j^k + b_n X_n^L + b_0,$$ where

$\hat{Y}$ is the predicted dependent variable,

$b_1\ b_2\ b_3\ \ldots\ b_i\ \ldots\ b_n$ are the regression coefficients for the respective independent variables,

$X_1\ X_1^2\ X_1^3\ \ldots\ X_j^k\ \ldots\ X_n^L$ are the independent variables in linear and various polynomial forms, and

b_0 is the Y-intercept.

A4-39

Category: Regression Analysis

Statistic: Stepwise Multiple Regression

Description: A procedure in multiple regression in which the independent variables are entered into the regression equation one at a time to build a regression equation in steps. As each independent variable is added to the equation, the new equation is evaluated by comparing it to the equation in the previous step in terms of how well the two equations account for variation in the dependent variable.

Application: A powerful technique commonly used to make predictions from several independent variables in a way that examines the power of the different independent variables to predict the dependent measure. Stepwise regression is often used to study theoretical constructs about relationships among variables. Independent variables that provide redundant information about the dependent measure are easily detected and eliminated by using stepwise regression.

Data: The dependent and independent variables are interval or ratio.

Limitations: The variables involved are assumed to be normally distributed.

Comments: A variety of criteria may be used for building equations in a stepwise regression analysis. The common stepwise procedure involves starting with the independent variable that has the highest correlation with the dependent measure and then building up the equation one independent variable at a time. The independent variable added at each step is the one that has the highest correlation with the residuals in the previous step.

Formula: Stepwise multiple regression equations are of the form:

$$\hat{Y} = b_1 X_1 + b_2 X_2 + \ldots + b_i X_i + \ldots + b_n X_n + b_0,$$ where

$\hat{Y}$ is the predicted dependent variable,

$b_1\ b_2\ \ldots\ b_i\ \ldots\ b_n$ are the regression coefficients for the respective independent variables,

$X_1\ X_2\ \ldots\ X_i\ \ldots\ X_n$ are the independent variables, and

b_0 is the Y-intercept.

A4-40

COMPARISONS: ONE SAMPLE

(A4-41) Sample Mean Compared to Some Value
(A4-42) Sample Variance Compared to Some Value
(A4-43) Sample Proportion Compared to Some Value
(A4-44) Sample Correlation Compared to 0
(A4-45) Sample Correlation Compared to Some Value
(A4-46) Chi-square Test for Goodness of Fit
(A4-47) Kolmogorov-Smirnov One-Sample Test

A4-41

Category: Comparison: One Sample

Statistic: Sample Mean Compared to Some Value, t-test and z-test

Description: A t-statistic or z-statistic is used to determine whether a sample mean differs from some hypothesized value by an amount that exceeds a specified significance level.

Application: Commonly used to test whether some sample mean differs significantly from the population mean on some variable.

Data: Used with interval or ratio scales.

Limitations: Assumes the sample has been randomly selected, and that the dependent variable is normally distributed.

Comments: The t-statistic is used if the population variance must be estimated from the sample variance. The z-statistic is used if the population variance is known.

Formula:

$$t = \frac{\overline{X} - \mu}{s/\sqrt{n}}, \quad (n-1) \text{ df}$$

$$Z = \frac{\overline{X} - \mu}{\sigma_{\overline{x}}}, \quad \sigma_{\overline{x}} = \sigma/\sqrt{n}$$

A4-42

Category: Comparisons: One Sample

Statistic: Sample Variance Compared to Some Value, Chi-square Test

Description: A Chi-square statistic is used to test whether a sample variance differs from some hypothesized value by an amount that exceeds some specified level of significance.

Application: Used to determine whether a sample variance differs significantly from some expected or hypothesized value.

Data: Interval or ratio.

Limitations: Like all Chi-square tests, this test is sensitive to sample size.

Comments: Helpful in educational settings to determine whether students are more or less homogeneous than some value of interest.

Formula:

$$\chi^2 = \frac{(n-1)S^2}{a}, \quad (n-1) \text{ df}$$

S^2 = variance
a = hypothesized value

A4-43

Category: Comparisons: One Sample

Statistic: Sample Proportion Compared to Some Value, z-statistic

Description: A z-statistic used to determine whether a sample proportion differs from some hypothesized value by an amount that exceeds a specified level of significance.

Application: Often used to check if a sample differs significantly from the population on some demographic characteristic.

Data: Used with interval or ratio scales.

Limitations: Sensitive to sample size. This test may be overly powerful with very large samples.

Comments: Helpful in determining if the proportional representation of some group in a sample is equivalent to that group's proportional representation in the population.

Formula:

$$z = \frac{p - P}{\sqrt{\frac{P(1-P)}{n}}}$$

p = sample proportion
P = hypothesized value

A4-44

Category: Comparisons: One Sample

Statistic: Sample Correlation Compared to 0, t-statistic

Description: A t-statistic used to determine whether a sample correlation differs from 0 by an amount that exceeds a specified significance level.

Application: This test is used to answer the basic question involved in determining whether a correlation is statistically significant.

Data: Pearson correlation coefficient

Limitations: Direct calculation is rarely needed since critical values for correlations are found in tables in most statistics books.

Comments: It is important to note that testing the correlation coefficient against 0 is not always the relevant question.

68 RESEARCH PROCESSES

Formula:

$$t = r\sqrt{\frac{n-2}{1-r^2}}, \quad (n-2) \text{ degrees of freedom}$$

A4-45

Category: Comparisons: One Sample

Statistic: Sample Correlation Compared to Some Value, z-statistic

Description: A z-statistic used to determine whether a sample correlation coefficient differs from some hypothesized value by an amount that exceeds a specified significance level. The test statistic is applied to the correlation coefficients after they have been transformed using Fisher's Z-transformation.

Application: Often used to test whether a sample correlation differs significantly (1) from the correlation in the population or (2), from some value suggested by a review of the literature.

Data: Applies to Pearson product-moment correlation coefficient.

Limitations: This procedure is sensitive to sample size.

Comments: Based on Fisher's Z-transformation. Fisher's Z-transformation is:
$Z = \frac{1}{2}\{\log(1+r)\} - \frac{1}{2}\{\log(1-r)\}$

Formula:

$$z = \frac{Z_r - Z_p}{\sqrt{\frac{1}{n-3}}}$$

z = the z-statistic, unit normal deviate

Z_r = the Fisher Z-transformation of r

Z_p = the Fisher Z-transformation of ρ, the hypothesized value

A4-46

Category: Comparisons: One Sample

Statistic: Chi-square Test for Goodness of Fit

Description: A procedure used with one sample to describe whether the observed frequencies on some variable with C categories is statistically equivalent to the frequencies expected based on theoretical or empirical criteria.

Application: Often used to determine if the frequency distribution for a sample across levels of some variable of interest is statistically equivalent to known population values.

Data: Categories are discrete and defined on a nominal, ordinal, interval or ratio variable.

Limitations: Like all Chi-square tests, the goodness-of-fit application of the Chi-square is sensitive to sample size. With small samples, the test is not very powerful, but with large samples the test can be overly powerful.

Comments: The Chi-square goodness-of-fit test is very useful in describing the extent to which the frequency distribution of some sample on a nominal variable resembles population values.

Formula:

$$\chi^2 = \sum_{i=1}^{c} \frac{(O_i - E_i)^2}{E_i} \quad \text{where}$$

O = observed frequency
E = expected frequency
For C-categories, there are
(C − 1) degrees of freedom

A4-47

Category: Comparisons: One Sample

Statistic: Kolmogorov-Smirnov One-Sample Test Comparing Distributions

Description: The Kolmogorov-Smirnov one-sample test is used to determine whether the distribution of some variable for a given sample is statistically equivalent to some particular distribution of interest. Often the criterion distribution is a theoretical distribution such as the normal distribution. The statistical test is based on the largest difference between the proportional cumulative frequencies for the observed and criterion distributions evaluated across all points on the distributions. The sampling distribution of the maximum difference in proportional cumulative frequencies is known and critical values at various significance levels can be found in nonparametric tables.

Application: The Kolmogorov-Smirnov one-sample test is a goodness-of-fit test (**A4-46**) used when a researcher wants to test the hypothesis that the distribution of some variable for a sample of interest follows some particular theoretical distribution. This test is often used to test whether a variable is distributed normally.

Data: Groups are classified on a discrete variable and measured on a continuous variable that may be ordinal but is typically interval or ratio.

Limitations: The Kolmogorov-Smirnov test is not a statistically powerful test.

Comments: The Kolmogorov-Smirnov test is broadly applicable because it requires few assumptions, but at the same time it is not a statistically powerful test.

Formula:

$$D = \text{Maximum} \, | F_c(X) - F_s(X) | \quad \text{where}$$

$F_c(X)$ = cumulative frequency distribution for the criterion group at each value of X
$F_s(X)$ = cumulative frequency distribution for the sample at each value of X

A4-48

COMPARISONS: TWO SAMPLES (PARAMETRIC)

(A4-49) Comparing Two Sample Means—Independent Samples
(A4-50) Comparing Two Sample Means—Dependent or Correlated Samples
(A4-51) Comparing Two Sample Variances—Independent Samples
(A4-52) Comparing Two Sample Variances—Dependent or Correlated Samples
(A4-53) Comparing Two Sample Proportions—Independent Samples
(A4-54) Comparing Two Sample Proportions—Dependent or Correlated Samples
(A4-55) Comparing Two Sample Correlations—Independent Samples
(A4-56) Comparing Two Sample Correlations—Dependent or Correlated Samples

A4-49

Category: Comparisons: Two Samples (Parametric)

Statistic: Comparison of Two Independent Sample Means, t-test

Description: A t-statistic used to determine if two sample means are significantly different from each other. More generally, a test of whether the difference between two means differs significantly from some value.

Application: Commonly used to determine if two sample means represent samples from the same population, that is, the difference between the two sample means is not different from 0.

Data: Dependent variable is interval or ratio.

Limitations: Assumes that the variables are distributed normally in the population and that the variances for the two populations are equal.

Comments: Fairly robust with respect to the assumptions. Does not require equal sample sizes, but the sample sizes cannot be widely different because of the effect on the assumptions.

Formula:

For equal sample sizes:

$$t = \frac{\overline{X}_1 - \overline{X}_2}{\sqrt{\frac{S_1^2 + S_2^2}{n}}}$$

with $(2n - 2)$ or
$2(n - 1)$ degrees of freedom

For unequal sample sizes:

$$t = \frac{\overline{X}_1 - \overline{X}_2}{\sqrt{\frac{(n_1 - 1) S_1^2 + (n_2 - 1) S_2^2}{n_1 + n_2 - 2} \left[\frac{1}{n_1} + \frac{1}{n_2}\right]}}$$

with $n_1 + n_2 - 2$ degrees of freedom

A4-50

Category: Comparisons: Two Samples (Parametric)

Statistic: Comparison of Two Sample Means—Dependent or Correlated Samples, t-test

Description: A t-statistic used to determine if the means of two dependent or correlated samples are significantly different from each other. More generally, a test of whether the difference between the means of two dependent or correlated samples differ significantly from some value.

Application: Commonly used to determine if the means from two correlated samples are drawn from the same population. A standard procedure in one-group pretest-posttest designs and in matching designs.

Data: Dependent variable is interval or ratio.

Limitations: Assumes that the variables are normally distributed in the population. Assumes that the variances of the two populations are equal (homogeneity of variance).

Comments: A more powerful statistical test than the independent sample t-test **(A4-49)** because it removes between-subject variance from the estimates of error variance.

Formula:

$$t = \frac{\overline{d}}{s_d/\sqrt{n}}, \quad (n - 1) \text{ degrees of freedom}$$

$$\overline{d} = \frac{\sum_{i=1}^{n} (X_{1i} - X_{2i})}{n} \quad \text{and}$$

$$S_d = \sqrt{\frac{\sum_{i=1}^{n} (d_i - \overline{d})^2}{n - 1}}$$

A4-51

Category: Comparisons: Two Samples (Parametric)

Statistic: Comparison of Two Independent Sample Variances, F-statistic

Description: An F-statistic used to determine if the variances from two independent samples are equivalent.

Application: Commonly used to test the assumption of homogeneity of variance.

Data: Variances from interval or ratio scales.

Limitations: Assumes normal distributions in the populations.

Comments: Can be used to test the effect of some independent variable on the variance of a dependent variable measured on an experimental group and a control group.

Formula:

$$F = \frac{S_1^2}{S_2^2}$$

with $(n_1 - 1)$ and $(n_2 - 1)$ degrees of freedom

A4-52

Category: Comparisons: Two Samples (Parametric)

Statistic: Comparison of Two Sample Variances—Dependent or Correlated Samples, t-statistic

Description: A t-statistic used to determine if the variances from two dependent or correlated samples are equivalent.

Application: Used to test whether the variance on a pretest is equivalent to the variance on the posttest.

Data: Variances from interval or ratio data and the correlation between the matched scores.

Limitations: The comparison of variances for two samples is confounded by any differences in the scales of the two variables.

Comments: Can be used to test the effect of some independent variable on the variance of the dependent measure in a pretest-posttest design.

Formula:

$$t = \frac{S_1^2 - S_2^2}{\sqrt{\frac{4S_1^2 S_2^2}{n - 2}(1 - r_{12}^2)}}$$

with $(n - 2)$ degrees of freedom

A4-53

Category: Comparisons: Two Samples (Parametric)

Statistic: Comparison of Two Independent Sample Proportions, z-statistic

Description: A z-statistic used to compare proportions from two independent samples.

Application: Used to determine if two samples are drawn from populations with similar proportions with respect to some variable. Commonly used to determine if two samples were proportionally equivalent with respect to some demographic variable.

Data: Used with proportions.

Limitations: The sample sizes (n) must be large enough so that two quantities (1) (n × proportion) and (2) [n × (1-proportion)] are greater than 5 for both groups.

Comments: If the requirements for sample size are not met, a Chi-square test for independence (**A4-61**) must be used.

Formula:

$$Z = \frac{(p_1 - p_2)}{\sqrt{p(1-p)\left(\frac{1}{n_1} + \frac{1}{n_2}\right)}}$$

$$\text{where } p = \frac{f_1 + f_2}{n_1 + n_2},$$

f_1 = frequency in sample 1
f_2 = frequency in sample 2

A4-54

Category: Comparisons: Two Samples (Parametric)

Statistic: Comparison of Two Sample Proportions—Dependent or Correlated Samples, z-statistic

Description: A z-statistic used to compare proportions from two dependent or correlated samples. Calculated from a 2 × 2 contingency table with the form:

		Posttest	
		−	+	
Pretest	+	A	B	A+B
	−	C	D	C+D
		A+C	B+D	N=A+B+C+D

Application: Used to compare proportions before and after some treatment in a pretest-posttest design. In the contingency table shown, for example, subjects indicate whether they are positive (+) or negative (−) about some stimulus before (pretest) and after (posttest) some intervention program. Often used to compare preferences of one group to two stimuli.

Data: Proportions or frequencies in a 2 × 2 contingency table.

Limitations: In the 2 × 2 contingency table, the sum of the frequencies in the cells of both diagonals must be greater than 10. If the frequency limitation is not met, a Chi-square test may be used.

Comments: When used in a pretest-posttest design, this test is the same as McNemar's test for change with dependent samples **(A4-63)**.

Formula:

$$z = \frac{A - D}{\sqrt{A + D}}$$

A and D as shown in the description

A4-55

Category: Comparisons: Two Samples (Parametric)

Statistic: Comparison of Two Independent Sample Correlations, z-statistic

Description: A z-statistic used to determine if correlation coefficients from two independent samples are significantly different.

Application: Used to test whether the relationship between two variables is statistically equivalent for two different samples.

Data: Applies to Pearson correlation coefficients.

Limitations: The test can be overly powerful with large samples.

Comments: Based on the Fisher's Z-transformation of the correlation coefficients. Fisher's Z-transformation is:
$Z = \frac{1}{2} \{\log(1 + r)\} - \frac{1}{2} \{\log(1 - r)\}$

Formula:

$$z = \frac{Z_{r_1} - Z_{r_2}}{\sqrt{\frac{1}{n_1 - 3} + \frac{1}{n_2 - 3}}}$$

where
Z_{r_1} = Fisher Z-transformation of r_1
Z_{r_2} = Fisher Z-transformation of r_2

A4-56

Category: Comparisons: Two Samples (Parametric)

Statistic: Comparisons of Two Sample Correlation Coefficients—Correlated or Dependent Samples, t-statistic

Description: A t-statistic used to determine if correlations between pairs of variables from two dependent samples are significantly different.

Application: Often used to determine if pairs of correlations for different variables for the same sample are statistically equivalent.

Data: Used with Pearson product-moment correlation coefficients.

Limitations: Sensitive to sample size.

Comments: Commonly used in the case of three variables, X, Y, Z, to determine if the correlations of each of two variables with the third are equivalent, for example, is r_{xy} equal to r_{xz}.

Formula:

$$t = \frac{(r_{xy} - r_{xz})\sqrt{(n-3)(1+r_{yz})}}{\sqrt{2(1 - r_{xy}^2 - r_{xz}^2 - r_{yz}^2 + 2r_{xy}r_{xz}r_{yz})}}$$

with $(n-3)$ degrees of freedom

A4-57

COMPARISONS: TWO SAMPLES (NONPARAMETRIC)

- **(A4-58)** Median Test—Sign Test for Two Independent Samples
- **(A4-59)** Wilcoxon Rank Sum Test—Rank Test for Independent Samples
- **(A4-60)** Mann-Whitney U-Test—Signed Rank Test for Independent Samples
- **(A4-61)** Chi-square Test of Independence (Homogeneity)
- **(A4-62)** Kolmogorov-Smirnov Two-Sample Test Comparing Distributions
- **(A4-63)** McNemar's Test for Significance of Change—Dependent or Correlated Samples
- **(A4-64)** Sign Test for Dependent Samples
- **(A4-65)** Wilcoxon Matched-Pairs Signed-Rank Test for Dependent Samples

A4-58

Category: Comparisons: Two Samples (Nonparametric)

Statistic: Median Test—Sign Test for Two Independent Samples, Chi-square Statistic

Description: A test that compares the median for two independent samples of sizes n_1 and n_2 respectively. The median for the two samples combined is calculated and then the number of subjects above and below the combined median is determined within each of the two separate samples. A 2×2 Chi-square table composed of samples (1, 2) by position relative to the combined median (above, below) is constructed to test the hypothesis that there is no significant difference between the medians of the population from which the samples are drawn. The 2×2 table is:

Position Relative to Combined Groups Median

		Below	Above	
Sample 1	1	A	B	A+B
Sample 2	2	C	D	C+D
		A+C	B+D	N=A+B+C+D

Application: Used to test differences between the central tendency of two samples when the assumptions required for an independent sample t-test (**A4-49**) are not met. Used when the dependent variable is ordinal.

Data: Dependent variable is at least ordinal but could be interval or ratio. The independent variable on the basis of which groups are defined is discrete.

Limitations: The sign test is not as statistically powerful as the parametric t-test.

Comments: A useful test when assumptions needed for the t-test have not been met. With small samples, Yates' correction for continuity may be required. Yates' correction for continuity increases by .5 observed frequencies that are less than expected and decreases by .5 observed frequencies that are greater than expected.

Formula:

$$\chi^2 = \frac{N(AD - BC)^2}{(A + B)(C + D)(B + D)(A + C)}$$

with 1 degree of freedom

A4-59

Category: Comparisons: Two Samples (Nonparametric)

Statistic: Wilcoxon Rank Sum Test—Rank Test for Independent Samples

Description: The Wilcoxon rank sum test is a nonparametric equivalent to the independent sample t-test (**A4-49**). It is very close to being as statistically powerful as the t-test for normal and rectangular distributions. For samples of size n_1 and n_2, all $N = n_1 + n_2$ observations are combined and rank ordered from 1 to N. The test statistic is R_1 which is the sum of the ranks for the smaller of the two samples, or of either sample if they are the same size.- The exact distribution of R_1 is known, and for situations in which both samples are less than 25, the appropriate critical value for R_1 can be found in tables.

There is a normal approximation for R_1 when n_1 and n_2 are greater than or equal 10. The expression for the z-statistic used in this approximation is shown in the Formula section.

Application: The Wilcoxon rank sum test is used when distributional assumptions needed for the appropriate application of the independent-sample t-test have not been met. This Wilcoxon test is used if the dependent variable is ordinal because the t-test assumes the dependent measure is interval or ratio.

Data: The dependent variable is at least ordinal. The independent variable, on the basis of which groups are classified, is discrete.

Limitations: The Wilcoxon rank sum test is slightly less powerful than the independent-sample t-test in some cases.

Comments: The Wilcoxon rank sum test is a useful and powerful alternative to the independent-sample t-test.

Formula:

$$Z = \frac{|R_1 - \overline{R}_1| - 1}{\sqrt{\dfrac{n_1 n_2 (n_1 + n_2 + 1)}{12}}}$$

where

$$\overline{R}_1 = \frac{n_1 (n_1 + n_2 + 1)}{2}$$

R_1 = sum of the ranks for the smaller of the two groups when ranked in the combined distribution

A4-60

Category: Comparisons: Two Samples (Nonparametric)

Statistic: Mann-Whitney U-Test—Signed Rank Test for Independent Samples

Description: A nonparametric procedure used to test differences between two independent samples. The Mann-Whitney U-test is a nonparametric alternative to the independent-sample t-test (**A4-49**) and is also statistically equivalent to the Wilcoxon rank sum test (**A4-59**). The test statistics U_1 and U_2 are calculated for samples 1 and 2, respectively. The calculations for U_1 and U_2 are shown in the Formula section. The smaller of these two test statistics is tested against tabled values for the U-statistic. The null hypothesis is rejected if the observed U-statistic is *less than the critical value.*

Application: The Mann-Whitney U-test is used when distributional assumptions needed for the appropriate application of the independent sample t-test have not been met. In addition, the Mann-Whitney U-test is used if the dependent variable is ordinal because the t-test assumes that the dependent measure is interval or ratio.

Data: The dependent variable is at least ordinal. The independent variable, on the basis of which groups are classified, is discrete.

Limitations: Like all nonparametric tests, the Mann-Whitney U-test is less powerful than corresponding parametric tests.

Comments: The Mann-Whitney U-test is a useful and powerful alternative to the independent sample t-test. The Mann-Whitney U-test yields results identical to those obtained by using the Wilcoxon rank sum test.

Formula:

$$U_1 = n_1 n_2 + \frac{n_1 (n_1 + 1)}{2} - R_1$$

$$U_2 = n_1 n_2 + \frac{n_2 (n_2 + 1)}{2} - R_2$$

where R_1 and R_2 are the sum of the ranks for samples 1 and 2 respectively in the combined ($n_1 + n_2$) distribution

A4-61

Category: Comparisons: Two Samples (Nonparametric)

Statistic: Chi-square Test of Independence (Homogeneity)

Description: A nonparametric test used to compare two groups on a nominal variable with two or more categories. If the nominal variable has only two categories, a Phi-coefficient (**A4-21**) can be used, or a test for the difference between proportions for two independent samples (**A4-53**) can be performed. Neither test is appropriate when the nominal variable has three or more categories. The Chi-square test for independence or homogeneity for two groups on a nominal variable involves a standard Chi-square calculation. It tests the hypothesis that the proportion of subjects in the two groups is equivalent across the categories of the nominal variable.

Application: The Chi-square test for independence or homogeneity is used to test the hypothesis that two or more groups are statistically equivalent with respect to some nominal variable that has two or more categories.

Data: Groups are defined on some discrete variable and measured on a nominal variable with two or more categories.

Limitations: The Chi-square test for independence is not a statistically powerful test of group differences.

Comments: This is a useful test when subjects in two groups are measured on a nominal level variable.

Formula:

For two variables with C categories and K categories

$$\chi^2 = \sum_{j=1}^{k} \sum_{i=1}^{c} \left(\frac{(O_{ij} - E_{ij})^2}{E_{ij}} \right)$$

with $(C - 1)(K - 1)$ degrees of freedom

A4-62

Category: Comparisons: Two Samples (Nonparametric)

Statistic: Kolmogorov-Smirnov Two-Sample Test Comparing Distributions

Description: The Kolmogorov-Smirnov two-sample test is used to determine whether two independent samples were drawn from the same population or from populations with the same distribution. The test is based on the largest difference between the proportional cumulative frequencies for the two distributions evaluated across all points on the distributions. The sampling distribution of the maximum difference in proportional cumulative frequencies and critical values at various significance levels can be found in nonparametric tables.

Application: This test is used when a researcher wants to test the hypothesis that two samples are equivalent with respect to their overall distributions.

Data: Groups are classified on a discrete variable and measured on a continuous variable that is typically interval or ratio.

Limitations: The Kolmogorov-Smirnov test is not a statistically powerful test.

Comments: This test is flexible in that it requires few assumptions but, again, it is not a statistically powerful test.

Formula:

$$D = \text{Maximum} \, | F_1(X) - F_2(X) |$$

where

$F_1(X)$ = cumulative frequency distribution for group 1 at the corresponding value of X

$F_2(X)$ = cumulative frequency distribution for group 2 at the corresponding value of X

A4-63

Category: Comparisons: Two Samples (Nonparametric)

Statistic: McNemar's Test for Significance of Change—Dependent or Correlated Samples, or Correlated Chi-square

Description: A nonparametric test used when two correlated samples are categorized on a dichotomous nominal variable. The test statistic is a Chi-square. The McNemar test is often used in pretest-posttest designs in which one group of subjects is classified into the two levels of a dichotomous nominal variable before and after some treatment. Such designs yield a 2 × 2 contingency table (pre, post × category 1, category 2). Only the two cells representing pretest-posttest change in classification are of interest. Such a table would be:

Posttest

		1	2	
Pretest	2	A	B	A+B
	1	C	D	C+D
		A+C	B+D	N=A+B+C+D

Application: The McNemar test is often used in pretest-posttest designs in which one group of subjects is classified into the two levels of a dichotomous nominal variable before and after some treatment. For example, the McNemar test would be used in a situation in which a group of subjects declared themselves "for" or "against" some stimulus (e.g., an issue, political position, candidate), on two different occasions. If "for" is defined as Category 1, and "against" is defined as Category 2, the design fits exactly the 2 × 2 contingency table shown above.

Data: Subjects are categorized on a nominal dichotomous variable.

Limitations: McNemar's test is not as statistically powerful as a parametric test like the dependent sample t-test **(A4-50)**.

Comments: This is a useful and informative test when a dependent measure at the ordinal, interval, or ratio level is not available.

Formula: In terms of the 2 × 2 table shown:

$$\chi^2 = \frac{(A - D)^2}{A + D}, \quad 1 \text{ degree of freedom}$$

A and D as shown in the description

A4-64

Category: Comparisons: Two Samples (Nonparametric)

Statistic: Sign Test for Dependent Samples

Description: This test is used to compare two correlated or dependent samples with N paired observations. The sign of the difference between subjects in each pair of observations is determined and tested against the null hypothesis that there should be an equal number of pairs with plus (+) and minus (−) signs. Pairs with a difference of zero are discarded. The exact probability of getting n pairs with a plus (+) sign and n pairs with a minus (−) sign out of N total pairs is calculated from the binomial expansion of the form $(.5 + .5)^N$, where N is the number of paired observations less the number of pairs with zero differences. With small N's, a Chi-square approximation to the binomial may be used with each sign having an expected value of (N/2). Yates' correction should be applied in such cases. A computationally convenient formula is shown in the Formula section.

Application: The sign test for dependent samples is used to test the difference between two correlated samples. Often, this would be used for the same sample measured twice in a pretest-posttest design. This test is analogous to the dependent sample t-test (**A4-50**), but does not require distributional assumptions. It is used when the dependent variable is ordinal.

Data: The dependent variable is at least ordinal. The independent variable, on the basis of which groups are defined, is discrete.

Limitations: The sign test for correlated samples is not as statistically powerful as the parametric dependent sample t-test.

Comments: The sign test for correlated samples is useful when the dependent variable is ordinal and when assumptions needed for the dependent sample t-test are not met. The sign test for dependent samples is not statistically very powerful compared with the dependent sample t-test (**A4-50**).

Formula:

$$z = \frac{|D| - 1}{\sqrt{n}}$$

where D = difference in the number of plus and number of minus signs

A4-65

Category: Comparisons: Two Samples (Nonparametric)

Statistic: Wilcoxon Matched-Pairs–Signed-Rank Test for Dependent Samples

Description: A nonparametric test used with matched pairs commonly encountered in pretest-posttest designs. This test is analogous to the dependent sample t-test **(A4-50)**. For small samples, n less than or equal to 25, the test statistic called "T" is used (this is not Student's t). T is determined by (1) calculating the difference between the matched scores for each subject or pair of subjects; (2) ranking the absolute value of the differences; (3) giving the average of the ranks occupied for tied rankings; (4) assigning the appropriate sign, plus (+) or minus (−), depending on the direction of the difference; and (5) summing the ranks of the less frequently occurring signed values, (+) or (−). T is the sum of the ranks with the less frequent sign. Critical values for T at various significance levels are found in nonparametric tables. With samples greater than 25, the sampling distribution of T has an approximation to the normal distribution. This approximation is given in the Formula section.

Application: The Wilcoxon matched-pairs–signed-rank test is used for correlated samples in much the same way that the dependent sample t-test is used. Such applications include pretest-posttest designs and designs in which pairs of subjects are matched on some variable the influence of which the researcher wants to control.

Data: The dependent variable is converted to ordinal differences. Groups are defined on a discrete independent variable.

Limitations: The Wilcoxon matched-pairs–signed-rank test is not as powerful as its parametric counterpart, the dependent-sample t-test.

Comments: In many situations, this test is a useful alternative to the dependent sample t-test, although it is somewhat less powerful than the parametric test. The lack of distributional assumptions makes this Wilcoxon test broadly applicable in situations where the assumptions of the dependent-sample t-test are questionable.

Formula:

$$Z = \frac{T - \frac{(n)(n+1)}{4}}{\sqrt{\frac{n(n+1)(2n+1)}{24}}}$$

where T is defined in the description

A4-66

COMPARISONS: TWO OR MORE SAMPLES (NONPARAMETRIC)

- **(A4-67)** Sign Test for K-Independent Samples
- **(A4-68)** Kruskal-Wallis Rank Test for K-Independent Samples
- **(A4-69)** Friedman Two-Way Analysis of Variance by Ranks for Dependent or Correlated Samples

A4-67

Category: Comparisons: Two or More Samples (Nonparametric)

Statistic: Sign Test for K-Independent Samples Chi-square

Description: An extension of the median test **(A4-58)** used to compare two or more independent samples.

The median for all samples combined is calculated and then the number of subjects above and below the combined median is determined within each of the separate samples. The (K × 2) Chi-square table composed of samples (K) by position relative to the combined median (above, below) is constructed to test the hypothesis that there is no significant difference between the medians of the populations from which the samples are drawn. The Chi-square is tested with K − 1 degrees of freedom.

The K × 2 table is of the form:

Position Relative to Combined Median

	Below 1	Above 2
Sample 1		
Sample 2		
Sample 3		
.		
Sample K		

Application: Used to test differences between two or more samples when the assumptions required for an independent sample t-test **(A4-49)** or analysis of variance **(A4-71)** are not met. Used when the dependent variable is ordinal.

Data: Dependent variable is at least ordinal, but could be interval or ratio. The independent variable, on the basis of which groups are defined, is discrete.

Limitations: The sign test is not as statistically powerful as the parametric t-test or analysis of variance.

Comments: A useful test when assumptions needed for the t-test or analysis of variance have not been met. With small samples, Yates' correction for continuity may be required.

Formula:

$$\chi^2 = \sum_{k=1}^{k} \sum_{i=1}^{2} \frac{(O_{ik} - E_{ik})^2}{E_{ik}}$$

with (K − 1) degrees of freedom

A4-68

Category: Comparisons: Two or More Samples (Nonparametric)

Statistic: Kruskal-Wallis Rank Test for K-Independent Samples

Description: The Kruskal-Wallis rank test for K-independent samples is a nonparametric equivalent to the one-way analysis of variance (**A4-71**) and is an extension of the Wilcoxon rank sum test (**A4-59**). The Kruskal-Wallis is calculated by combining all the observations from all the groups and then rank ordering them from 1 to N. The sum of the ranks for each of the K groups is then determined and used in calculating the test statistic "H," shown in the Formula section. For K=3 and n's less than or equal to 5, the exact distribution of H is known. In other cases, H approximates a Chi-square distribution with K − 1 degrees of freedom.

Application: The Kruskal-Wallis rank test is used when distributional assumptions needed for the appropriate application of an analysis of variance procedure have not been met. It is commonly applied when the dependent variable is ordinal.

Data: The dependent variable is at least ordinal. The independent variable, on the basis of which groups are classified, is discrete.

Limitations: The Kruskal-Wallis rank test for independent samples is somewhat less powerful than a one-way analysis of variance, but the difference in statistical power, in most cases, is slight.

Comments: The Kruskal-Wallis rank test is a useful and powerful alternative to the one-way analysis of variance when the assumptions required for the analysis of variance have not been met.

Formula:

$$H = \frac{12}{n(n+1)} \sum_{i=1}^{k} \left(\frac{R_i^2}{n_i}\right) - 3(n+1)$$

where R_i = Sum of the ranks for group i on the combined distribution

A4-69

Category: Comparisons: Two or More Samples (Nonparametric)

Statistic: Friedman Two-Way Analysis of Variance by Ranks for Dependent or Correlated Samples

Description: The Friedman two-way analysis of variance by ranks is a rank test for correlated samples. The Friedman test is commonly used when some sample of N subjects is measured on K different occasions. The Friedman test is performed by arranging the data into N rows with one row for each person, and K columns with one column for each occasion. The K entries in each row are then rank ordered from 1 to K and assigned rank-order values 1 through K. A test statistic S is then calculated as shown in the Formula section. For small values of K and N, the exact distribution of S is known. In general, the Friedman test statistic is a Chi-square statistic derived from S as shown in the Formula section. This statistic approximates a Chi-square with K − 1 degrees of freedom.

Application: The Friedman two-way analysis of variance by ranks is used to test hypotheses about differences among correlated samples when the assumptions needed to perform a repeated measures analysis of variance **(A4-79)** have not been met.

Data: The dependent variable is at least ordinal, while the independent variables are discrete.

Limitations: The Friedman test is not as statistically powerful as the parametric repeated-measures analysis of variance, but the Friedman test is a useful approach when the assumptions of the parametric procedure have not been met.

Comments: The Friedman test is the nonparametric equivalent to the two-way, mixed-model analysis of variance with subjects being the random factor and occasions being the fixed factor **(A4-74)**.

Formula:

$$S = \sum_{i=1}^{k} (R_i - \overline{R})^2$$

where

R_i = sum of the ranks for group i

$\overline{R}$ = mean of the R_i's

$$\chi^2 = \frac{12S}{NK(K-1)} \quad \text{with } (K-1) \text{ degrees of freedom}$$

A4-70

COMPARISONS: TWO OR MORE SAMPLES—ANALYSIS OF VARIANCE (ANOVA)

Factorial Analysis of Variance

(A4-71) One-Way Analysis of Variance

(A4-72) Two-Way Analysis of Variance—Fixed Effects Model

(A4-73) Two-Way Analysis of Variance—Random Effects Model

(A4-74) Two-Way Analysis of Variance—Mixed Effects Model

(A4-75) Three-Way Analysis of Variance—Fixed Effects Model

(A4-76) Three-Way Analysis of Variance—Mixed Model, One Factor Fixed, Two Factors Random

(A4-77) Three-Way Analysis of Variance—Mixed Model, Two Factors Fixed, One Factor Random

(A4-78) Three-Way Analysis of Variance—Random Model

Repeated-Measures Analysis of Variance/Split-Plot Analyses

(A4-79) One-Way Repeated Measures—Subjects by Occasions

(A4-80) Two-Way Split Plot, One Between-Subjects Factor and One Within-Subjects Factor

(A4-81) Three-Way Split Plot, One Between-Subjects Factor and Two Within-Subjects Factors

(A4-82) Three-Way Split Plot, Two Between-Subjects Factors and One Within-Subjects Factor

(A4-83) Four-Way Split Plot, Two Between-Subjects Factors and Two Within-Subjects Factors

Nested / Hierarchical Analysis of Variance
(A4-84) Two-Factor Design, B Nested in A
(A4-85) Three-Factor Designs, C Nested in B, B Nested in A
(A4-86) Three-Factor Designs, B Nested in A, C Crossed with B Nested in A

General Analysis of Variance Designs
(A4-87) Randomized-Blocks Designs
(A4-88) Latin Square Designs

Related Designs
(A4-89) Analysis of Covariance
(A4-90) Time-Series Analysis

A4-71

Category: Analysis of Variance (ANOVA)

Statistical Procedure: One-Way Analysis of Variance

Description: A procedure for examining the variance in some dependent variable in terms of the portion of variance that can be attributed to certain factors. In a one-way analysis of variance total variance is partitioned into (1) variation due to different levels of the independent variable, called between-group variance, and (2) variation within the levels of the independent variable, called within-group or random variance.

The basic arrangement for a one-way analysis of variance is shown below.

```
Factor A, A Levels
 1  2  3  .  .  A
┌──┬──┬──┬──┬──┬──┐
│  │  │  │  │  │  │
└──┴──┴──┴──┴──┴──┘
```

Application: Commonly used to test hypotheses about different groups. The groups may be naturally occurring or represent different levels of some treatment variable.

Data: The dependent variable is interval or ratio. The independent variable is discrete and may be nominal, ordinal, or, in some cases, interval.

Limitations: Assumes homogeneity of variance within group, normal distributions in the populations, and independence among observations (e.g., the subjects are not correlated).

Comments: Robust with respect to the assumption of homogeneity of variance. Often used with *a priori* or *post hoc* contrasts.

Formula: One-Way ANOVA

Source of Variation	Degrees of Freedom	Mean Square	F-Ratio
Between Groups	$A - 1$	MSB	MSB/MSW
Within Groups	$N - A$	MSW	
TOTAL	$N - 1$		

A4-72

Category: Analysis of Variance (ANOVA)

Statistical Procedure: Two-Way Analysis of Variance—Fixed Effects Model

Description: A procedure for examining the variance in some dependent measure in terms of the portion of variance that can be attributed to certain factors. In the two-way analysis of variance, total variation is partitioned into variation due to independent variables (1) A (with A levels) and (2) B (with B levels), (3) the interaction of the independent variables (A*B), and (4) within-cell or random variation. In the fixed effects model, all possible levels of the independent variables are represented in the design and analysis.

The basic arrangements for a two-way analysis of variance follow; Factor A is shown with A levels and Factor B is shown with 3 levels.

```
                    Factor A, A Levels
                    1  2  3  . . A
              1  ┌──┬──┬──┬──┬──┐
   Factor B   2  ├──┼──┼──┼──┼──┤
              3  └──┴──┴──┴──┴──┘
```

Application: Standard procedure for designs with two independent variables. All levels of the independent variables are represented in the design. One of the independent variables may be a blocking variable used to reduce within-cell variance as it occurs in the context of a one-way design, **(A4-71).**

Data: The dependent variable is interval or ratio. The independent variable is discrete and may be nominal, ordinal, or, in some cases, interval.

Limitations: Assumes homogeneity of variance within groups, normal distributions in the populations, and independence among observations (e.g., the subjects are not correlated).

Comments: Robust with respect to the assumption of homogeneity of variance. Often used with *a priori* or *post hoc* contrasts. One of the independent variables may be a blocking variable designed to increase statistical power by reducing within-cell variance, the MSW or error term for hypothesis testing. Main effects cannot be interpreted directly if there is a significant interaction effect.

Formula: Two-Way Analysis of Variance, Fixed Effects Model

Source of Variation	Degrees of Freedom	Mean Square	F-Ratio
Between A	A − 1	MSA	MSA/MSW
Between B	B − 1	MSB	MSB/MSW
Interaction of A*B	(A − 1)(B − 1)	MSAB	MSAB/MSW
Within	N − (A × B)	MSW	
TOTAL	N − 1		

A4-73

Category: Analysis of Variance (ANOVA)

Statistical Procedure: Two-Way Analysis of Variance—Random Effects Model

Description: A procedure for examining the variance in some dependent measure in terms of the portion of variance that can be attributed to certain factors. In the two-way analysis of variance, total variation is partitioned into variation due to independent variables (1) A (with A levels) and (2) B (with B levels), (3) the interaction of the independent variables (A*B), and (4) within-cell or random variation. In the random effects model, the levels of both independent variables are sampled from larger sets of possible levels.

Application: A procedure used with designs that have two independent variables, when levels of the independent variables are sampled. One of the independent variables may be a blocking variable that is used to reduce within-cell variance as it occurs in the context of a one-way design.

Data: The dependent variable is interval or ratio. The independent variable is discrete and may be nominal, ordinal, or, in some cases, interval.

Limitations: Assumes homogeneity of variance within groups, normal distributions in the populations, and independence among observations (e.g., the subjects are not correlated).

Comments: Robust with respect to the assumption of homogeneity of variance. Often used with *a priori* or *post hoc* contrasts **(A4-91)**. One of the independent variables may be a blocking variable designed to increase statistical power by reducing error variance for hypothesis testing. Main effects cannot be interpreted directly in the presence of a significant interaction effect.

Formula: Two-Way Analysis of Variance, Random Effects Model

Source of Variation	Degrees of Freedom	Mean Square	F-Ratio
Between A	A − 1	MSA	MSA/MSAB
Between B	B − 1	MSB	MSB/MSAB
Interaction of A∗B	(A − 1)(B − 1)	MSAB	MSAB/MSW
Within	N − (A × B)	MSW	
TOTAL	N − 1		

A4-74

Category: Analysis of Variance (ANOVA)

Statistical Procedure: Two-Way Analysis of Variance—Mixed Effects Model

Description: A procedure for examining the variance in some dependent measure in terms of the portion of variance that can be attributed to certain factors. In the two-way mixed effects model analysis of variance, there is a fixed independent variable and a random independent variable. All levels of the fixed variable are reflected in the design. The levels of the random independent variable are sampled from a larger set of levels. Total variation is partitioned into variation due to (1) a fixed independent variable A (with A levels) and (2) random independent variable B (with B levels), (3) the interaction of the independent variables (A∗B), and (4) within-cell or random variation.

Application: A common application of the two-way mixed effects model is the repeated measures design (**A4-79**) in which subjects are measured on multiple occasions or trials. "Subjects" is the random factor, while occasions, or trials, is the fixed factor. (There is no within-cell variance in the mixed model repeated measures analysis.) The two-way mixed-effects model is also used in randomized block designs when levels of the blocking variable are randomly selected and used to reduce error variance as it occurs in the context of a one-way design.

Data: The dependent variable is interval or ratio. The independent variable is discrete and may be nominal, ordinal, or, in some cases, interval.

Limitations: Assumes homogeneity of variance within groups, normal distributions in the populations, and independence among observations (e.g., the subjects are not correlated).

Comments: Robust with respect to the assumption of homogeneity of variance. Often used with *a priori* or *post hoc* contrasts. One of the independent variables may be a blocking variable that is designed to increase statistical power. (See Randomized-Blocks Designs, **A4-87**.) The repeated measures analysis is the multiple-group equivalent to the two-group dependent t-test (**A4-50**). Main effects cannot be interpreted directly in the presence of a significant interaction effect.

Formula: Two-Way Analysis of Variance, Mixed Effects Model, Factor A Fixed, Factor B Random

Source of Variation	*Degrees of Freedom*	*Mean Square*	*F-Ratio*
Between A	A − 1	MSA	MSA/MSAB
Between B	B − 1	MSB	MSB/MSW
Interaction of A∗B	(A − 1)(B − 1)	MSAB	MSAB/MSW
Within	N − (A × B)	MSW	
TOTAL	N − 1		

A4-75

Category: Analysis of Variance (ANOVA)

Statistical Procedure: Three-Way Analysis of Variance—Fixed Effects Model

Description: A procedure for examining the variance in some dependent measure in terms of the portion of variance that can be attributed to certain factors. In the three-way ANOVA, total variation is partitioned into variation due to independent variables (1) A (with A levels), (2) B (with B levels), and (3) C (with C levels); the two-way interactions (4) A∗B, (5) A∗C, and (6) B∗C; the (7) A∗B∗C three-way interaction and (8) the within-group variance. In the fixed effects model, all levels of each independent variable are fixed and represented in the design and analysis.

The basic arrangements for a three-way analysis of variance are shown. Factor A is shown with A levels, Factor B is shown with 3 levels, and Factor C is shown with 3 levels.

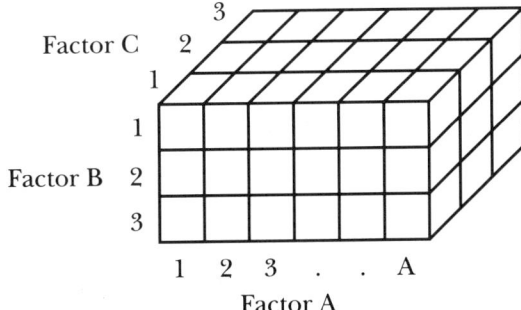

Application: The appropriate procedure when the design involves three independent variables and all levels of each variable are represented in the design. One

of the independent variables may be a blocking variable used to reduce within-cell variance used as the error term in a two-way analysis of variance without the blocking variable. (See Randomized-Blocks Designs, **A4-87**.)

Data: The dependent variable is interval or ratio. The independent variables are discrete and may be nominal, ordinal, or, in some cases, interval.

Limitations: Assumes homogeneity of variance within groups (cells), normal distributions in the populations, and independence among observations (e.g., the subjects are not correlated).

Comments: Robust with respect to the assumption of homogeneity of variance. Often used with *a priori* or *post hoc* contrasts. One of the independent variables may be a blocking variable designed to increase statistical power by reducing error variance for hypothesis testing. Main effects cannot be interpreted directly in the presence of a significant interaction effect.

Formula: Three-Way Analysis of Variance, Fixed Effects Model

Source of Variation	Degrees of Freedom	Mean Square	F-Ratio
Between A	A − 1	MSA	MSA/MSW
Between B	B − 1	MSB	MSB/MSW
Between C	C − 1	MSC	MSC/MSW
Interaction of A*B	(A − 1)(B − 1)	MSAB	MSAB/MSW
Interaction of A*C	(A − 1)(C − 1)	MSAC	MSAC/MSW
Interaction of B*C	(B − 1)(C − 1)	MSBC	MSBC/MSW
Interaction of A*B*C	(A − 1)(B − 1)(C − 1)	MSABC	MSABC/MSW
Within	N − (A × B × C)	MSW	
TOTAL	N − 1		

A4-76

Category: Analysis of Variance (ANOVA)

Statistical Procedure: Three-Way Analysis of Variance
　　　　　　　　　　　Mixed Effects Model
　　　　　　　　　　　One Fixed Independent Variable
　　　　　　　　　　　Two Random Independent Variables

Description: A procedure for examining the variance in some dependent measure in terms of the portion of variance that can be attributed to certain factors. In the three-way analysis of variance, total variation is partitioned into variation due to independent variables (1) A (with A levels), (2) B (with B levels), and (3) C (with C levels); the two-way interactions (4) A*B, (5) A*C, and (6) B*C; the (7) A*B*C three-way interaction and (8) the within-group variance. In this mixed effects model, all levels of one independent variable (A) are fixed and represented in the design, and the levels of two of the independent variables (B and C) are sampled from a larger set of possible levels.

Application: The appropriate procedure when the design involves three independent variables with one variable being fixed and the other two being random. One of the independent variables may be a blocking variable used to reduce within-cell variance (used as the error term) in a two-way analysis of variance without the blocking variable.

Data: The dependent variable is interval or ratio. The independent variables are discrete and may be nominal, ordinal, or, in some cases, interval.

Limitations: Assumes homogeneity of variance within groups (cells), normal distributions in the populations, and independence among observations (e.g., the subjects are not correlated).

Comments: Robust with respect to the assumption of homogeneity of variance. Often used with *a priori* or *post hoc* contrasts. One of the independent variables may be a blocking variable designed to increase statistical power by reducing error variance for hypothesis testing. Main effects cannot be interpreted directly in the presence of a significant interaction effect.

Formula: Three-Way Analysis of Variance, Mixed Model, One Factor Fixed (A), Two Factors Random (B, C)

Source of Variation	Degrees of Freedom	Mean Square	F-Ratio
Between A	A − 1	MSA	MSA / (MSAB + MSAC − MSABC)
Between B	B − 1	MSB	MSB/MSBC
Between C	C − 1	MSC	MSC/MSBC
Interaction of A*B	(A − 1)(B − 1)	MSAB	MSAB/MSABC
Interaction of A*C	(A − 1)(C − 1)	MSAC	MSAC/MSABC
Interaction of B*C	(B − 1)(C − 1)	MSBC	MSBC/MSW
Interaction of A*B*C	(A − 1)(B − 1)(C − 1)	MSABC	MSABC/MSW
Within	N − (A × B × C)	MSW	
TOTAL	N − 1		

A4-77

Category: Analysis of Variance (ANOVA)

Statistical Procedure: Three-Way Analysis of Variance
Mixed Effects Model
Two Independent Variables Fixed
One Independent Variable Random

Description: A procedure for examining the variance in some dependent measure in terms of the portion of variance that can be attributed to certain factors. In the three-way ANOVA, total variation is partitioned into variation due to independent variables (1) A (with A levels), (2) B (with B levels), and (3) C (with C levels); the two-way interactions (4) A*B, (5) A*C, and (6) B*C; the (7) A*B*C three-way interaction and (8) the within-group variance. In this mixed effects model, all levels of two independent variables (A and B) are fixed and represented in the design, and the levels of one of the independent variables (C) are sampled from a larger set of possible levels.

Application: The appropriate procedure when the design involves three independent variables with the levels of two variables being fixed and the levels of the other variable being sampled in the design. One of the independent variables may be a blocking variable used to reduce within-cell variance (used as the error term) in a two-way analysis of variance without the blocking variable.

Data: The dependent variable is interval or ratio. The independent variables are discrete and may be nominal, ordinal, or, in some cases, interval.

Limitations: Assumes homogeneity of variance within groups (cells), normal distributions in the populations, and independence among observations (e.g., the subjects are not correlated).

Comments: Robust with respect to the assumption of homogeneity of variance. Often used with *a priori* or *post hoc* contrasts **(A4-91).** One of the independent variables may be a blocking variable designed to increase statistical power by reducing error variance for hypothesis testing. Main effects cannot be interpreted directly in the presence of a significant interaction effect.

Formula: Three-Way Analysis of Variance, Mixed Model, Two Factors Fixed (A, B), One Factor Random (C)

Source of Variation	Degrees of Freedom	Mean Square	F-Ratio
Between A	A − 1	MSA	MSA/MSAC
Between B	B − 1	MSB	MSB/MSBC
Between C	C − 1	MSC	MSC/MSW
Interaction of A∗B	(A − 1)(B − 1)	MSAB	MSAB/MSABC
Interaction of A∗C	(A − 1)(C − 1)	MSAC	MSAC/MSW
Interaction of B∗C	(B − 1)(C − 1)	MSBC	MSBC/MSW
Interaction of A∗B∗C	(A − 1)(B − 1)(C − 1)	MSABC	MSABC/MSW
Within	N − (A × B × C)	MSW	
TOTAL	N − 1		

A4-78

Category: Analysis of Variance (ANOVA)

Statistical Procedure: Three-Way Analysis of Variance—Random Effects Model

Description: A procedure for examining the variance in some dependent measure in terms of the portion of variance that can be attributed to certain factors. In the three-way ANOVA, total variation is partitioned into variation due to independent variables (1) A (with A levels), (2) B (with B levels), and (3) C (with C levels); the two-way interactions (4) A∗B, (5) A∗C, and (6) B∗C; the (7) A∗B∗C three-way interaction and (8) the within-group variance. In the random effects model all levels of each independent variable are randomly sampled from a larger set of levels.

Application: The appropriate procedure when the design involves three independent variables and the levels of each variable are sampled from a larger set of levels. One of the independent variables may be a blocking variable used to reduce within-cell variance (the error term) in a two-way analysis of variance without the blocking variable.

Data: The dependent variable is interval or ratio. The independent variables are discrete and may be nominal, ordinal, or, in some cases, interval.

Limitations: Assumes homogeneity of variance within groups (cells), normal distributions in the populations, and independence among observations (e.g., the subjects are not correlated).

Comments: Robust with respect to the assumption of homogeneity of variance. Often used with *a priori* or *post hoc* contrasts **(A4-91)**. One of the independent variables may be a blocking variable designed to increase statistical power by reducing error variance for hypothesis testing. Main effects cannot be interpreted directly in the presence of a significant interaction effect.

Formula: Three-Way Analysis of Variance, Random Effects Model

Source of Variation	Degrees of Freedom	Mean Square	F-Ratio
Between A	A − 1	MSA	MSA / (MSAB + MSAC − MSABC)
Between B	B − 1	MSB	MSB / (MSAB + MSBC − MSABC)
Between C	C − 1	MSC	MSC / (MSAC + MSBC − MSABC)
Interaction of A∗B	(A − 1)(B − 1)	MSAB	MSAB/MSABC
Interaction of A∗C	(A − 1)(C − 1)	MSAC	MSAC/MSABC
Interaction of B∗C	(B − 1)(C − 1)	MSBC	MSBC/MSABC
Interaction of A∗B∗C	(A − 1)(B − 1)(C − 1)	MSABC	MSABC/MSW
Within	N − (A × B × C)	MSW	
TOTAL	N − 1		

A4-79

Category: Analysis of Variance (ANOVA)

Statistical Procedure: Repeated Measures Analysis of Variance—Subjects by Occasions

Description: An analysis of variance procedure used to test hypotheses about one group of subjects measured on two or more occasions. The procedure examines the total variance in some dependent measure in terms of variance attributable to between-subject variance (variation between different subjects), and within-subject variance (variation for the same subject across occasions). The procedure is described as a one-way repeated measures analysis. Nonetheless, it is equivalent to the two-way mixed model analysis of variance **(A4-74)** when subjects are viewed as a random variable and occasions as a fixed variable.

The basic arrangement for a one-way repeated measures analysis of variance is illustrated below with Factor O, Occasions, consisting of 4 levels, and n subjects (not shown) measured on each occasion.

Factor A, Occasions (4 Levels)

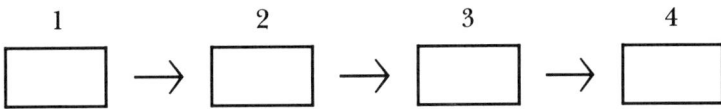

Application: The appropriate analysis when subjects are measured on multiple occasions. Commonly used in longitudinal designs. Appropriate analysis for randomized-blocks designs when A subjects in each block are randomly assigned to one of A levels of the fixed independent variable (see **A4-87**).

Data: Dependent variable is interval or ratio. Independent variables are discrete, nominal, ordinal and, sometimes, interval.

Limitations: Assumes homogeneity of variance within group, homogeneity of covariance between groups, normal distributions in the population, and the scale on which the dependent variable is measured is equivalent across occasions.

Comments: A powerful technique in which subjects act as their own controls. Analogous to the dependent sample t-test (**A4-50**) but appropriate for two or more occasions. Orthogonal polynomial contrasts are often used to test trends across occasions.

Formula: One-Way Repeated Measures Analysis of Variance for n Subjects on O Occasions.

Source of Variation	Degrees of Freedom	Mean Square	F-Ratio
Between Subjects	n − 1	MSsub	
Within Subjects			
Occasions	O − 1	MSO	$\dfrac{\text{MSO}}{\text{MSSub*Occ}}$
Sub*Occs	(n − 1)(O − 1)	MSSub*Occ	
TOTAL	(n × O) − 1		

A4-80

Category: Analysis of Variance (ANOVA)

Statistical Procedure: Two-Way Split-Pilot Analysis of Variance
  One Between-Subjects Factor
  One Within-Subjects Factor (Repeated Measures on Within-Subjects Factor)

Description: An analysis of variance procedure that examines the variance in some dependent measure in terms of the portion of variance attributable to a between-subjects Factor A (with A levels) and a within-subjects Factor B (with B levels). Subjects within the levels of Factor A are measured repeatedly across the levels of Factor B. Factor A may be different treatment conditions, experimental and control, and Factor B might be measures taken before, during, immediately after, and long after the experimental treatment. Factors A and B are generally considered fixed factors and n subjects are randomly assigned to the levels of Factor A.

The basic arrangement for this design is illustrated below with 4 levels of Factor B (Occasions) and 2 levels of Factor A (Groups).

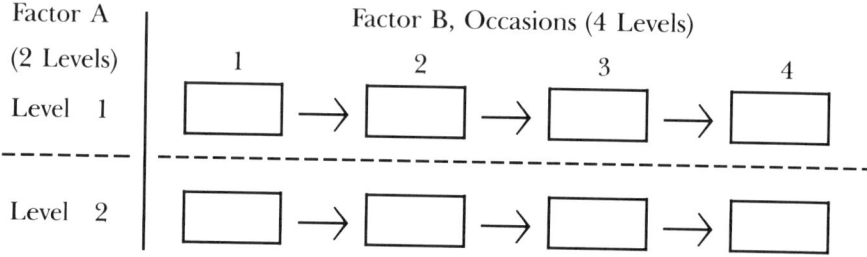

Application: The appropriate analysis when subjects in different groups or categories are measured across multiple occasions. For example, Factor A could be a classification variable such as gender, and Factor B might be successive treatment conditions. Commonly used in longitudinal designs when subjects in different groups are measured across time on several occasions. The appropriate analysis with a randomized-blocks design when subjects are grouped into blocks and measured across successive treatment conditions.

Data: Dependent variable is interval or ratio. Independent variables are discrete, nominal, ordinal, and, sometimes, interval.

Limitations: Assumes homogeneity of variance within group, homogeneity of covariance between groups, normal distributions in the population, and the scale on which the dependent variable is measured is equivalent across occasions.

Comments: A powerful technique in which variation between subjects is removed as a source of variation in error terms that are used to test the hypotheses of interest. Orthogonal polynomial contrasts are often used to test trends across occasions.

Formula: Two-Way Split-Plot Analysis of Variance
Factor A, Between Subjects, has A levels.
There are n subjects (S's) in each level of A.
Factor B, Within Subjects, has B levels.

Source of Variation	Degrees of Freedom	Mean Square	F-Ratio
Between Subjects			
A	A − 1	MSA	$\dfrac{\text{MSA}}{\text{MS(S's in A)}}$
S's in A	A (n − 1)	MS(S's in A)	
Within Subjects			
B	B − 1	MSB	$\dfrac{\text{MSB}}{\text{MSB*(S's in A)}}$
A*B	(A − 1)(B − 1)	MSAB	$\dfrac{\text{MSAB}}{\text{MSB*(S's in A)}}$
B*(S's in A)	(B − 1)[A (n − 1)]	MSB*(S's in A)	
TOTAL	(A × B × n) − 1		

A4-81

Category: Analysis of Variance (ANOVA)

Statistical Procedure: Three-Way Split-Plot Analysis of Variance
One Between-Subjects Factor
Two Within-Subjects Factors (Repeated Measures on Within-Subjects Factors)

Description: An ANOVA procedure that examines the variance in some dependent measure in terms of the portion of variance attributable to a between-subjects Factor A (with A levels) and within-subjects Factor B (with B levels) and Factor C (with C levels). Subjects within the levels of Factor A are measured repeatedly across the levels of Factor B and across the levels of Factor C. Factor A (between-subjects) may be different treatment conditions. Factor B (within-subjects) might be repeated trials on some task. The trials of Factor B are attempted on two or more occasions, and these occasions represent Factor C. Factor A could be a classification variable such as gender. Factor B might be a set of trials on an experimental task, and Factor C might be successive treatment conditions under which the trials of Factor B are attempted. Factors A, B, and C are generally considered fixed factors, and n subjects are randomly assigned to the levels of Factor A.

The basic arrangements for this design are illustrated. Factor A, between-subjects, has 2 levels; Factor B, within-subjects, has 3 levels; and Factor C, within-subjects, has 4 levels.

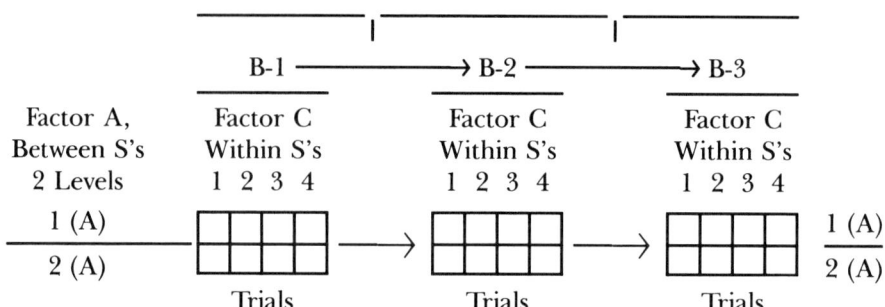

Application: The appropriate analysis when subjects in different groups or categories (Factor A) are measured repeatedly (Factor C) across multiple occasions (Factor B). Commonly used in longitudinal designs when subjects in different groups are measured under different conditions across time on several occasions. This analysis could be used with a randomized blocks design (A4-87) when subjects are grouped into blocks and measured repeatedly across successive treatment conditions.

Data: Dependent variable is interval or ratio. Independent variables are discrete, nominal, ordinal and, sometimes, interval.

Limitations: Assumes homogeneity of variance within groups, homogeneity of covariance between groups, normal distributions in the population, and the scale on which the dependent variable is measured is equivalent across occasions.

Comments: A powerful technique in which variation between subjects is removed as a source of variation in error terms used to test the hypotheses of interest. Orthogonal polynomial contrasts are often used to test trends across occasions.

Formula: Three-Way Split-Plot Analysis of Variance
Factor A, Between Subjects, has A levels.
There are n subjects (S's) in each levels of A.
Factor B, Within Subjects, has B levels.
Factor C, Within Subjects, has C levels.
Factors A, B, and C are fixed.

Source of Variation	Degrees of Freedom	Mean Square	F-Ratio
Between Subjects			
A	A − 1	MSA	$\dfrac{\text{MSA}}{\text{MS(S's in A)}}$
S's in A	A(n − 1)	MS(S's in A)	
Within Subjects			
B	B − 1	MSB	$\dfrac{\text{MSB}}{\text{MSB*(S's in A)}}$
B*A	(B − 1)(A − 1)	MSAB	$\dfrac{\text{MSAB}}{\text{MSB*(S's in A)}}$
B*(S's in A)	(B − 1)[A(n − 1)]	MSB*(S's in A)	
C	C − 1	MSC	$\dfrac{\text{MSC}}{\text{MSC*(S's in A)}}$
C*A	(C − 1)(A − 1)	MSAC	$\dfrac{\text{MSAC}}{\text{MSC*(S's in A)}}$
C*(S's in A)	(C − 1)[A(n − 1)]	MSC*(S's in A)	
C*B	(C − 1)(B − 1)	MSCB	$\dfrac{\text{MSCB}}{\text{MSCB*(S's in A)}}$
C*B*A	(C − 1)(B − 1)(A − 1)	MSABC	$\dfrac{\text{MSABC}}{\text{MSCB*(S's in A)}}$
C*B*(S's in A)	(C − 1)(B − 1)[A(n − 1)]	MSCB*(S's in A)	
TOTAL	(A × B × C × n) − 1		

A4-82

Category: Analysis of Variance (ANOVA)

Statistical Procedure: Three-Way Split-Plot Analysis of Variance
Two Between-Subjects Factors

One Within-Subjects Factor (Repeated Measures on Within-Subjects Factor)

Description: An analysis of variance procedure that examines the variance in some dependent measure in terms of the portion of variance attributable to between-subjects Factor A (with A levels), and Factor B (with B levels), and within-subjects Factor C (with C levels). Subjects within the levels of Factor B are within the levels of Factor A and are measured repeatedly across the levels of Factor C. Factor A (between subjects) might be a classification variable such as gender, Factor B might be a set of treatment conditions, and Factor C (within subjects) might be successive measurements of the subjects before, during, and after the treatments are applied. Factors A, B, and C are generally considered fixed factors and n subjects in the different levels of Factor A are randomly assigned to the levels of Factor B.

The basic arrangements for this design are illustrated. Factor A, between subjects, has 2 levels; Factor B, between subjects, has 3 levels; and Factor C, within subjects, has 3 levels.

Application: The appropriate analysis when subjects in different groups or categories (Factor A) are assigned to different levels of a second Factor B, and measured repeatedly across multiple occasions (Factor C). Commonly used in longitudinal designs where subjects in different groups are assigned to different treatment conditions and measured on several occasions.

Data: Dependent variable is interval or ratio. Independent variables are discrete, nominal, ordinal and, sometimes, interval.

Limitations: Assumes homogeneity of variance within group, homogeneity of covariance between groups, normal distributions in the population, and the scale on which the dependent variable is measured is equivalent across occasions.

Comments: A powerful technique in which variation between subjects is removed as a source of variation in error terms that are used to test the hypotheses of interest. Orthogonal polynomial contrasts are often used to test trends across occasions.

Formula: Three-Way Split-Plot Analysis of Variance
Factor A, between subjects, has A levels.
Factor B, between subjects, has B levels.
There are n subjects (S's) in each AB level.
Factor C, within subjects, has C levels.
Factors A, B, and C are fixed.

Source of Variation	Degrees of Freedom	Mean Square	F-Ratio
Between Subjects			
A	A − 1	MSA	$\dfrac{\text{MSA}}{\text{MS(S's in AB)}}$
B	B − 1	MSB	$\dfrac{\text{MSB}}{\text{MS(S's in AB)}}$
AB	(A − 1)(B − 1)	MSAB	$\dfrac{\text{MSAB}}{\text{MS(S's in AB)}}$
S's in AB	AB(n − 1)	MS(S's in AB)	
Within Subjects			
C	C − 1	MSC	$\dfrac{\text{MSC}}{\text{MSC*(S's in AB)}}$
C*A	(C − 1)(A − 1)	MSCA	$\dfrac{\text{MSCA}}{\text{MSC*(S's in AB)}}$
C*B	(C − 1)(B − 1)	MSCB	$\dfrac{\text{MSCB}}{\text{MSC*(S's in AB)}}$
C*AB	(C − 1)(A − 1)(B − 1)	MSCAB	$\dfrac{\text{MSCAB}}{\text{MSC*(S's in AB)}}$
C*(S's in AB)	(C − 1)[AB(n − 1)]	MSC*(S's in AB)	
TOTAL	(A × B × C × n) − 1		

A4-83

Category: Analysis of Variance

Statistical Procedure: Four-Way Split-Plot Analysis of Variance
Two Between Subjects Factors
Two Within-Subjects Factors (Repeated Measures on Within-Subjects Factors)

Description: An ANOVA procedure that examines the variance in some dependent measure in terms of the portion of variance attributable to between-subjects

Factor A (with A levels), Factor B (with B levels), and within-subjects Factor C (with C levels) and Factor D (with D levels). Subjects within the levels of Factor A are randomly assigned to the levels of Factor B and are measured repeatedly across the levels of Factor C on multiple occasions, which are the levels of Factor D. Factor A (between subjects) might be a classification variable such as gender. Factor B might be a set of treatment conditions, and Factor C (within subjects) might be repeated trials on some tasks attempted before and after (Factor D) the treatments are applied. Factors A, B, C, and D are generally considered fixed factors and n subjects in the different levels of Factor A are randomly assigned to the levels of Factor B.

Application: The appropriate analysis when subjects in different groups or categories (Factor A) are assigned to different levels of a second Factor B, and are measured repeatedly (Factor C) on multiple occasions (Factor D). This procedure is used in longitudinal designs when subjects in different groups are assigned to different treatment conditions and are measured several times on several occasions.

Data: Dependent variable is interval or ratio. Independent variables are discrete, nominal, ordinal, or, sometimes, interval.

Limitations: Assumes homogeneity of variance within groups, homogeneity of covariance between groups, normal distributions in the population, and the scale on which the dependent variable is measured is equivalent across occasions.

Comments: A powerful technique in which variation between subjects is removed as a source of variation in error terms that are used to test the hypotheses of interest. Orthogonal polynomial contrasts are often used to test trends across occasions.

Formula: Four-Way Split-Plot Analysis of Variance
Factor A, Between Subjects, has A levels.
Factor B, Between Subjects, has B levels.
There are n subjects (S's) in each AB level.
Factor C, Within Subjects, has C levels.
Factor D, Within Subjects, has D levels.
Factors A, B, C, and D are fixed.

Source of Variation	Degrees of Freedom	Mean Square	F-Ratio
Between Subjects			
A	A − 1	MSA	$\dfrac{\text{MSA}}{\text{MS(S's in AB)}}$
B	B − 1	MSB	$\dfrac{\text{MSB}}{\text{MS(S's in AB)}}$
AB	(A − 1)(B − 1)	MSAB	$\dfrac{\text{MSAB}}{\text{MS(S's in AB)}}$
S's in AB	AB(n − 1)	MS(S's in AB)	
Within Subjects			
C	C − 1	MSC	$\dfrac{\text{MSC}}{\text{MSC*(S's in AB)}}$
C*A	(C − 1)(A − 1)	MSCA	$\dfrac{\text{MSCA}}{\text{MSC*(S's in AB)}}$
C*B	(C − 1)(B − 1)	MSCB	$\dfrac{\text{MSCB}}{\text{MSC*(S's in AB)}}$
C*AB	(C − 1)(A − 1)(B − 1)	MSCAB	$\dfrac{\text{MSCAB}}{\text{MSC*(S's in AB)}}$
C*(S's in AB)	(C − 1)[AB(n − 1)]	MSC*(S's in AB)	
D	D − 1	MSD	$\dfrac{\text{MSD}}{\text{MSD*(S's in AB)}}$
D*A	(D − 1)(A − 1)	MSDA	$\dfrac{\text{MSDA}}{\text{MSD*(S's in AB)}}$
D*B	(D − 1)(B − 1)	MSDB	$\dfrac{\text{MSDB}}{\text{MSD*(S's in AB)}}$
D*AB	(D − 1)(A − 1)(B − 1)	MSDAB	$\dfrac{\text{MSDAB}}{\text{MSD*(S's in AB)}}$
D*(S's in AB)	(D − 1)[AB(n − 1)]	MSD*(S's in AB)	
CD	(C − 1)(D − 1)	MSCD	$\dfrac{\text{MSCD}}{\text{MSCD*(S's in AB)}}$
CD*A	(C − 1)(D − 1)(A − 1)	MSCDA	$\dfrac{\text{MSCDA}}{\text{MSCD*(S's in AB)}}$
CD*B	(C − 1)(D − 1)(B − 1)	MSCDB	$\dfrac{\text{MSCDB}}{\text{MSCD*(S's in AB)}}$
CD*AB	(C − 1)(D − 1)(A − 1)(B − 1)	MSCDAB	$\dfrac{\text{MSCDAB}}{\text{MSCD*(S's in AB)}}$
CD*(S's in AB)	(C − 1)(D − 1)[AB(n − 1)]	MSCD*(S's in AB)	
TOTAL	(ABCDn) − 1		

A4-84

Category: Analysis of Variance (ANOVA)

Statistical Procedure: Two-Factor Nested (Hierarchical) Design
One Factor Nested in Another Factor

Description: An analysis of variance procedure that examines the variance in some dependent measure that can be attributed to different levels of some Factor B, (with B levels), which are nested within the levels of Factor A (with A levels). Not all levels of Factor B appear across all levels of Factor A. Rather a subset of the levels of Factor B appear under the different levels of Factor A.

For example, Factor B might be teachers and Factor A might be two different methods of instruction. If half the teachers use one method of instruction and half use the other, then teachers are nested under treatment.

The general arrangement of a two-factor nested design is illustrated for Factor A with 2 levels and Factor B with 10 levels.

Factor A	
Level 1	Level 2
Factor B	Factor B
Level 1	Level 6
Level 2	Level 7
Level 3	Level 8
Level 4	Level 9
Level 5	Level 10

Application: The appropriate analysis in two-factor designs when only some of the levels of one Factor (B) appear under each of the levels of the other Factor (A). Applicable when teachers or classrooms (Factor B) are assigned to use one of a set of different teaching methods (Factor A). Teachers/classrooms are nested under treatment in such designs. The appropriate procedure when intact groups such as classrooms are assigned to levels of some treatment factor.

Data: Dependent variable is interval or ratio. Independent variables are nominal, ordinal, or, sometimes, interval.

Limitations: Assumes homogeneity of variance within the nested cells, normal distributions in the populations. The subjects in the nested cells are not independent but the levels of the nested factor are independent. Assumes no interaction between factors B and A.

Comments: The appropriate but often ignored analysis when different teachers use different instructional methods in studies designed to assess differences in instructional methods. Often analyzed with a one-way analysis of variance **(A4-71)**, with teaching method as the only independent variable. Such analyses violate the assumption of subject independence since sets of subjects within treatment conditions are related by reason of being in the same classroom with the same teacher. Furthermore, the one-way analysis of variance incorporates the variation between teacher within teaching method into the within-cell variance used to test hypotheses about treatment effects.

Formula: Two-Factor Nested Analysis of Variance (ANOVA)

Factor A with A levels

Factor B with B levels nested in each level of A

Factor B is random.

There are n subjects in each level of B nested in A.

Source of Variation	Degrees of Freedom	Mean Square	F-Ratio
A	A − 1	MSA	$\dfrac{\text{MSA}^a}{\text{MSB (in A)}}$
B (in A)	A(B − 1)	MSB (in A)	$\dfrac{\text{MSB (in A)}}{\text{MSE}}$
Error within cell	AB(n − 1)	MSE	
TOTAL	(ABn) − 1		

[a]If Factor B is fixed, $F = \dfrac{\text{MSA}}{\text{MSE}}$

A4-85

Category: Analysis of Variance (ANOVA)

Statistical Procedure: Nested (Hierarchical) Design
　　　　　　　　　　　Three Factors, Two Levels of Nesting
　　　　　　　　　　　Factor C Nested in Factor B
　　　　　　　　　　　Factor B Nested in Factor A

Description: An ANOVA procedure that examines the variance in some dependent measure that can be attributed to different levels of some Factor C (with C levels) nested within the levels of Factor B (with B levels). The levels of Factor B are themselves nested within the levels of Factor A (with A levels). Not all levels of Factor C appear across all levels of Factor B and not all levels of Factor B appear across all levels of Factor A. Rather, a subset of the levels of Factor C appear under different levels of Factor B and a subset of the levels of Factor B appear under the different levels of Factor A. For example, teachers might be Factor C, nested in different schools (Factor B), using different instructional methods (Factor A). The basic arrangement for such a design follows. Factor A has 2 levels, Factor B has 4 levels, and Factor C has 12 levels. Factor C, levels 1, 2, and 3, are nested under level 1 of Factor B. Factor C, levels 4, 5, and 6, are nested under level 2 of Factor B. Factor C, levels 7, 8, and 9, are nested under Factor B, level 3. Factor C, levels 10, 11, and 12, are nested under Factor B, level 4. Factor B, levels 1 and 2, are nested under Factor A, level 1. Factor B, levels 3 and 4, are nested under Factor A, level 2.

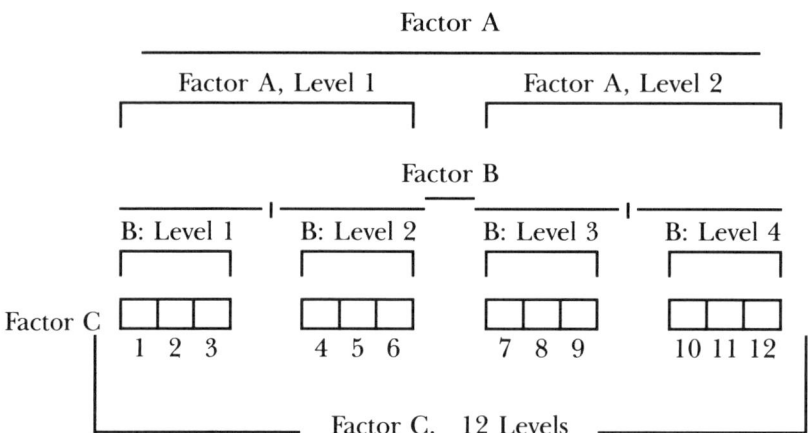

Application: The appropriate analysis in three-factor designs where only some of the levels of one Factor (C) appear under one of the levels of another Factor (B), which itself is nested under a third Factor (A). Applicable when teachers/classrooms (Factor C), within different schools (Factor B) are assigned to use one of a set of different teaching methods (Factor A). Teachers/classrooms are nested under schools and under treatment in such designs. The appropriate procedure when intact groups such as classrooms are assigned to levels of some treatment factor.

Data: Dependent variable is interval or ratio. Independent variables are nominal, ordinal, or, sometimes, interval.

Limitations: Assumes homogeneity of variance within the nested cells, normal distributions in the populations. The subjects in the nested cells are not independent, but the levels of the nested factor are independent. Assumes no interactions between or among the factors.

Comments: The appropriate but often ignored analysis when different teachers in different schools use different instructional methods in studies designed to assess differences in instructional methods. Schools and methods are not independent. Often analyzed (inappropriately) with a one-way analysis of variance (teaching method). Such analyses violate the assumption of subject independence since sets of subjects within treatment conditions are related by reason of being in the same classroom with the same teacher.

Formula: Three-Factor Nested Analysis of Variance
  Factor A with A levels
  Factor B with B levels nested in each level of A
  Factor C with C levels nested in each level of B
  Factors B and C are random.
  There are n subjects in each level of C nested in B nested in A.

Source of Variation	Degrees of Freedom	Mean Square	F-Ratio
A	A − 1	MSA	$\dfrac{\text{MSA}^a}{\text{MSB (in A)}}$
B (in A)	A(B − 1)	MSB (in A)	$\dfrac{\text{MSB (in A)}^a}{\text{MSC (in B in A)}}$
C (in B in A)	AB(C − 1)	MSC (in B in A)	$\dfrac{\text{MSC (in B in A)}^a}{\text{MSE}}$
Error within cell	ABC(n − 1)	MSE	
TOTAL	(ABCn) − 1		

[a] If all factors are fixed, the MSE is the denominator in all the F-ratios.

A4-86

Category: Analysis of Variance (ANOVA)

Statistical Procedure: Three-Factor Design
One Factor Nested in Another
Third Factor Crossed with the Nested Factors

Description: An analysis of variance procedure that examines the variance in some dependent measure in terms of the portion of variance that can be attributed to (1) Factor B (with B levels), which is nested in (2) Factor A (with A levels) and (3) Factor C (with C levels), which is crossed with Factor B nested in A. Appropriate in nested designs that have a third factor crossing all levels of the nested factors.

The basic arrangement for this type of design is shown. In this illustration, Factor C, with 2 levels, is crossed with Factor B, with 6 levels. These factors are crossed because all levels of Factor C appear under all levels of Factor B. Factor B is nested in Factor A, with 2 levels. These factors are nested since only some of the levels of Factor B appear under some of the levels of Factor A. Levels 1, 2, and 3 of Factor B are nested under level 1 of Factor A. Levels 4, 5, and 6 of Factor B are nested under level 2 of Factor A.

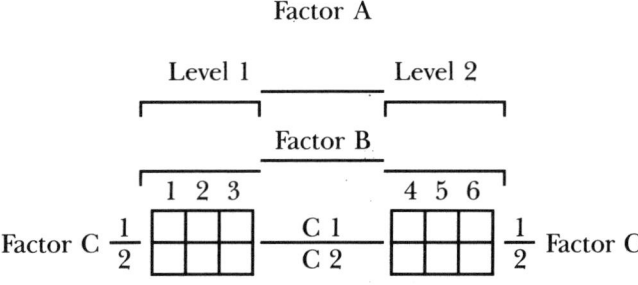

Application: Used when some classification or categorical Factor C crosses the levels of some nested Factor B within the levels of some third Factor A. For example, this analysis would be used in a design in which teachers (Factor B) were nested within experimental teaching method (Factor A) and students' gender (Factor C) was crossed with teachers nested in method.

Data: The dependent variable is interval or ratio. The independent variables are discrete, nominal, ordinal, and, sometimes, interval.

Limitations: Assumes homogeneity of variance within groups, normal distributions in the populations, and no interactions between the nested factors (e.g., Factor B and Factor A).

Comments: An appropriate analysis when teaching methods (Factor A) are assigned to schools and taught by different teachers (Factor B). Factor C might be gender of students or some aptitude measure employed to assess aptitude-treatment interaction (C × A) effects.

Formula: Crossed and Nested Analysis of Variance
Factor A with A levels
Factor B with B levels nested in each level of Factor A
Factor C with C levels crossed with B in A
Factors A and C are fixed, Factor B is random.
There are n subjects in each level of C crossed with B in A.

Source of Variation	Degrees of Freedom	Mean Square	F-Ratio
A	A − 1	MSA	$\dfrac{\text{MSA}^a}{\text{MSB (in A)}}$
B (in A)	A(B − 1)	MSB (in A)	$\dfrac{\text{MSB (in A)}}{\text{MSE}}$
C	C − 1	MSC	$\dfrac{\text{MSC}^a}{\text{MSC*[B (in A)]}}$
C*A	(C − 1)(A − 1)	MSCA	$\dfrac{\text{MSCA}^a}{\text{MSC*[B (in A)]}}$
C*[B (in A)]	(C − 1)[A(B − 1)]	MSC*B (in A)	$\dfrac{\text{MSC*[B (in A)]}}{\text{MSE}}$
Error within cell	ABC (n − 1)	MSE	
TOTAL	(ABCn) − 1		

[a] If all effects are fixed, the MSE is in the denominator for all F-ratios.

A4-87

Category: Analysis of Variance (ANOVA)

Statistical Procedure: Randomized-Blocks Designs

Description: A large class of research designs and analysis of variance procedures in which a blocking variable is used in the design and analysis to reduce error variance. The blocking variable is an independent variable added to the design and analysis. Hypotheses about the blocking factor are not usually of any particular interest. There are always statistically significant differences in the levels of the blocking factor or the factor is of no value in reducing error variance. The analysis with the blocking factor, in contrast to the same analysis without the blocking factor, partitions the error term into a new source of systematic variation, namely variation due to the blocking factor.

Application: Randomized-blocks designs can be used with virtually any analysis of variance model. The levels of the blocking factor may contain one subject in each cell in which case a repeated measures design is appropriate with replications on the blocking variable. Such a design would be treated as a two-way mixed model **(A4-74),** with treatments fixed and subjects (the blocking factor) random. In all applications, the researcher must determine whether the levels of the blocking factor are fixed or random, and whether each cell will contain one or more subjects.

Data: The dependent variable is interval or ratio. Independent variables are discrete and may be nominal, ordinal, or, in some cases, interval.

Limitations: Assumes homogeneity of variance within groups, normal distributions in the populations, and other assumptions depending on the specific nature of the design.

Comments: Randomized-blocks designs with appropriate analyses are a class of very powerful error-reducing procedures. Occasionally, a factor can be both a blocking variable and a variable of interest, as in the case of aptitude-treatment interaction research. Levels of the aptitude factor reduce within-group variance and thus increase the power to detect treatment effects and, at the same time, enable the researcher to check for aptitude-treatment interactions.

Care must be exercised in using randomized-blocks designs. If the blocking variable does not account for a significant proportion of variation, the analysis becomes less powerful in detecting effects on the other factors because degrees of freedom for error are reduced when the blocking factor is partitioned out of the error term. In addition, it is essential that all blocking factors in the design be incorporated into the analysis because, otherwise, within-cell error is maximized and degrees of freedom for error are reduced.

Formula: Randomized-blocks designs can be incorporated into virtually all of the previously described ANOVA procedures (**A4-71** to **A4-83**) with the blocking factor simply becoming one of the factors in the design and analysis.

A4-88

Category: Analysis of Variance (ANOVA)

Statistical Procedure: Latin Square Designs

Description: A large class of research designs and analysis of variance procedures in which two sources of potential extraneous variation can be controlled by including factors in the design and analysis which represent them. Very often used to control for effects due to the order in which treatments are presented. Also used to control for practice effects that are related to the order of the treatments.

Latin squares will be described using a 3 × 3 model to assess the effects of three treatments, A, B, and C. The design is constructed by first selecting a Latin square of the appropriate size from a table of Latin square designs. In the 3 × 3 case with treatments A, B, and C, one such design would have the form:

| | Columns | | |
	1	2	3
Row 1	A	B	C
Row 2	B	C	A
Row 3	C	A	B

Each treatment appears once in each row and column. This is referred to as the standard form. The three rows are then randomly repositioned and the three columns are randomly repositioned. Each treatment still appears only once in each row and column.

Application: Latin square designs are commonly used to control variation due to order, sequence, practice effects, and other possible confounding effects in experimental research. In this example of a 3 × 3 Latin square design, suppose A, B and C were experimental treatments and n subjects were assigned to each of the 9 cells in the design. The row and column effects would partition out from the error term variation due to order (ABC; BCA; CAB) and ordinal position (first, second, third).

Latin square designs are commonly used in repeated measures analyses in which a subject or group of subjects are assigned to each row and are measured repeatedly across the columns.

Lastly, Latin square designs are used to simplify higher order factorial designs by using the Latin square as an incomplete factorial design. With a 3 × 3 × 3 design for example, two of the factors could be used as row and column control variables in the Latin square used to study the third factor.

Data: The dependent variable is interval or ratio. The independent variables are discrete and are nominal, ordinal, or, sometimes, interval.

Limitations: Assumes homogeneity of variance within groups, normal distributions in the respective populations, and, with repeated measures designs, the assumption of homogeneity of covariance is made.

A major assumption made in the analysis of Latin square designs is that row-by-column interaction effects are negligible or zero.

Comments: Latin square designs are useful and powerful in controlling sources of confounding variance in experimental research. The application of Latin square designs requires considerable control in conducting an experiment, and thus such designs often cannot be used in applied educational research where rigorous control is not possible.

Formula: For an L × L Latin Square Design
With Rows (R), Columns (C), and Treatments (T) each with L levels, and n subjects in each cell.

Source of Variation	Degrees of Freedom	Mean Square	F-Ratio
Rows	L − 1	MSR	MSR/MSW
Columns	L − 1	MSC	MSC/MSW
Treatment	L − 1	MSTrt	MSTrt/MSW
Residual	(L − 1)(L − 2)	MSRes	MSRes/MSW
Within	L × L × (n − 1)	MSW	
TOTAL	(L × L × n) − 1		

In Latin squares with repeated measures such that n = 1, the denominator for all F-ratios is MSRs.

A4-89

Category: Analysis of Variance (ANOVA)
   Related Designs
   Analysis of Covariance (ANCOVA)

Statistical Procedure: Analysis of Covariance

Description: A large class of designs and associated analyses in which a continuous independent variable known as a covariate is introduced into an analysis of variance design to increase the statistical power of the analysis. The covariate is sometimes called a concomitant variable, and more than one covariate can be used. The covariate is systematically related to the dependent variable and is used to reduce the estimate of random or error variance in the dependent measure. The reduction of error variance is accomplished by regressing the dependent variable onto the covariate and extracting the sum of squares due to regression from the sum of squares due to error. Covariates can be used in virtually all analysis of variance designs.

Application: Analysis of covariance is commonly used in pretest-posttest designs with the pretest as a covariate. Alternate forms of tests can be used as covariate and dependent measures. Aptitude measures are often used as covariates in analyses that have achievement measures as dependent variables.
   Analysis of covariance is sometimes used to correct for sampling bias or error. This application of analysis of covariance can be misleading.

Data: The dependent variable is interval or ratio. Independent variables are discrete and may be nominal, ordinal, or, sometimes, interval. The covariate is continuous and is interval or ratio.

Limitations: All assumptions of the corresponding analysis of variance procedure apply to the analysis of covariance. In addition, analysis of covariance assumes that the relationships between the covariates and the dependent measure are statistically equivalent within all groups or cells in the design. With a one-way analysis of covariance, this assumption states that the regression lines of the dependent variable that has been regressed onto the covariate within each group are all parallel. In general, this assumption is referred to as parallelism of regression. Violations of this assumption seriously jeopardize the validity of inferences drawn from the analysis.

Comments: The value of analysis of covariance is directly related to the strength of the relationship between the dependent variable and the covariate. The stronger the relationship, the greater the gain in power and precision. Using a covariate that is weakly related to the dependent variable can be counterproductive and actually reduce power and precision. This is because the covariate reduces the error degrees of freedom without reducing the sum of squares due to error.

Using analysis of covariance to correct for bias or errors in sampling, or for the actual difference between groups, can be problematic. Such applications create statistical conditions that may have no real-life counterpart. For example, consider an experiment in which a remedial-education treatment group is compared to a control group composed of children in regular classes, using a one-way analysis of covariance with a pretest as covariate and posttest as the dependent measure. The analysis of covariance leads to inferences about group differences on the posttest under the statistical condition that the two groups were equivalent at the beginning of the study. Such inferences are valid and generalize to other situations only if the same conditions exist—that is, only if the groups are equivalent at the beginning of instruction. In real school settings, however, students referred for remedial instruction and students in regular classes are not equivalent prior to instruction. Thus the conditional inference from the analysis of covariance could be misleading since it does not apply to the context of greatest interest.

Formula: The analysis of covariance is quite similar to the analysis of variance into which the covariate is introduced. The analysis of covariance has an additional source of variation due to the covariate(s). This source of variation has one degree of freedom for each covariate, and the degrees of freedom are extracted from the error degrees of freedom.

A one-way analysis of covariance is shown as an illustration.

The number of covariates = I

Factor A, between groups, has A levels.

There are n subjects in each group.

Source of Variation	Degrees of Freedom	Mean Square	F-Ratio
Covariate	I	MSCOV	MSCOV/I
Between Groups	A − 1	MSB	MSB/MSW
Within Groups	N − A − I	MSW	
TOTAL	N − 1		

A4-90

Category: Analysis of Variance (ANOVA)
  Related Designs
  Time-Series Analysis

Statistical Procedure: Time-Series Analysis
  Interrupted Time-Series Analysis

Description: Time-series analysis and interrupted time-series analysis refer to a large class of procedures and analyses used to examine trends or changes over time. These procedures can be performed using various regression procedures or a specialized set of procedures called Auto Regressive Integrated Moving Average (ARIMA) models. The ARIMA models are useful in a wide variety of applications. ARIMA models are designed to deal with three sources of "noise" that confound the study of change over time. These are: (1) trend—gradual but consistent drifts upward or downward over time, (2) seasonality—cyclic trends repeating at certain times, and (3) random error. ARIMA models have three structural parameters, namely P, D, and Q. ARIMA models are specified in terms of these parameters as ARIMA (P, D, Q) models. P refers to the autoregressive relationship. D refers to the series being "differenced," for example, subtracting successive values in the series. And Q refers to the number of moving average structures in the model.

  For additional information, the interested reader should consult Ostrom (1978), McDowall, McCleary, Meidinger, and Hay (1980) (both cited at the end of this chapter), and Cook and Campbell, (1979), (cited at the end of chapter 3 above).

Application: Time-series analysis can be applied to any situation where there are periodic observations or measurements over some period of time. Time-series analysis is used in a wide range of disciplines including economics, education, history, psychology, and sociology. Interrupted time-series analysis is used to assess the impact of some event or treatment on some dependent measure. In education, for example, these procedures could be used to assess the impact of changes in school policy on dropout rates. Dropout rates for a period of years before and after the policy change would be examined using the appropriate ARIMA. The ARIMA model would be used to determine if there had been a change in the dropout rate associated with the policy change.

Data: A wide range of data can be used with these procedures. Many data points are required for their application.

Limitations: Time-series analysis is not applicable to many educational situations because a large number of observations or measurements must be made over time.

Comments: Time-series procedures are generally applied in quasi-experimental settings where the researcher does not have control over the selection and assignment of subjects. The application of these procedures can, nonetheless, be quite informative.

  The number of periodic observations needed to apply these procedures is difficult to specify; 20 observations would constitute a relatively short series of

114   RESEARCH PROCESSES

observation. With a small number of observations, Repeated Measures and Split-Plot Analysis of Variance (**A4-79** to **A4-83**) procedures need to be considered if the necessary assumption are met.

Formula: The references in the Description section provide the various formulas for time-series analysis.

A4-91
MULTIPLE-COMPARISON PROCEDURES

A Priori Planned Comparisons
(**A4-92**) Orthogonal Contrasts
(**A4-93**) Orthogonal Polynomial Contrasts

Post Hoc Comparisons
(**A4-94**) Scheffé Contrasts
(**A4-95**) Tukey's Honestly Significant Difference
(**A4-96**) Newman-Keuls Test
(**A4-97**) Duncan Multiple-Range Test

A4-92

Category: Multiple Comparison Procedures—*A Priori* Planned Comparisons

Statistical Procedure: Orthogonal Contrasts

Description: A general procedure for independent tests of particular hypotheses about specific means or combinations of means in the context of analysis of variance when the overall test of the null hypothesis is not of interest. The use of orthogonal contrasts maintains the Type I error rate for the individual comparisons and for the total set of analyses. Orthogonal contrasts test particular hypotheses by partitioning the sum of squares and degrees of freedom for a given factor into separate and independent sources of variation. For Factor A (with A levels), there are $A - 1$ degrees of freedom for orthogonal contrasts. There can be only $A - 1$ orthogonal contrasts. Contrasts are orthogonal if the sum of the cross products of their corresponding coefficients is equal to zero. These contrasts do not require equal cell sizes.

Application: Orthogonal contrasts are used to test particular hypotheses about the difference between specific means or combinations of means. In such cases, the overall F-test for the null hypothesis may be of no particular interest. Studies involving experimental treatment conditions and a control group often use orthogonal contrasts to compare different experimental conditions to each other and to compare the combined experimental treatment conditions with the control group.

Data: The same data are used in the orthogonal contrasts as are used in the analysis of variance.

Limitations: All assumptions of the analysis of variance procedure apply. By definition, the orthogonal contrasts are independent of each other, which means that the sum of the cross products of the corresponding coefficients is zero. There can be only as many orthogonal contrasts as there are degrees of freedom for the factor in question.

Comments: Orthogonal contrasts are used when the researcher has specific hypotheses within the overall analysis of variance design. Often used in experimental research in which theoretical considerations lead to specific hypotheses designed to test theoretical claims.

Formula: Orthogonal contrasts are tested with an F-statistic using 1 and the degrees of freedom within from the analysis of variance procedure, for example, the degrees of freedom for contrast = (1, df within).

$$F = \frac{\left(\sum_{j=1}^{A} C_j \overline{X}_j\right)^2}{\text{MSW}\left[\sum_{j=1}^{A} (C_j^2/n_j)\right]}$$

C_j = Contrast coefficient for Group j
MSW = Mean square within from ANOVA

A4-93

Category: Multiple Comparisons Procedures—*A Priori* Planned Comparisons

Statistical Procedure: Orthogonal Polynomial Contrasts

Description: A special type of orthogonal contrast that tests geometric trends in the differences among means. Orthogonal polynomial contrasts have all the characteristics of general orthogonal contrasts. In addition, they specifically test whether differences among group means are linear, quadratic, cubic, quartic, or some higher order polynomial.

Application: Orthogonal polynomial contrasts are very commonly used in longitudinal research with repeated measures designs to test the nature of change across developmental periods. Similar use is made of these procedures in cross-sectional research designs. Orthogonal polynomials are often used when the levels of some independent variable differ quantitatively not qualitatively. In such applications, the procedure describes trends in the outcome measure across groups (levels of the independent variable). These trends might be described as linear, quadratic, cubic, or whatever order polynomial is appropriate.

Data: The same data are used in the orthogonal polynomial contrasts as are used in the analysis of variance.

Limitations: All assumptions of the analysis of variance procedure apply. By definition, the orthogonal polynomial contrasts are independent of each other. There are as many orthogonal polynomial contrasts as there are degrees of freedom for the factor in question.

Comments: Orthogonal polynomial contrasts are used when the researcher has specific hypotheses about the nature of the differences among groups in the overall analysis of variance design. Often used in developmental research with repeated measures designs to examine the nature of growth that occurs across developmental periods.

Formula: Orthogonal polynomial contrasts are tested with an F-statistic with 1 and the degrees of freedom within from the analysis of variance.

Orthogonal polynomial coefficients for factors with 3, 4 and 5 levels are:

Levels	Trend	Coefficients
3	Linear	−1 0 1
3	Quadratic	1 −2 1
4	Linear	−3 −1 1 3
4	Quadratic	1 −1 −1 1
4	Cubic	−1 3 −3 1
5	Linear	−2 −1 0 1 2
5	Quadratic	2 −1 −2 −1 2
5	Cubic	−1 2 0 −2 1
5	Quartic	1 −4 6 −4 1

$$F = \frac{\left(\sum_{j=1}^{A} C_j \overline{X}_j\right)^2}{MSW \left[\sum_{j=1}^{A} (C_j^2/n_j)\right]}$$

C_j = Contrast coefficients for group j
MSW = Mean square within

A4-94

Category: Multiple Comparison Procedures—*Post Hoc* Comparisons

Statistical Procedure: Scheffé Contrasts

Description: A general procedure for comparing two or more means in the context of analysis of variance. Specific hypotheses about pairs of means or other combinations of more than two means can be tested with Scheffé contrasts. The calculation of the Scheffé contrasts is identical to the calculation of orthogonal contrasts (**A4-92**).

The critical value for testing the significance of Scheffé contrasts is the critical value from the analysis of variance for the factor involving the contrasts, multiplied by the degrees of freedom for that factor. The critical value is thus inflated, making this a conservative test.

Application: Used when a significant overall F-test in an analysis of variance is obtained. Scheffé contrasts are used to determine exactly which means or sets of means are different from each other at a statistically significant level. Often used in experiments with multiple-treatment conditions and a control group to test differences between specific treatments, and the difference between all treatments combined and the control group.

Data: Dependent variable is interval or ratio. The independent variable defining groups is nominal, ordinal or, sometimes, interval.

Limitations: Scheffé contrasts are not independent (orthogonal). Scheffé contrasts are conservative and result in fewer significant differences than are observed using Tukey **(A4-95)** or Newman-Keuls **(A4-96)**. Because of the conservative nature of Scheffé contrasts, they may not be desirable for pairwise contrasts with equal n's. Scheffé contrasts should be used for complex contrasts or situations with unequal cells.

Comments: The use of Scheffé contrasts maintains the experiment-wise Type I error rate. Scheffé contrasts are very versatile but conservative in testing hypotheses.

Formula:

$$F = \frac{\left(\sum_{j=1}^{A} C_j \overline{X}_j\right)^2}{\text{MSW}\left[\sum_{j=1}^{A}(C_j^2/n_j)\right]}$$

C_j = Contrast coefficients for group j
MSW = Mean square within from ANOVA
n_j = number of S's in group j

Tested against $(A - 1) \times F$,
where F is the critical value from
the ANOVA for Factor A with A Levels

A4-95

Category: Multiple Comparison Procedures—*Post Hoc* Comparisons

Statistical Procedure: Tukey's Honestly Significant Difference Using the Q-statistic

Description: A procedure for comparing all pairs of means in the context of analysis of variance to test for significant differences between pairs of means. The Tukey procedure maintains the overall experiment-wise error rate at the predetermined alpha level. The Q-statistic is used to test whether differences between pairs of means are statistically significant. The Q-statistic is the difference between the means divided by the square root of the ANOVA mean square after it is divided by the number of subjects in each cell. The Tukey procedure is not as powerful as the Newman-Keuls procedure (bA4-96) but the Tukey method maintains the Type I error rate. The degrees of freedom for the critical value of Q is (r, df within), where r is the number of means, and df-within is the degrees of freedom within group from the ANOVA.

Application: A standard procedure for examining pairwise differences after a significant overall F-ratio has been found in an analysis of variance. Often tests of all pairwise differences are made when the researcher does not have conceptually based hypotheses that focus attention on specific pairs of means or combinations of means.

Data: Dependent variable is interval or ratio. The independent variable defining groups is nominal, ordinal, or, sometimes, interval.

Limitations: The Tukey method is appropriate only when there are equal cell sizes in all levels of the independent variable in question. The Tukey method is not as statistically powerful as some other methods, notably Newman-Keuls, but Tukey maintains the Type I error rate at the predetermined level.

Comments: The Tukey method is somewhat conservative, but pairwise differences detected using this method would be detected using any other method.

Although the procedure requires equal cell sizes for the levels of the independent variable in question, an approximation of the Tukey method can be made for unequal cell sizes by using the harmonic $\hat{n}$ for the cell size. The harmonic $\hat{n}$ is of the form:

$$\hat{n} = \frac{A}{\frac{1}{n_1} + \frac{1}{n_2} + \ldots + \frac{1}{n_A}} \quad A = \text{Number of groups}$$

Formula:

$$Q = \frac{\overline{X}_i - \overline{X}_j}{\sqrt{MSW/n}}$$

where MSW = Mean square within from ANOVA

A4-96

Category: Multiple Comparison Procedures—*Post Hoc* Comparisons

Statistical Procedure: Newman-Keuls Test for Pairwise Differences Using the Q-statistic

Description: A procedure for comparing all pairs of means in the context of analysis of variance to test for significant differences between pairs of means. The procedure requires that the means be ordered for testing, generally from lowest to highest, left to right. Pairwise differences between all means are calculated and, since the means are ordered, adjacent pairs are closer in value than other pairs. The Q-statistic is used to test whether differences between pairs of means are statistically significant. The Q-statistic, calculated as in the Tukey method, is the difference between the means divided by the square root of the ANOVA mean square after it has been divided by the number of subjects in each cell. The Newman-Keuls method is more powerful in detecting significant pairwise differences than the Tukey procedure (A4-95). The Newman-Keuls method, however, inflates the experiment-wise Type I error rate. The critical values of Q used in the Newman-Keuls tests are determined sequentially as a function of r and the error degrees of freedom, where r is the number of steps between ordered means tested plus 1, and the error degrees of freedom are the degrees of freedom within from the ANOVA.

Application: A standard procedure for examining pairwise differences after a significant overall F-ratio has been found in an analysis of variance. Often, tests of all pairwise differences are made when the researcher does not have conceptually based hypotheses that focus attention on specific pairs of means or combinations of means.

Data: Dependent variable is interval or ratio. The independent variable defining groups is nominal, ordinal, or, sometimes, interval.

Limitations: The Newman-Keuls method is appropriate only when there are equal cell sizes in all levels of the independent variable in question. The Newman-Keuls method is statistically powerful but inflates the overall experiment-wise Type I error rate.

Comments: The Newman-Keuls method is statistically powerful and hence detects pairwise differences that might not be detected using other methods. Although the procedure requires equal cell sizes for the levels of the independent variable in question, an approximation to the Newman-Keuls method can be made for unequal cell sizes by using the harmonic $\hat{n}$ for the cell size. The harmonic $\hat{n}$ is of the form:

$$\hat{n} = \frac{A}{\frac{1}{n_1} + \frac{1}{n_2} + \ldots + \frac{1}{n_A}} \quad A = \text{Number of groups}$$

Formula:

$$Q = \frac{\overline{X}_i - \overline{X}_j}{\sqrt{MSW/n}}$$

where MSW = Mean square within from ANOVA

A4-97

Category: Multiple Comparison Procedures—*Post Hoc* Comparisons

Statistical Procedure: Duncan Multiple-Range Test for Pairwise Differences Using the Q-Statistic

Description: A procedure for comparing all pairs of means in the context of analysis of variance to test for significant differences between pairs of means. The procedure requires that the means be ordered for testing, generally from lowest to highest, left to right. Pairwise differences between all means are calculated and, since the means are ordered, adjacent pairs are closer in value than other pairs. The Q-statistic is used to test whether differences between pairs of means are statistically significant. The Q-statistic, calculated as in the Tukey and Newman-Keuls methods, is the difference between two means divided by the square root of the analysis of variance mean square after it has been divided by the number of subjects in each cell. The Duncan Multiple Range Test detects more significant differences than the Newman-Keuls method (**A4-96**) which in turn is more powerful in detecting significant pairwise differences than the Tukey procedure (**A4-95**). However, the Duncan method, like the Newman-Keuls method, inflates the experiment-wise Type I error rate. Like the Newman-Keuls method, the critical values of Q used in the Duncan method are determined sequentially as a function of r and the error degrees of freedom, where r is the number of steps between ordered means being tested plus 1, and the error degrees of freedom are the degrees of freedom within from the analysis of variance. The Duncan method differs substantially from the Newman-Keuls method in that it inflates the significance level for each paired comparison depending on how many other means fall between the two means being compared. The farther apart two means are in the ordering of means, the more lenient the significance level used to test the difference. The adjustment is of the form $1 - (1 - \alpha)^{r-1}$, where α is the Type I error rate and r is the number of means between the two ordered means being compared plus 1.

Application: A common procedure for examining pairwise differences after a significant overall F-ratio is found in an analysis of variance. Often, tests of all pairwise differences are made when the researcher does not have conceptually based hypotheses focusing attention on specific pairs of means or combinations of means.

Data: Dependent variable is interval or ratio. The independent variable defining groups is nominal, ordinal, or, sometimes, interval.

Limitations: The Duncan Multiple Range Test is appropriate only when there are equal cell sizes in all levels of the independent variable in question. The Duncan test is statistically powerful but inflates the overall experiment-wise Type I error rate.

Comments: The Duncan test is statistically powerful and, hence, detects pairwise differences that might not be detected using other methods. Although the procedure requires equal cell sizes for the levels of the independent variable in question, an approximation to the Duncan method can be made for unequal cell sizes by using the harmonic n̂ for the cell size. The harmonic n̂ is of the form:

$$\hat{n} = \frac{A}{\frac{1}{n_1} + \frac{1}{n_2} + \ldots + \frac{1}{n_A}} \quad A = \text{Number of groups}$$

Formula:

$$Q = \frac{\overline{X}_i - \overline{X}_j}{\sqrt{MSW/n}}$$

where MSW = Mean square within from ANOVA

References and Resources

Barcikowski, R. S., ed. *Computer Packages and Research Designs, Volume 1: BMDP.* Lanham, MD: University Press of America, 1983.

Barcikowski, R. S., ed. *Computer Packages and Research Designs, Volume 2: SAS.* Lanham, MD: University Press of America, 1983.

Barcikowski, R. S., ed. *Computer Packages and Research Designs, Volume 3: SPSS.* Lanham, MD: University Press of America, 1983.

Cody, R. P., and Smith, J. K. *Applied Statistics and the SAS Programming Language.* 2d edition. New York: North Holland, 1987.

Dixon, W. J., ed. *BMDP Statistical Software, 1981 Edition.* Los Angeles: University of California Press, 1981.

Ferguson, G. A. *Statistical Analysis in Education and Psychology.* 4th edition. New York: McGraw-Hill, 1976.

Glass, G. V., and Stanley, J. C. *Statistical Methods in Education and Psychology.* Englewood Cliffs, NJ: Prentice-Hall, 1970.

Hays, W. L. *Statistics for the Social Sciences.* 2d edition. New York: Holt, Rinehart and Winston, 1973.

Hinkle, D. E., Wiersma, W., and Jurs, S. G. *Applied Statistics for the Behavioral Sciences.* Boston: Houghton Mifflin, 1979.

Hollander, M., and Wolfe, D. A. *Nonparametric Statistical Methods.* New York: Wiley & Sons, 1973.

Kim, J., and Mueller, C. W. *Introduction to Factor Analysis: What It Is and How to Do It.* Sage University Paper series on Quantitative Applications in the Social Sciences, 07-013. Beverly Hills and London: Sage Publications, 1978.

Marascuilo, L.A., and McSweeney, M. *Nonparametric and Distribution Free Methods for the Social Sciences.* Monterey, CA: Brooks/Cole Publishing Company, 1977.

McDowall, D., Mcleary, R., Meidinger, E. E., and Hay, R. Jr., *Interrupted Time Series Analysis.* Sage University Paper series on Quantitative Applications in the Social Sciences, 07-021. Beverly Hills and London: Sage Publications, 1980.

Mulaik, S. A. *The Foundations of Factor Analysis.* New York: McGraw-Hill, 1972.

Nie, N. N., Hull, C. H., Jenkins, J. G., Steinbrenner, K., and Brent, D. H. *SPSS: Statistical Package for the Social Sciences.* 2d edition. New York: McGraw-Hill, 1975.

Ostrom, C. W., Jr. *Time Series Analysis: Regression Techniques.* Sage University Paper series on Quantitative Applications in the Social Sciences, 07–009. Beverly Hills and London: Sage Publications, 1978.

SAS Institute, Inc. *Introductory Guide, Version 5 Edition.* Cary, NC: SAS Institute Inc., 1982.

SAS Institute, Inc. *SAS User's Guide: Basics, Version 5 Edition.* Cary, NC: SAS Institute, 1982.

SAS Institute, Inc. *SAS User's Guide: Statistics, Version 5 Edition.* Cary, NC: SAS Institute, 1982.

Seigel, S. *Nonparametric Statistics for the Behavioral Sciences.* New York: McGraw-Hill, 1956.

Snedecor, G. W., and Cochran, W. G. *Statistical Methods.* 6th edition. Ames, IA: Iowa State University Press, 1967.

Winer, B. J. *Statistical Principals in Experimental Design.* 2d edition. New York: McGraw-Hill, 1971.

CHAPTER 5

A Summary of Sampling Techniques

Introduction

The issues involved in obtaining a useful sample of subjects from some population are many and complex. Choices have to be made in a process that contains numerous steps, each with several options. Kish (1965) suggests four criteria for assessing sampling designs: (1) goal orientation, (2) measurability, (3) practicality, and (4) economy. "Goal orientation" refers to the need to consider the overall purpose of the research project when making decisions about sampling issues. "Measurability" means that the sampling design allows for the computation of valid estimates of sampling variability or sampling error. "Practicality" means that the procedures required by the design can actually be carried out—in practice, not just in theory. "Economy" refers to the productive use of the often limited resources available for collecting the data needed to complete the research project. All four of these criteria should be carefully considered by the researcher as decisions are made about a sampling design.

The goal-orientation criterion focuses the researcher's attention on the purpose of the research project. Broadly speaking, sampling designs can be used in two types of research settings, each with a different research purpose or orientation. These differences have implications for the sampling procedures.

First, sampling designs are critical in survey research and are generally referred to as survey sampling. In this setting, samples are selected from a population and used to estimate directly characteristics of that population, such as the mean, the sum of a variable, or a proportion of subjects in some category. Survey sampling has two integral components: the selection process and the estimation process. These components are not considered separately but are interactive. Often, the researcher determines which type of estimation is desired and this dictates the selection process. In other situations, the criteria of practicality and economy dictate the selection process which in turn determines the estimation process.

Second, sampling designs are critical in experimental research. In this application, a sample is selected from a population and then assigned to groups receiving different experimental treatment conditions to test hypotheses about the effects of those various treatments. In this setting, careful sampling procedures are employed to improve the generalizability of the experimental research. The generalizability of an experimental study is also called the study's external validity.

This chapter is concerned primarily with sampling as it relates to the second of these two applications, namely, experimental research and hypothesis testing. Major sampling concepts and procedures used to improve the external validity of experimental studies will be reviewed. The technical issues involved in the estimation processes of survey sampling will not be examined in detail. Kish (1965) and Jaeger (1984) are suggested for information on survey-sampling selection and estimation.

Selection and Assignment

It is important to differentiate two separate steps, sample selection and sample assignment, in the use of samples in experimental research. Sample selection refers to the procedures used to obtain a sample from a population. Sample assignment refers to the procedures used to allocate subjects from the sample to different treatment conditions or groups. The role of sample selection and assignment is illustrated in the following figure:

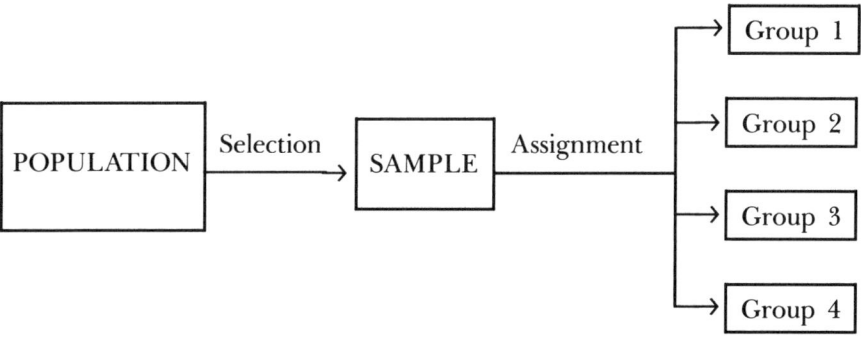

Sample selection, the process of obtaining a sample from a population, is concerned primarily with generalizability or external validity. Sample assignment procedures, used for allocating subjects from the sample to treatment groups, are concerned primarily with internal validity. The example shows four groups, but any number of groups might be involved.

In this chapter entries are organized according to three major categories, namely, (1) sampling concepts, (2) sampling procedures, and (3) assignment procedures.

Sampling Terms and Procedures

Sampling Concepts
(A5-1) Population
(A5-2) Sampling Frame

(A5-3) Sampling Unit
(A5-4) Model Sampling (Nonrandom Sampling)
(A5-5) Probability Sampling (Random Sampling)
(A5-6) Estimators
(A5-7) Sampling Without Replacement
(A5-8) Multistage Sampling
(A5-9) Multiple Matrix Sampling

Sampling Procedures
(A5-10) Simple Random Sampling
(A5-11) Stratified Sampling
(A5-12) Homogeneous Sampling
(A5-13) Systematic Sampling
(A5-14) Cluster Sampling

Assignment Procedures
(A5-15) Assigning Subjects to Groups

A5-1

Category: Sampling Concepts

Concept: Population

Description: The aggregation of elements or subjects for which the results of a study are to be generalized. A precise definition of a population would describe it in terms of (1) content, (2) units, (3) extent, and (4) time (Kish, 1965, p. 7). For example, (1) all students, (2) in third-grade classrooms, (3) in South Carolina, (4) in spring 1988. The population of interest is called the target population. For practical reasons, samples cannot always be obtained from the target population as defined. In such cases, the definition of the target population is modified to define the accessible or operational population.

Applications: Carefully defining the population is a critical and often overlooked step in educational research. The difference between the target population and accessible population should be noted with care. When nonrandom samples are used in educational research, hypothetical populations are sometimes defined to reflect the characteristics of the available sample. The sample is then alleged to represent the theoretical population.

Comments: The external validity of a study is not enhanced by defining a hypothetical population for which research results, based on nonrandom samples, can be generalized. If the researcher wishes to generalize the results of a study, a well-defined population must be identified from which some type of random sample can be drawn.

A5-2

Category: Sampling Concepts

Concept: Sampling Frame

Description: The sampling frame is the actual listing of members of the population from which the sample is drawn. In a sense, the sampling frame is the authoritative operational definition of the accessible population. The ideal sampling frame lists every element of the target population only once, and lists no other extraneous elements. In practice, sampling frames rarely have a perfect one-to-one matching with the target population. In such cases, the target population might be redefined to match the sampling frame. Alternatively, the sampling frame might be edited or augmented with other lists to increase the congruence between the target population and sampling frame.

Applications: All sampling designs involve the use of a sampling frame at some point. In educational research, the sampling frame might be a list of students, teachers, schools, or school districts. In practice, sampling frames are often used because they are conveniently available and little care is taken to assess the quality of the sampling frame. For example, a common problem is that lists of students and teachers often do not reflect recent changes. In such cases, the sampling frames do not reflect transfer students, students who drop out, new teachers, teachers on leave, and teachers who retire.

Comments: The careful selection, preparation, or documentation of a sampling frame is a critical step in the sampling process because it is the operational definition of the accessible population.

A5-3

Category: Sampling Concepts

Concept: Sampling Unit

Description: Populations are defined in terms of the elements or units for which the researcher seeks information. The sampling unit refers to the basic elements that are sampled to represent the accessible population.

Applications: In educational research, the sampling unit can be any one of a number of basic elements. For example, the sample unit might be individual students, classrooms of students, school districts, or entire states.

Comments: It is important to define the sampling unit clearly because the sampling unit, in most cases, must be the unit used in the statistical analysis, and generalization can be made only to these units. For example, if classrooms are the sampling unit, data must be analyzed at the classroom level, not at the student level.

A5-4

Category: Sampling Concepts

Concept: Model Sampling or Nonrandom Sampling

Description: Model sampling refers to nonrandom sampling procedures that are based on an implicit "model." The model makes general assumptions about the distribution of the relevant variable or variables in the population. The assumption implicit in the use of the model is that the variable or variables of interest are distributed in the population in such a way that the researcher can obtain representative samples without random selection.

Applications: Many types of model sampling or nonrandom sampling are well known. These include:

VOLUNTEER SAMPLES. Subjects are invited to participate in a study in which they volunteer.

CONVENIENCE SAMPLES. Subjects are chosen because they are readily available.

PAID SAMPLES. Subjects are encouraged to participate in a study with some form of payment.

JUDGMENT SAMPELS. Subjects are selected based on expert judgment of their representativeness.

QUOTA SAMPLES. Subjects are selected to match the population proportion on certain demographic variables.

Comments: Model sampling is based on the general assumption (or hope) that the variable or variables of interest will be distributed in the population in a way that will yield a representative sample, using the techniques described. This assumption is rarely appropriate. For example, judgment sampling would assume that a school principal could identify "typical classrooms" for use in an experiment. Samples obtained using these procedures cannot be considered representative. Such samples may be useful for pilot studies and for field testing measuring instruments; however, these sampling procedures do not support the external validity of research studies.

A5-5

Category: Sampling Concepts

Concept: Probability Sampling or Random Sampling

Description: In general, probability sampling, referred to more generally as random sampling, has three defining attributes. First, all sampling units in the population have some probability of being selected that is greater than zero. It is not necessary that these probabilities be equal. Second, the probability that any sampling unit might be selected must be specifiable prior to selection. Third, the potential sampling units that comprise the population and those that are excluded from the population are explicitly identified.

Applications: Four major types of random sampling are generally applicable in educational research. These are: Simple Random Sampling (**A5-10**); Stratified Sampling (**A5-11**); Systematic Sampling (**A5-13**); and Cluster Sampling (**A5-14**). The major purpose of these procedures is to obtain a sample that is representative of the population and thus improve external validity.

Comments: It is important to recognize that it is not always possible to draw a random sample in applied educational research. The sampling unit (**A5-3**) of interest, whether it is student, teacher, classroom, or school, cannot always be selected in a way that meets the three criteria for a random sample. In such cases, the researcher must address the issue of generalizability by seeking to replicate the research in different settings with different samples.

A5-6

Category: Sampling Concepts

Concept: Estimators

Description: An estimator is a formula for calculating sample estimates for population parameters. An estimator is a property of a sample design that includes both the selection procedure and the procedure for estimation. The sample design defines an estimator and can be used to generate the sampling distribution of the estimator. Important properties of estimators include:

> BIAS. An estimator is biased if, for all possible samples of a particular size, the average estimate for all the samples is larger or smaller than the population value. An estimator is unbiased if the average estimate for all samples is equal to the population parameter.
>
> ERROR OF ESTIMATION. The estimation error is the difference between the sample estimate and the population parameter. The variance of an unbiased estimator is the average squared error of estimation over all samples of a particular size.
>
> MEAN SQUARE ERROR. The sum of the estimator variance and the square of the estimator bias equals the mean square error.
>
> EFFICIENCY. One estimator is said to be more efficient than another if, for a given sample size, it has a smaller mean square error.
>
> CONSISTENCY. An estimator is consistent if its mean square error decreases as the sample size increases.

Applications: The careful study of estimators and their properties is critical in survey sampling because the purpose of survey sampling is estimating population values from sample statistics.

Comments: It is important to emphasize that the properties of estimators are determined by the sample design, which includes both the selection procedure and the estimation procedure.

"Bias," as a property of an estimator, should not be confused with "bias" used in a more general sense. "Bias" is sometimes used to mean that a sample is not representative.

A5-7

Category: Sampling Concepts

Concept: Sampling Without Replacement

Description: Sampling without replacement is a characteristic of sampling procedures which means that sampling units are removed from the accessible population after they have been selected for the sample. Sampling units can only appear in the sample once using this procedure. Sampling *with* replacement describes sampling procedures in which sampling units remain in the accessible population even if they are selected to be in the sample. Sampling units can appear in the sample more than once using sampling with replacement.

Applications: In most educational research settings, sampling is done without replacement. Sampling with replacement might be used in multistage sampling **(A5-8)**.

Comments: Sampling without replacement is used in most educational research studies. In selecting a random sample of students from a school, for example, students are removed from the accessible population after they have been selected to be in the sample.

A5-8

Category: Sampling Concepts

Concept: Multistage Sampling

Description: Multistage sampling refers to sampling procedures carried out in stages or steps. Large units are sampled first, then smaller units are sampled from these larger units. The large units are called primary sampling units; the small units are elements or elementary units. The number of stages depends on the sampling situation and a variety of other criteria (e.g., practicality, economy).

Applications: Multistage sampling is often used in educational research. In the study of statewide student achievement, for example, school districts might be sampled first as the primary sampling unit. Individual schools within those school districts would then be selected as the elementary sampling unit. The National Assessment of Educational Progress (NAEP) uses a three-stage sampling process. First, counties, or groups of counties, are sampled. Second, schools are sampled from the counties chosen in stage one. Third, students are selected from the schools selected in stage two.

Comments: Multistage sampling can be a practical necessity in many situations. The procedure facilitates drawing a representative sample of individual subjects from very large populations when the subjects are contained in units, within units that are within even larger units.

A5-9

Category: Sampling Concepts

Concept: Multiple Matrix Sampling

Description: Multiple matrix sampling refers to situations in which sampling is used for both subjects and material/conditions to which the subjects are assigned.

Applications: Multiple matrix sampling is used in large-scale assessment programs, such as the National Assessment of Education Progress (NAEP). In the NAEP studies, a sample of subjects is selected (see **A5-8,** under Applications), then the subjects are randomly assigned to different test booklets. The test booklets each contain a sample of the test questions from the population of questions of interest. No student answers all the test questions, and no test question is answered by all students. By using careful sampling procedures, however, very informative estimates of how the student population would do on all test questions are obtained.

Comments: Multiple matrix sampling is very useful for field testing large numbers of items in test-development activities. The procedure maximizes the number of items that can be employed without burdening children with excessively long tests.

A5-10

Category: Sampling Procedures

Procedure: Simple Random Sampling

Description: Simple random sampling (SRS) refers to a sampling procedure in which all elements of the population have an equal probability of being selected. This is equivalent to saying that all possible samples that could be drawn from the population have an equal probability of being selected. Simple random sampling is performed without replacement **(A5-7)**.

Applications: The application of SRS generally involves the use of a table of random numbers. To use this procedure, the N elements in the population are each assigned an individual listing number. For a sample of size n, n equally probable random numbers are then selected from the table of random numbers. The random sample is composed of the n elements in the population whose listing numbers match the n randomly selected numbers from the table.

Comments: Simple random sampling is useful and practical in many small scale research settings if the mechanics of the procedure are followed carefully. SRS is not very useful in large-scale research projects.

A5-11

Category: Sampling Procedures

Procedure: Stratified Sampling

Description: Stratified sampling involves both the sampling variable (i.e., the variable of interest) and a classification or stratification variable or variables. The classification variable is used to sort population elements into categories or strata.

Subjects are selected to be included in the sample from the strata, not directly from the population at large. It is quite common to select subjects from each strata in proportion to the size of the strata in the population. This procedure for allocating subjects is called proportional stratified random sampling. Optimal allocation, used in survey sampling, is another procedure for allocating subjects. With this procedure, the number of subjects from each stratum is proportional to the product of the stratum size and standard deviation.

The basic arrangement for a proportional stratified random sample is illustrated in the following figure:

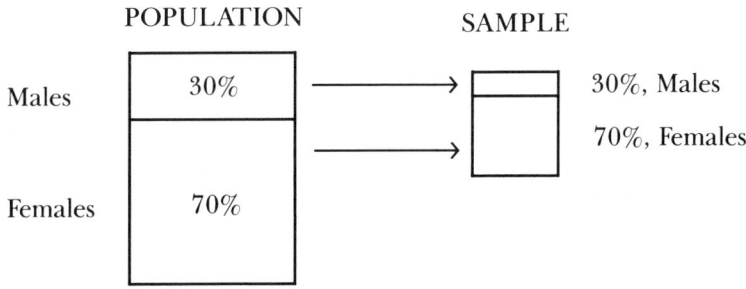

In this example, the stratification variable is gender. In the population, 30% of the elements are males and 70% are females. If a sample of size n is to be drawn, the number of males will be (.3 × n), and the number of females will be (.7 × n). SRS **(A5-10)**, or any other method, can be used to select the appropriate number of males and females from the population.

Applications: Stratified sampling has many applications in educational research. Classification variables are often selected because they are correlated with the dependent measure and thus are potential extraneous variables. Stratified samples provide a mechanism for controlling the influence of the extraneous variable, whereas simple random selection would, at best, randomly distribute the extraneous variable. When stratified sampling is used in experimental research, the stratification variables are sometimes used in the assignment of subjects to groups **(A5-15),** and as independent variables in the statistical analysis.

Comments: Stratification variables have different roles in survey sampling and in experimental research. In survey sampling, stratification variables are used to improve the properties of the estimators of the sampling variable of interest. In experimental research, designed to test hypotheses, stratification variables are used to improve the representativeness of the sample and, also, as variables of interest themselves. The stratification variables may be independent variables for which the researcher has specific hypotheses, or potential extraneous variables that the researcher wishes to control.

A5-12

Category: Sampling Procedures

Procedure: Homogeneous Sampling

Description: Homogeneous sampling is a special adaptation of stratified selection in which elements or subjects from only one stratum are selected for the sample. With this procedure, the sample is homogeneous with respect to the stratification variable or variables. This approach is generally not identified as a specific sampling procedure.

Homogeneous sampling is a design procedure used in experimental research to control variation due to the stratification variable. This approach can increase the internal validity of a study. It always decreases a study's external validity.

Applications: Homogeneous sampling is used when the researcher chooses to improve the internal validity of a study at the expense of the study's external validity. Often this choice is made when the researcher recognizes potentially serious confounding variables, stratifies the population on these variables, but does not have the resources to sample from all strata created by this stratification.

Comments: Homogeneous sampling is a practical necessity in many educational research situations. It is critical to recognize, of course, that generalizations cannot be made if homogeneous sampling has been employed. In such cases, the research must be replicated using subjects selected from different strata on the various replications.

A5-13

Category: Sampling Procedures

Procedure: Systematic Sampling

Description: Systematic sampling refers to a set of procedures in which (1) elements of the population are arranged in an ordered sequence, and (2) the elements are selected at a fixed interval from the ordered sequence. In general, the basis for ordering the elements can be arbitrary as long as each element in the sampling frame can be assigned a unique position in the sequence. In practice, however, the ordering is often systematic and should be examined with care. For example, the elements can be ordered alphabetically, grouped by classrooms, or arranged along an achievement scale from low to high. The nature of the ordering has a major effect on the characteristics of the sample.

The sampling interval, k, is a function of N, the number of elements in the sampling frame, and n, the number of elements required for the sample. The sampling interval k is equal to N/n, or the integer that is closest to N/n. A random number r, between 1 and k, is selected from a table of random numbers as the first case. Every k-th element after this first case is then selected as part of the sample.

Applications: Systematic sampling is a very practical and useful procedure in many educational research settings. Lists of schools, teachers, and students are often readily available and can be used as sampling frames (**A5-2**). The procedures for determining the sampling interval, k, and the first case, the r-th element, are straightforward. Selecting every k-th element after the first case basically involves counting.

Systematic sampling is especially convenient when elements of the sampling frame are stored on magnetic tape. Once r and k are determined, the researcher basically needs a program that counts and stores selected cases.

Comments: Systematic sampling is very useful; however, several problems can arise in its application. The most serious problems arise when subjects are systematically ordered within groups, such as classrooms or schools, and groups are listed sequentially in the sampling frame. For example, consider a situation in which students are ordered within classrooms from low to high on some achievement measure, and the classrooms have approximately the same number of students. In this case, the systematic selection of every k-th student will yield a sample of students who have approximately the same level of achievement. Problems of this sort can occur with other, far less obvious, systematic orderings within groups.

If the entire population of elements across the whole sampling frame is ordered from low to high on some achievement measure, then systematic sampling will resemble stratified sampling. One subject will be selected from each interval where intervals function as achievement strata.

A5-14

Category: Sampling Procedures

Procedure: Cluster Sampling

Description: Cluster sampling refers to a set of procedures in which groups of elements, rather than individual elements, are sampled. These groups, or clusters, are often naturally occurring, for example, residents of a neighborhood, town, or city. In educational research, the clusters are often classrooms, schools, or school districts. Clusters can be selected using a variety of procedures. Two common procedures for selecting clusters of elements are:

SIMPLE RANDOM SELECTION. All clusters are equally likely to be selected.

PROBABILITY PROPORTIONAL TO SIZE (PPS). The probability that a cluster will be selected is directly proportional to its size. For example, clusters with large numbers of elements are more likely to be selected than clusters with a smaller number of elements.

Applications: Cluster sampling is commonly used in educational research with classrooms, schools, or school districts serving as the cluster and is often the only realistic or practical approach. Cluster sampling is often used in a multistage sample design. When the clusters—for example, classrooms—are to be randomly assigned to treatment conditions in experimental research, cluster sampling is required.

Comments: Cluster sampling is often a practical necessity and can be very useful in many educational applications. Special care must be taken in the analysis of data obtained from a cluster sample. If clusters are randomly sampled and then randomly assigned to treatment conditions, the clusters are the unit of analysis for the statistical procedure used for hypothesis testing. For example, if classrooms are sampled and randomly assigned to treatment conditions, it would be incorrect to analyze the performance of individual students to test hypotheses about treatment effects. A nested analysis of variance **(A4-83)**, with classrooms nested in treatment, would be the correct approach.

A5-15

Category: Assignment Procedures

Procedure: Assigning Subjects to Groups

Description: In experimental research, subjects in the sample must be assigned to different treatment conditions. The relationship between sample selection and assignment is:

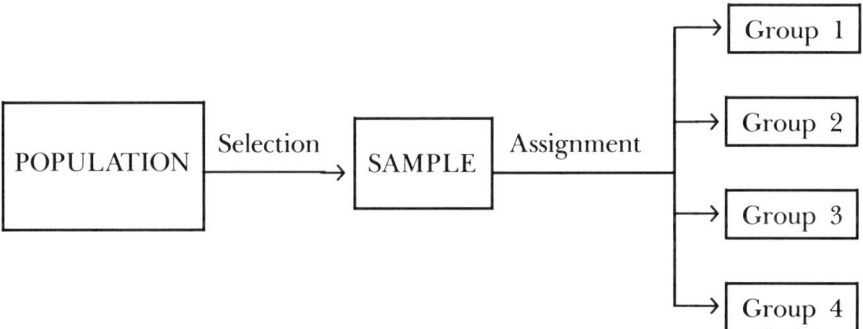

The assignment of subjects to groups can be performed in ways that are directly analogous to the procedures used to select samples from the population. For example, nonrandom or random assignment can be employed. In some cases, nonrandom assignment is necessary because subjects must consent to being assigned to a treatment condition. In general, nonrandom assignment diminishes internal validity.

The random assignment of subjects from a sample to treatment groups is generally called randomization. Randomization can be performed using a variety of procedures of which simple random assignment is most common. Stratifying the sample on a variable of interest, and then randomly assigning from strata to treatment groups, can be very useful. In such cases, the stratification variable is generally used in the sample selection process. The stratification variable may be an independent variable of interest or an extraneous variable, the influence of which the researcher wishes to control. In either case, the stratification variable is used as an independent variable in the statistical analysis phase of the research.

Applications: The assignment of subjects from the sample to different treatment groups is a critical step in experimental research. Random assignment of subjects is a major safeguard against threats to internal validity. Characteristics of the subjects that might influence the dependent measure are spread evenly across all treatment groups with random assignment.

Procedures that include the stratification of subjects in the sample, followed by random assignment from each stratum to groups, are called randomized-blocks designs **(A4-86)**.

Comments: Random assignment is critical in experimental research if the researcher wishes to make causal inferences. Random assignment does not assure the validity of such inferences, that is, the internal validity of the study, but the absence of randomization virtually prohibits causal inferences.

Randomization is not practical in many applied educational research settings. In the absence of randomization, other procedures for controlling threats to internal validity should be employed and attempts at causal inferences should be qualified.

References and Resources

Kish, L. *Survey Sampling.* New York: Wiley & Sons, 1965.
Jaeger, R. *Sampling in Education and the Social Sciences.* New York: Longman, 1984.

CHAPTER

6

A Research Process Checklist

Introduction

This Research Process Checklist is designed to provide general guidelines for conducting educational research. It can be used as a planning guide before a research study begins, as a review procedure for a research study in progress, or as a criterion for evaluating completed research.

The checklist is organized around five major steps in the research process. These steps are:

I. DEFINING THE RESEARCH QUESTION AND THE NATURE OF THE RESEARCH

Information related to this step is contained in Chapter 2, *Linking Research Questions to Research Designs and Statistical Procedures.*

II. DEFINING VARIABLES, SUBJECTS, AND THE RESEARCH DESIGN

Information related to this step is contained in Chapter 5, *A Summary of Sampling Techniques,* and Chapter 3, *A Summary of Research Designs.*

III. VERIFYING THE OBJECTIVITY, RELIABILITY, AND VALIDITY OF OBSERVATION INSTRUMENTS AND PROCEDURES

Information related to this step is contained in books on educational measurement and evaluation.

IV. ANALYZING DATA

Information related to this step is contained in Chapter 4, *A Summary of Statistical Procedures.*

V. INTERPRETING RESEARCH RESULTS

Information related to this step is contained throughout this section (A) of the book.

The major issues that must be addressed in each step of the research process are enumerated in the checklist. The listing of issues is appropriate for a large variety of research studies but may not be appropriate in all cases. Depending on the nature of a particular study, certain of these issues might be ignored while other issues, not listed in this general checklist, might be addressed.

The Checklist

I. DEFINING THE RESEARCH QUESTION AND THE NATURE OF THE RESEARCH

 _____ 1. Specify the research problem in clear and explicit terms.
 2. Review the literature related to the research problem.
 _____ A. Review substantive/conceptual issues.
 _____ B. Review technical/methodological issues.
 _____ C. Enumerate recommendations for improving research offered by those who have conducted previous research.
 3. Determine whether the research question deals with:
 _____ A. Description (See A2, Part 1)
 _____ B. Relationships or Correlation (See A2, Part 2)
 _____ C. Differences (See A2, Part 3)
 _____ 4. Specify each research hypothesis to be explored in the study.
 5. Classify the research approach as:
 _____ A. Descriptive
 _____ B. Correlational
 _____ C. Quasi-experimental
 _____ D. Experimental
 _____ 6. Specify delimitations of the study.
 _____ 7. Specify operating assumptions.

II. DEFINING VARIABLES, SUBJECTS, AND THE RESEARCH DESIGN

Variables

 8. Specify a conceptual definition of each of the following:
 _____ A. All independent variables
 _____ B. All dependent variables
 _____ C. All controlled extraneous variables
 _____ D. All other extraneous variables
 9. Specify an operational definition of each of the following:
 _____ A. All independent variables
 _____ B. All dependent variables

    _____   C. All controlled extraneous variables
    _____   D. All other extraneous variables

Subjects

_____ 10. Specify the target population.
_____ 11. Specify the accessible population.
_____ 12. Specify the procedures that will be used for sample selection.
_____ 13. Identify any limitations to generalizability due to the sample-selection procedure.
_____ 14. When appropriate, specify procedures for assigning subjects to treatment groups.
_____ 15. Specify review procedures for assuring compliance with guidelines for ethical use of human subjects.
_____ 16. Specify procedures for obtaining the subjects' informed consent to participate in the study.

Research Design

        17. Specify procedures for controlling extraneous variables:
_____   A. Design procedures
_____   B. Statistical procedures
        18. Summarize the overall research design, specifying:
_____   A. The nature of the research (See steps 3 and 5 above.)
_____   B. The number and nature of any groups (i.e., experimental, control)
_____   C. The number and sequence of observations made on the groups (e.g., pretests, posttests)

III. VERIFYING THE OBJECTIVITY, RELIABILITY, AND VALIDITY OF OBSERVATION INSTRUMENTS AND PROCEDURES

        19. Will the independent, dependent, control, and extraneous variables be observed using:
_____   A. Currently available instruments/procedures?
_____   B. Instruments/procedures developed for this research?
        20. If instruments/procedures for observing variables will be developed (19B), specify:
_____   A. Procedures for developing the instrument(s).
_____   B. Procedures for pilot testing the instrument(s).
_____ 21. Specify evidence that the instruments and procedures used to observe the independent, dependent, control, and extraneous variables (steps 19 and 20 above) are objective.
_____ 22. Specify evidence that the instruments and procedures used to observe the independent, dependent, control, and extraneous variables (steps 9, 19, and 20 above) are reliable.
_____   A. For each variable, specify the index of reliability used.
_____ 23. Specify evidence that the instruments and procedures used to observe the independent, dependent, control, and extraneous variables (steps 9, 19, and 20 above) are valid.
_____   A. For each variable, specify the index of validity used.

IV. ANALYZING DATA

      24. Specify the procedures that will be used to provide basic descriptive information:
_____   A. Visual displays of data (e.g., graphs, charts)
_____   B. Descriptive statistics
_____   C. Indices of relationships
_____ 25. Specify the statistical hypothesis that corresponds to each research hypothesis described in step 4 above.
_____ 26. Specify the appropriate statistical procedure for testing each hypothesis stated in step 25.
_____   A. Indicate the significance level for inferential tests.
_____   B. Indicate possible violations of assumptions or other limitations of the procedures.

V. INTERPRETING RESEARCH RESULTS

Internal Validity
      27. Determine the internal validity of the study:
_____   A. Enumerate threats to internal validity that must be assessed in interpreting the results of the study.
_____   B. Evaluate the internal validity of the study.

External Validity
      28. Determine the external validity of the study:
_____   A. Enumerate threats to external validity that must be assessed in interpreting the results of the study.
_____   B. Evaluate the external validity of the study.

Findings, Conclusions, and Recommendations
_____ 29. Summarize the study's findings.
_____ 30. Provide conclusions based on the study.
_____ 31. Describe how the results of the study related to the findings from previous research on the problem.
      32. Provide recommendations based on the study.
_____   A. Substantial recommendations
_____   B. Methodological recommendations

SECTION B

Microcomputer Software Profiles

CHAPTER

1

Nonstatistical Packages for Microcomputers

Introduction

When one explores the history of computers, it is quite evident that computers were developed initially to perform large and tedious calculations. Their evolution into the systems now represented by the new supercomputer generation has not suppressed this early purpose. Computers are still the fastest and most accurate method available for calculating large sets of data. Today packages for mainframe computers, minicomputers, and microcomputers are available to the educational researcher. The choice of a package is limited mostly by the quality of the computer system and the expertise of the user.

The range of available computer statistical packages runs from the simple, menu-driven, to the complex, statistical model–driven. Most menu-driven packages are designed for ease of use and application. They are intended more for the novice user than the experienced user. In the final consideration, the type of package chosen must be dictated more by the sophistication of the user than by the limits of the machine.

The type and quality of the available packages is vast. One may choose from single-task packages (e.g., to do only regression) or from the fully integrated systems (i.e., performing most, if not all, descriptive and inferential analyses). The integrated packages are also divided into complete systems (you buy the package and you get all of the available analyses) or package sets where you may purchase various combinations of programs to perform only specific analyses. Again, the type of package chosen must be mandated by the task to be performed, the size of the data, and the skill of the user.

Some of the best programs for descriptive data analysis are spreadsheets, database managers, and the multiprogram integrated packages. These are all, to some extent, effective for input, storage, and organization of data. While it is true that these programs are not designed for intensive and in-depth analyses, the selective use of such a program can increase speed and efficiency for the user.

Please note that the forthcoming descriptions of each of the packages include the suggested list prices of the manufacturers. The reader is advised that these were the current prices at the time of research, and they are subject to change. Furthermore, be aware that new versions of software packages are constantly being released.

Spreadsheets

The fundamental purpose of an electronic spreadsheet is the collection, analysis, and reporting of numerical information. In other words, spreadsheets are designed to handle the type of information typically used by educational researchers; namely, numbers. Spreadsheets range from the style designed for the home-budget user to the corporate-level financial planner—that is, from very limited applications to very complex uses. This discussion, while limited to several programs, will focus upon programs at the mid- or advanced application level.

A spreadsheet is an electronic worksheet that utilizes a series of cells to store numerical information and contains the formulas used to analyze the data. The information is stored in a table format of rows and columns. Typically, the type of data coded by a researcher can be categorized into rows (data values or observations) and columns (data labels or data types). This format is perhaps the major strength of a spreadsheet, especially as far as the educational researcher is concerned. The spreadsheet permits the calculations of counts, sums, averages, standard deviations, and variances for the data sets.

In addition to the descriptive type of calculations permitted by spreadsheets, many of the more advanced programs also enable the user to create graphic representations of the data values. The types of graphs available differ from program to program, but most possess bar graphs, line graphs, and pie graphs. Thus, a relatively advanced spreadsheet program will calculate descriptive statistics and produce visual images of the data. The limitation of spreadsheets is in their primary purpose: they are designed for financial analysis, not statistical analysis.

Profiles of Spreadsheets

LOTUS 1-2-3
Lotus Development Corporation
55 Cambridge Parkway
Cambridge, MA 02142

Cost: List $495

System: IBM PC and compatibles (MS-DOS)

Memory: Requires a minimum of 256K (640K recommended); will make use of extended memory boards.

Version: 2.01

Functions: Descriptive statistics; data-file manipulations (including sorts and subsetting); regression statistics; matrix algebra; and bar, line, and pie graphs

Comments: LOTUS 1-2-3 is the best-selling program in the history of microcomputers. In fact, much of the proliferation of microcomputers may be attributed to *LOTUS 1-2-3* and its predecessor, *VISI-CALC*. *LOTUS 1-2-3* will produce a multitude of statistical data analyses. For example, *LOTUS 1-2-3* will calculate standard deviation and variance, but the two functions (@STD and @VAR) produce population-based results. To calculate a sample-based standard deviation statistic using *LOTUS 1-2-3*, the following formula must be used:

@SQRT((@COUNT(list)/(@COUNT(list)−1))*@VAR(list))

To calculate a sample variance this formula is used:

(((@COUNT(list)/(@COUNT(list)−1))*@VAR(list))

LOTUS 1-2-3 (version 2.00 and above) also calculates regression and correlation statistics. Using a special feature of the DATA functions (/DATA REGRESSION), *LOTUS 1-2-3* will produce regression analysis for up to 16 variables (one dependent). The function will handle up to 8,192 observations per variable, but the number of observations per variable must be equal. The output of this function will produce the following: (1) constant, (2) standard error of estimate, (3) R squared, (4) degrees of freedom, (5) x coefficients, and (6) standard error of coefficients. The resulting information can be used to calculate predicted values, residuals, and do estimates of best-fits.

Two other functions of *LOTUS 1-2-3* that are of interest to educational researchers are /DATA DISTRIBUTION and /DATA MATRIX. The /DATA DISTRIBUTION function permits the user to obtain distribution counts and to present graphs of the distribution. The /DATA MATRIX function permits the user to utilize matrix algebra operations (including inversion) to perform a multitude of calculations. In fact, using the matrix function, a mathematically skilled user could execute many very sophisticated statistics.

Overall, *LOTUS 1-2-3* is useful for the general calculation of noninferential statistics. The ease of use for a beginner can be significantly enhanced with the addition of *HAL*, a user-interface system for operating *LOTUS 1-2-3*. Version 2.00 and above of *LOTUS 1-2-3* will import *dbase III, dbase II,* DIF and ASCII files. In addition, *LOTUS* will export DIF and ASCII files. The ability to export ASCII data means that exported files can be read by any program capable of reading ASCII files for further analysis.

Multiplan
Microsoft Corporation
16011 N.E. 36th Way,
P.O. Box 97017
Redmond, WA 98073

Cost: List $195; List $100 (for Apple II and Commodore systems)

System: IBM PC and compatibles (MS-DOS); Apple II series; Commodore C-64 series

Memory: 256K for IBM PC and compatibles (MS-DOS); 64K for Apple and Commodore

Version: 3.0

Functions: Descriptive statistics and data-file manipulation, as well as regression analysis and matrix operations

Comments: *Multiplan* is one of the oldest spreadsheet programs on the market and its key strokes and operations are common across all systems. In other words, the key-stroke operations that work on the IBM PC and compatibles (MS-DOS) systems work on the Apple II series as well. None of the data are cross-compatible between systems.

Multiplan performs all of the common descriptive statistics but requires a formula approach to sample statistics (similar to *LOTUS 1-2-3*), including regression analysis and matrix operations. *Multiplan's* ability to perform regression analysis is excellent, but the output from the analysis is rather limited. The matrix operations are not well-documented in the manual, but do perform well under skilled hands.

Multiplan's graphic capabilities are limited to bar graphs. However, the data produced by *Multiplan* can be read by several of the graphics-only packages, such as *Chart-Star, Chart-Master, Freelance,* and others.

Multiplan is able to export and import *LOTUS 1-2-3,* DIF, and ASCII files.

SuperCalc 3
Computer Associates
2195 Fortune Dr.
San Jose, CA 95131

Cost: List $200

System: Apple II

Memory: 64K (will use 128K)

Version: 3, release 2.0

Functions: Descriptive statistics and data-file manipulations (including sorts and subsetting). Graphics include area, bar, line, and pie graphs.

Comments: *SuperCalc 3* is the only version of *SuperCalc* currently available for the Apple II family. The program provides excellent descriptive functions as well as data-file manipulation. Command structures are very similar to those of *SuperCalc 4*. The system is useful on the Apple II family; data can be exported to other Apple II programs, but not to non-Apple systems.

SuperCalc 3 does not offer a function for standard deviation and variance, but these may be calculated through the use of a formula.

SuperCalc 5
Computer Associates
2195 Fortune Dr.
San Jose, CA 95131

Cost: List $495 (version 5)

System: SuperCalc 5—IBM PC and compatibles (MS-DOS)

Memory: 512K (640K recommended)

Version: 5, release 1.0

Functions: Descriptive statistics and data-file manipulations (including sorts and subsetting). Graphics include area, bar, line, pie, and high-low graphs.

Comments: SuperCalc 5 is an alternative to *LOTUS 1-2-3*. With regard to general spreadsheet operations, *SuperCalc 5* literally does all that *LOTUS 1-2-3* does and can be used in the production of complete, flexible graphic representations of data. The high-low graph feature of *SuperCalc 5* is especially useful when graphing high-low values of data around a central point. The statistical functions of *SuperCalc 5* are very similar to *LOTUS 1-2-3*s. As with *LOTUS 1-2-3*, *SuperCalc 5* lacks a function for sample data and a corrective formula must be used. *SuperCalc 5* also does handle regression analysis and matrix operations.

Overall, *SuperCalc 5* is useful for calculating general descriptive statistics and producing graphs of the data. *SuperCalc 5* exports data in ASCII form so that programs using ASCII can read the data. In addition, the program can import *LOTUS 1-2-3, dbase II, dbase III*, and DIF files, and readily imports straight ASCII files from nonspreadsheet programs.

VP-Planner

Paperback Software International
2830 Ninth St.
Berkeley, CA 94710

Cost: List $99.95 ($109.95 for non–copy protected version)

System: IBM PC and compatibles (MS-DOS)

Memory: 256K

Version: 1.21

Functions: Descriptive statistics and data-file manipulations (sorts and subsetting). Graphics include bar, line, and pie graphs.

Comments: VP-Planner was developed initially as an inexpensive alternative to *LOTUS 1-2-3*, version 1.1A. While it does not possess the enhanced functions of version 2.0 (and above) of *LOTUS 1-2-3*, it does operate precisely as version 1.1A. It does not offer regression, distribution, and matrix functions, but is useful for tabulating data, calculating general descriptive statistics, and presenting bar or line graphs. While standard deviation and variance are calculated for a population rather than a sample, the same conversion formula noted for *LOTUS 1-2-3* will work with *VP-Planner*.

A major addition of version 1.21 of *VP-Planner* is its multidimensional capability (i.e., ability to link different dimensions or spreadsheets). Multidimensions permit the user to make changes in one spreadsheet (or dimension), and the changes will automatically carry over to the other spreadsheets (or dimensions) linked to the first. *VP-Planner* exports data in ASCII form so that programs using ASCII can read the data. In addition, the program can import *LOTUS 1-2-3, dbase II, dbase III*, and DIF files and readily imports straight ASCII files from nonspreadsheet programs.

Additional Spreadsheet Programs

The following programs were not available or not provided for review. Some limited information is offered to aid the researcher in identifying additional packages on the market. The reader is advised to seek other published reviews concerning these programs.

Boeing Calc
Boeing Computer Services
Software and Education Products Group
P. O. Box 24346, M.S.
Seattle, WA 98124-0346

Cost: List $399

System: IBM PC and compatibles (MS-DOS)

Memory: 384K

Version: 3.00

Functions: Descriptive statistics, data-file manipulation (sorting), and file exporting and importing. Permits multidimensional linking of spreadsheets.

Math Plan
WordPerfect Corporation
288 West Center
Orem, UT 84057

Cost: List $395

System: IBM PC and compatibles (MS-DOS)

Memory: 256K

Version: 3.0

Functions: Descriptive statistics and data-file manipulations (including sorts and subsetting). Graphics include area, bar, line, pie, and high-low graphs. Advertisements report the capacity of the program to do regression analysis and matrix operations. Integrates with *WordPerfect* word-processing program.

PFS Professional Plan
Software Publishing Corporation
1901 Landings Dr.
P. O. Box 7210
Mountain View, CA 94039

Cost: List $249

System: IBM PC and compatibles (MS-DOS)

Memory: 384K

Version: 1.0

Functions: Descriptive statistics and data-file manipulations (including sorts and subsetting). Graphics include area, bar, line, and pie graphs.

Comments: Versions of the first generation of this program, *PFS Plan,* are available for Apple II series. The program is an excellent stand-alone spreadsheet for Apple II systems, and it also passes data between the PFS series, *Write, File,* and *Graph. Plan* and *Graph* can be used together for descriptive analysis.

Database Managers

The management of data is one of the aspects of research that often gets lost in the haste of research. Researchers are ready to compile information into a computer program and to analyze the results, but before an analysis can take place the data must be organized in some manner. The practice of organizing the data can be tedious and time consuming, yet effective organization can not only increase the speed of analysis, but may lead to the discovery of unanticipated contingencies within the data itself.

Database managers have been a dominant force in the use of microcomputers since the inception of the microcomputer. A database manager allows the researcher to develop a format to structure the data and to input the data into the structure. A well-developed database permits the user to sort, categorize, subset, and review the information contained within the database.

The type and quality of database selected by a researcher should reflect several aspects: (1) the expertise of the user; (2) the amount of information (data); (3) the degree of manipulation (i.e., sorting, categorizing) required; and (4) the ability of the database manager to export the data to external programs. A second level of concern associated with selecting a database manager is the extent of the monetary resources available to the researcher. Certainly one of the top-of-the-line database managers would accomplish just about everything necessary, but would a less expensive model meet the needs just as effectively?

PC Magazine* reviewed more than 100 of the top relational and programmable databases and still did not scratch the surface of the total number available. The editors at *PC Magazine* divided the types of available database managers into three divisions:

1. Programmable-relational databases
2. Relational databases
3. Flat-field databases

The type of database selected depends, to a large extent, upon the size of the dataset and the degree of data manipulation required by the user. For example, providing the dataset is not exceptionally large and requires minimal sorting, a spreadsheet with a sorting feature such as that possessed by *LOTUS 1-2-3* or *SuperCalc 4* would be adequate. If multiple levels of sorts are required, then

---

*"Project Database II." *PC Magazine* 5, no. 12 (June 24, 1986), pp. 108–227.

perhaps the databases integrated into *Symphony* or *Framework II* would meet the needs of the user, and the purchase of an additional program simply to store and sort data would be unnecessary.

The database development industry is producing database programs nearly as fast as reviewers can write about them. It seems that every issue of the industry's trade magazines touts one or more new database managers or revisions to current ones, and the user can find any number of database managers on the shelves of any software store. Each is a good product in its own right; the plight facing the user is the selection.

For example, imagine the following scenario. Educator Jones is conducting a survey of student needs for her school and wants to use a database manager to organize her data. She goes to XYZ Computers to purchase a database manager. The salesperson persuades Jones to purchase the *Walk-on-the-Water Manager,* a super-deluxe programmable-relational database. Jones carts the database and its 12 manuals home to set up the system. When Jones gets home she discovers that she needs two graduate students, a manual to explain the manuals, and a series of courses in programming. The choice was correct; the database manager will do the job, but so would a less complicated and involved program. Because the salesperson assumed expertise on the part of the customer, Jones was prevented from finding the proper database manager to meet her needs.

The quality of database managers has increased immensely since the early days of the microcomputer. The user is now faced with almost too many database managers from which to choose. Perhaps the most effective way to present a limited selection of programs is to identify them in the following categories:

1. System required
2. Memory requirements, including need for a hard disk
3. Type of database: programmable, relational, or flat-field
4. Record size, if known
5. Report capability
6. File import and export capabilities

Profiles of Database Managers

DataEase
DataEase International Ltd.
12 Cambridge Dr.
Trumbull, CT 06611

Cost: List $600

System: IBM PC and compatibles (MS-DOS)

Memory: 512K; hard disk recommended

Version: 2.5

Type: Programmable-relational

Record Size: 65,000

Reports: Sorts are performed during report generation.

File Import/Export: dbase III, dbase II, LOTUS, DIF, and ASCII

Comments: DataEase is a relatively new entry into the high-end database manager field.

dBase III Plus
Ashton-Tate
20101 Hamilton Ave.
Torrance, CA 90502

Cost: List $695

System: IBM PC and compatibles (MS-DOS)

Memory: 256K (384K recommended); hard disk recommended

Version: 1.1 (*dBase IV* now available)

Type: Programmable-relational

Record Size: 1 billion

Reports: Report generation, but no statistical functions

File Import/Export: dbase III, dbase II, LOTUS, DIF, and ASCII

Comments: dBase III Plus is the industry leader. There is very little that cannot be done with the program, except statistical operations. A wealth of third-party support materials is available.

Paradox
ANSA Software
1301 Shoreway Rd.
Belmont, CA 94002

Cost: $495

System: IBM PC and compatibles (MS-DOS)

Memory: 512K; hard disk recommended

Version: 2.0

Type: Programmable-relational

Record Size: 65,000

Reports: Ability to layer reports

File Import/Export: dbase III, dbase II, LOTUS, DIF, and ASCII

Comments: Paradox has features very similar to *dbase III* and *R-Base*. The report generation is flexible to meet user needs, and permits layered reports. The user interface can be mastered quickly. The program permits most descriptive statistics (mean, standard deviation, variance) and columnar math operations.

PC-File/R
ButtonWare, Inc.
P. O. Box 5786
Bellevue, WA 98006

Cost: $149

System: IBM PC and compatibles (MS-DOS)

Memory: 256K

Version: 1.0

Type: Relational

Record Size: 32,000

Reports: Reports can be generated through various form construction features.

File Import/Export: Limited, does work well with ASCII files.

Comments: PC-File/R is the non-shareware version of the very popular *PC-File+*, a shareware program available directly from ButtonWare or through many bulletin boards. The newest version incorporates features of some of the best relational database managers. *PC-File+* is a possibility for users wishing to use a database with a dataset of reasonable size.

PFS: File
Software Publishing Company
1901 Landings Dr.
P. O. Box 7210
Mountain View, CA 94043

Cost: $139

System: IBM PC and compatibles (MS-DOS); Apple IIe and IIc

Memory: 128K for IBM; 64K for Apple

Version: B

Type: Flat-field

Record Size: 2,200

Reports: Requires the additional purchase of *PFS: Reports* ($125) to generate reports. Together they produce an acceptable range of reports.

File Import/Export: None

Comments: *PFS: File* was one of the earliest database managers available for more than one microcomputer system. A new series for IBM and compatibles is called *PFS: First Choice,* an integrated package that includes a new flat-field database called *PFS: Professional File,* which lists separately for $249.

R-Base 5000
Microrim
3380 14th Place, SE
Bellevue, WA 98007

Cost: $495

System: IBM PC and compatibles (MS-DOS)

Memory: 256K

Version: 1.1

Type: Programmable-relational

Record Size: Limited by system memory.

Reports: Now has a companion graphics program.

File Import/Export: dbase II, dbase III, LOTUS, DIF, and ASCII

Comments: R-Base 5000 is the version of the *R-Base* series that operates on non-hard-disk-based systems. It is capable of multiple manipulations of a large set of data. There is a large number of third-party manuals to aid the new user.

R-Base for DOS
Microrim
3380 14th Place, SE
Bellevue, WA 98007

Cost: List $795

System: IBM PC and compatibles (MS-DOS)

Memory: 512K (640K recommended); hard drive required

Version: 1.1

Type: Programmable-relational

Record Size: Limited by system memory.

Reports: Report generator with a companion graphics program available

File Import/Export: Imports and exports all major types, reads *R-Base System V*, and translates *R-Base 4000* records.

Comments: R-Base for DOS is an advanced version of *System V*. The program fully integrates all program operations and applications. The various Express Features (e.g., reports, forms) permit the rapid creation of files, input forms, and report forms. Full arithmetical and descriptive statistics are available. The program is able to handle large data sets and the fast creation of reports.

R-Base System V
Microrim
3380 14th Place, SE
Bellevue, WA 98007

Cost: List $700

System: IBM PC and compatibles (MS-DOS)

Memory: 512K; hard disk required

Version: 1.1

Type: Programmable-relational

Record Size: Limited by system memory.

Reports: Now has a companion graphics program.

File Import/Export: dbase III, dbase II, LOTUS, DIF, and ASCII

Comments: R-Base System V is the real challenger to the *dBase III Plus*'s supremacy in the database-manager field. This new release adds a program generator that removes the difficulty of programming from the user. Flexible and with a large memory, it is useful if the user has a large set of data requiring multiple manipulations. The program now has descriptive statistical functions.

Reflex
Borland International, Inc.
4585 Scotts Valley Dr.
Scotts Valley, CA 95066

Cost: $149.95

System: IBM PC and compatibles (MS-DOS); Macintosh version available.

Memory: 384K

Version: 1.1

Type: Relational (flat-field). Hard to classify, since it does possess some of the operations of a relational database.

Record Size: 65,000

Reports: Not as elaborate as *R-Base System V* or *dBase III Plus*.

File Import/Export: Exports only ASCII files; imports all major types.

Comments: Reflex is useful for the researcher who needs data storage and organization. The program interfaces well with other programs, especially *LOTUS 1-2-3*. *Reflex* offers on-screen graphics as the data is manipulated.

The following database managers were not available for review but are listed in an effort to provide additional sources for the user.

ClearCut
Menlosoft/Business Day & Software
5 Cheryl Place
Menlo Park, CA 94205

Cost: List $49.95

System: IBM PC and compatibles (MS-DOS)

Memory: 128K

Version: 1.1

Type: Flat-field

dBase II
Ashton-Tate
20101 Hamilton Ave.
Torrance, CA 90502

Cost: List $295

System: IBM PC and compatibles (MS-DOS); Apple IIe version available.

Memory: 128K

Version: 2.4

Type: Relational

File
Microsoft Corporation
16011 N.E. 36th Way
P. O. Box 97017
Redmond, WA 98703

Cost: List $195

System: Macintosh

Memory: 512K

Version: 1.04

Type: Flat-field

Comments: Interfaces with the Macintosh version of *Microsoft Word,* version 3.0.

Helix
Odesta Corporation
4084 Commercial Ave.
Northbrook, IL 60062

Cost: List $175

System: Macintosh

Memory: 512K

Version: 2.0

Type: Relational

Comments: Also available in a custom version, *Double Helix,* for $495, and a multiuser version, *Multiuser Helix,* for $695.

Infoscope
Microstuf, Inc.
1000 Holcomb Woods Parkway, Suite 440
Roswell, GA 30076

Cost: List $79

System: IBM PC and compatibles (MS-DOS)

Memory: 192K

Version: 1.01

Type: Flat-field

KnowledgeMan
Micro Data Base Systems, Inc.
P. O. Box 248
Lafayette, IN 47902

Cost: List $525

System: IBM PC and compatibles (MS-DOS)

Memory: 320K

Version: 2.0

Type: Relational

Mac Base
Eqtron Corporation
330 Bay St., Suite 115
Toronto, Ontario, Canada M5H258

Cost: List $149

System: Macintosh

Memory: 512K

Version: 1.0

Type: Relational

McMax
Nantucket
12555 Jefferson Blvd.
Los Angeles, CA 90066

Cost: List $295

System: Macintosh

Memory: 512K

Version: 1.0

Type: Relational

Omnis 3 Plus
Blyth Software, Inc.
2929 Campus Dr., 4th Floor
San Mateo, CA 94403

Cost: List $495

System: Macintosh

Memory: 512K

Version: 3.24

Type: Relational

Practibase
PractiCorp International, Inc.
44 Oak St.
The Silk Mill
Newton Upper Falls, MA 02164

Cost: $99.95

System: IBM PC and compatibles (MS-DOS)

Memory: 256K

Version: 1.14

Type: Relational

Probase
Probase Group, Inc.
1738 W. LaPalma Ave.
Anaheim, CA 92801

Cost: List $395

System: IBM PC and compatibles (MS-DOS)

Memory: 192K

Version: 2.0

Type: Programmable-relational

The Sensible Solution
O'Hanlon Computer Systems
11508 Main St., Suite 10
Bellevue, WA 98004

Cost: $695

System: IBM PC and compatibles (MS-DOS)

Memory: 128K

Version: 2.0

Type: Relational

Integrated Packages

An integrated software package is a system of programs designed to perform a series of applications, generally consisting of a word processor, spreadsheet, graphics, data base, and, often but not always, a communication program. The quality and usefulness of integrated packages varies from extremely sophisticated to extremely sophomoric. The one obvious limitation of such a package is that no matter how good it is, a single dedicated program, i.e., one activity, is often better than its comparable program in an integrated package.

Another limitation of integrated packages is their memory requirements, although this is not as big an issue today as it was in 1984 or 1985. The availability of 640K (and larger, thanks to Extended Memory Systems) IBM PC and compatibles (MS-DOS) has reduced this concern somewhat, but the user must be willing to expand the system's memory to meet the memory requirements and needs of the package.

The major advantage of an integrated package is the ease by which information can be moved among applications. This is particularly useful when the database program is used as an input and storage system for the data and the data are transferred to the spreadsheet for analysis and back to the word processor for report writing. Another advantage is when the keystrokes are common among the applications, enabling the user to memorize a limited number of keystrokes. All in all the general usefulness of an integrated package must be weighed against the lack of full-powered operation available in dedicated programs and the memory requirements of the integrated package.

One final comment concerning integrated packages as a whole: If one does considerable work between microcomputers and mainframes, the added value of a communication program must be considered. The one caution worth reflecting upon is the added value of a communication program within the package. On the surface, the ability to move data between systems is very valuable; however, it should be remembered that the primary user of an integrated package is a business, rather than a researcher. The type of communication program needed by the former is vastly less polished than the program needed by the researcher. Yet if the transmission is limited to communication and ASCII file transfer, most of the communication programs available within the integrated packages are quite satisfactory.

Profiles of Integrated Packages

AppleWorks
Apple Computer
20525 Mariani Ave.
Cupertino, CA 95014

Cost: List $230

System: Apple IIe and IIc

Memory: 64K (minimum; works best at 128K)

Version: 2.0

Functions: Descriptive statistics and data-file manipulations (including sorts and subsetting). Includes no graphics or regression and matrix operations.

Programs: Word processor, database, and spreadsheet

Comments: AppleWorks is the only truly integrated package for the Apple II series of microcomputers. The program is easy to operate yet sufficiently flexible to provide skilled users with a solid storage system for their data. It takes very little time to learn to use the word processor, the spreadsheet, or the database manager. The new user can be up and operating within a few hours of beginning instruction.

The three programs included in the package can be considered superior to most stand-alones available for the Apple II systems, particularly the spreadsheet. The database is a flat-field system that permits movement of data back and forth to the word processor.

Full-time users of Apple systems will find the integration features of *AppleWorks* especially useful. The program permits the transfer of data from the spreadsheet and database to the word processor, but transfer between the spreadsheet and database takes a bit of file manipulation. The transfer of data requires the creation of an external file to temporarily hold the data before importing it back into the system.

Enable

The Software Group
Northway Ten, Executive Park
Ballston Lake, NY 12019

Cost: $695

System: IBM PC and compatibles (MS-DOS)

Memory: 320K

Version: 2.0

Functions: Descriptive statistics and data-file manipulations (including sorts and subsetting). Graphics include area, bar, line, pie, high-low graphs, and three-dimensional graphs. Will not perform regression or matrix operations.

Programs: Word processor, database, spreadsheet, communications, and graphics

Comments: Enable 2.0 is a package that meets the standard needs for descriptive analysis but lacks regression analysis and matrix operations. *Enable*'s spreadsheet and word processor are a bit more capable than its communication and database manager. The package also exports and imports a wide variety of files.

The package now comes bundled with *Perspective*, a versatile graphics package. The addition of *Perspective* changes *Enable* into a competitive graphics-generation program. The graphics available to the user include standard bar, line, pie, and area graphs as well as three-dimensional graphs. The graphic types can be mixed, to a degree, to produce precise representations of data. Graphics may be generated from either the spreadsheet or database-management program. A second feature of the package is its ability to interface with BASIC, C, and Pascal programming languages. This is of interest to the advanced user who wishes to program particular activities as part of data storage and analysis.

Framework II
Ashton-Tate
20101 Hamilton Ave.
Torrance, CA 90502

Cost: List $695

System: IBM PC and compatibles (MS-DOS)

Memory: 384K

Version: 1.1 (*Framework III* now available)

Functions: Descriptive statistics and data-file manipulations (including sorts and subsetting). Graphics include area, bar, line, pie, high-low graphs, and overlay graphs.

Programs: Word processor, database, spreadsheet, communications, and graphics

Comments: Framework II is a product of the same corporation that produces *dBase III Plus*. The database feature of *Framework II* is very capable. Unlike *dbase Framework* does not possess full programming features, but some user-developed programming is possible. Spreadsheet functions are similar to *VP-Planner's*, with the same caution for sample considerations noted earlier for *LOTUS 1-2-3* and *VP-Planner*. The program does offer matrix operations but no regression analysis. Because *Framework II* utilizes a window environment to integrate the programs, movement between any of the programs is simple.

The communications program lacks the ability to emulate various terminals. Data transfer is generally limited to ASCII text transfer unless the host system will accept XMODEM protocols for file transfer. The communication system and windowing environment of *Framework II* allows movement from spreadsheet (or database) to communication program to remote computer and back to the spreadsheet (or database). *Framework II* is useful to the researcher seeking a package that permits movement of raw data among the various programs of the package and to outside computers.

Jazz
Lotus Development Corporation
55 Cambridge Parkway
Cambridge, MA 02142

Cost: List $395

System: Macintosh

Memory: 512K

Version: 1.0

Functions: Descriptive statistics and data-file manipulations (including sorts and subsetting). Graphics include bar, line, pie, and high-low graphs.

Programs: Word processor, database, spreadsheet, communications, and graphics

Comments: *Jazz* on the Macintosh offers the features of *LOTUS 1-2-3,* version 1.1A, with the graphic clarity and speed of a 68000 microprocessor. The feature of the mouse-driven windows environment of the Macintosh enhances the accessibility of the package.

The same limitations of *LOTUS 1-2-3* apply, and regression analysis and matrix operations are absent as well. However, the integration of a database, word processor, and communications package within the Macintosh context provides a useful package for Macintosh users.

Smart Software System
Innovative Software, Inc.
9875 Widmer Rd.
Lenexa, KS 66215

Cost: List $895

System: IBM PC and compatibles (MS-DOS)

Memory: 384K

Version: 3.10

Functions: Descriptive statistics and data-file manipulations (including sorts and subsetting). Graphics include area, bar, line, pie, and high-low graphs.

Programs: Word processor, database, spreadsheet, communications, and graphics

Comments: Each program within the *Smart Software System* package is a stand-alone program that can be used independently. It is included in the integrated-systems category because of its ability to move swiftly among the programs within its system. Accompanying the system is documentation that is completely cross-referenced within each module of the system.

The spreadsheet is similar to that of *LOTUS 1-2-3*. The general functions typically necessary for a spreadsheet are all present, with the addition of a statistical function (STDEV) for calculating sample standard deviations. The spreadsheet possesses matrix operations and regression analysis, although the regression is a little less complete than the *LOTUS 1-2-3* function.

The *Smart Software System*'s graphics are comprehensive. Some 78 different types of possible graphs have been reported, produced in two and three dimensions (although the author must admit to having attempted only 32). Furthermore, the graphs may be directly imported into the word-processing program, an unusual feature useful for report writing.

The word processor, although not quite as capable as the top dedicated word-processor programs, offers more features than other word processors that are part of integrated packages.

The database handles data input, manipulation, and storage, and it is a relational database. Data can be readily moved from the database to any of the other programs with a couple of keystrokes.

The communication program, intended for business users, lacks any true terminal emulation but transmits and receives ASCII files. The program does make use of XMODEM protocols for file transfer.

Symphony
Lotus Development Corporation
55 Cambridge Parkway
Cambridge, MA 02142

Cost: List $695

System: IBM PC and compatibles (MS-DOS)

Memory: 384K

Version: 1.2

Functions: Descriptive statistics and data-file manipulations (including sorts and subsetting). Graphics include area, bar, line, and pie graphs. Includes regression analysis and matrix operations. Readily imports from and exports to *LOTUS 1-2-3*.

Programs: Word processor, spreadsheet, database, communications, and graphics

Comments: Symphony is the best-selling of the integrated packages. It offers a word processor, database, graphics, and communication program as well as a spreadsheet. The spreadsheet program is very similar to *LOTUS 1-2-3*'s program. The functions operate, for all practical purposes, exactly the same. If one is familiar with *LOTUS 1-2-3*, then *Symphony* operations are straightforward. *Symphony* is designed under the assumption that the data are based on populations, and adequate precautions must be taken not to bias one's estimations.

Symphony's database will handle data input and organization more effectively than *LOTUS 1-2-3*, but will not allow the type of data interrelationships that a relational database (such as *R-BASE—System V* or *dBase III Plus*) offers. The data can be readily moved between the database and the spreadsheet. Furthermore, data can be drawn from either the database or the spreadsheet to the graphics program. The graphics are nearly a mirror of *LOTUS 1-2-3*'s.

The user who wishes to perform data manipulations prior to exporting the information to a statistical package should find *Symphony*'s operations helpful. The package's general operations are very similar to *LOTUS 1-2-3*'s.

The following programs were not available or not provided for review. For further details, the reader is advised to seek other published reviews concerning these programs.

Ability
Migent, Inc.
P. O. Box 6062
Incline Village, NY 89450

Cost: Ability, list $99; *Ability Plus,* list $199

System: IBM PC and compatibles (MS-DOS)

Memory: 384K

Version: 1.2 (*Ability*)

Functions: Descriptive statistics and data-file manipulations (including sorts and subsetting). Good graphics including bar, line, pie, and high-low graphs. Does regression analysis and matrix operations.

Programs: Word processor, database, spreadsheet, communications, and graphics

Microsoft Works
Microsoft Corporation
16011 N.E. 36th Way
P. O. Box 97017
Redmond, WA 98073

Cost: List $295

System: Macintosh, IBM

Memory: 512K

Version: 1.0

Functions: Descriptive statistics and data file manipulations (including sorts and subsetting). Limited graphics including bar, line, pie, and high-low graphs. Does not perform regression analysis or matrix operations.

Open Access
Software Products International, Inc.
10240 Sorrento Valley Rd.
San Diego, CA 92121

Cost: List $595

System: IBM PC and compatibles (MS-DOS)

Memory: 256K

Version: 2.0

Functions: Descriptive statistics and data-file manipulations (including sorts and subsetting). Graphics, including bar, line, pie, and high-low graphs. Does include matrix operations but not regression analysis.

Programs: Word processor, database, spreadsheet, communications, and graphics

PFS First Choice
Software Publishing Company
1901 Landings Dr.
P. O. Box 7210
Mountain View, CA 94039

Cost: List $179

System: IBM PC and compatibles (MS-DOS)

Memory: 256K

Version: 1.03

Functions: Descriptive statistics and data-file manipulations (including sorts and subsetting). Graphics, including bar, line, pie, and high-low graphs. Does include matrix operations but not regression analysis.

Programs: Word processor, database, spreadsheet, and graphics

CHAPTER

2

Statistical Packages for Microcomputers

Introduction

Statistics and microcomputers: the two seem to be natural companions. The modern researcher, unless blessed with uncanny abilities, must rely upon computers to aid in the analysis of datasets. This is particularly true when one considers the degree of sophistication to which statistical analysis has advanced in the last 25 years.

Crunching numbers is no longer something done only by thick-lensed operators on large computers. The modern researcher is just as apt to be a practitioner as he or she is to be a researcher. Research has entered the realm of practical assessment and evaluation as well as theoretical development and construct testing. However, the analysis of data still requires an understanding of the underlying principles and models that impact upon the chosen analytic tool and the proposed question under study.

The tools of the modern researcher have evolved along with the strategies and techniques of research. It seems but a short time ago that we were dancing in the streets over the invention of the hand calculator to aid our inferential analyses; now hardly any analysis can be done without the aid of a computer or microcomputer. The choice of tools for a researcher has been made by the increasing demand for precision in the research field. The numerous packages available to aid the researcher range from the simple to the complex, from those designed only to provide single analysis to those designed to provide analytic tools for all types of problems. Many of the *full-system* packages available for microcomputers were derived from their mainframe big brothers and sisters; however, a good number of the available packages were developed for microcomputers alone.

The various packages included in this work will be limited to only full-system packages, i.e., those that perform more than three different applications. The rationale for limiting this review to those packages is that a researcher needs a complete set of tools, not merely a single-function program. A further comment concerns the method in which the packages were reviewed. The cost of most of the packages prohibited their purchase simply for a tryout; hence, publishers were initially contacted about supplying review copies. Several, most notably BMDP,

Inc., SYSTAT, Minitab, and SAS provided actual demonstration packages or complete sets. Others were tried out through the kind aid of co-workers who owned the packages or through company representatives at national conventions. However, many packages were not available for an actual tryout. Some of the companies sent only printed documentation, most often sales information, while others did not respond at all. As a consequence, the information for some packages may not be as precise as would have been preferred.

The information included here focuses primarily on several selected features, namely:

1. Type of system(s) the package runs on
2. Various features of the package, including editors, statistical functions, and data import and export
3. Memory requirements of the package
4. Expertise required of the user

Profiles of Statistical Packages

BMDP PC

BMDP Statistical Software, Inc.
1440 Sepulveda Blvd., Suite 316
Los Angeles, CA 90025

Cost: Initial package $495; complete package $1,800

System: IBM PC and compatibles (MS-DOS)

Memory: 640K and 8087 co-processor

Version: 1987 release

Format: Batch or interactive

Functions: Descriptive, correlations, ANOVA, ANCOVA, factor analysis, cross tables, multiple analysis of variance, plots and charts, nonparametric, regression

Comments: BMDP PC is a statistical package that absolutely mirrors the mainframe package. In fact, the manuals are the same regardless of which system you use, with the exception that a *PC User's Guide* is also included in the package. *BMDP PC* does what *BMDP* mainframe does, and it does it in precisely the same manner, with the exception of the interactive mode; however, the interactive mode might better be called a semi-interactive mode since the user types the program line by line, references the file containing the data, and submits the program to run. Former *BMDP* mainframe users will find this system familiar.

The user who wishes to perform a regression analysis on a data set uses the same program, 2R, that would be used on the mainframe. The program is constructed either through a batch level process or through the interactive line editor. Once the program is completed it is submitted. Users of the mainframe

BMDP will note that *BMDP PC* produces output almost exactly like the output of its parent program, including some features of output rarely found in other systems. Some of these features are offered, for example, in its computation of a multivariate analysis of variance.

One added feature of the microcomputer-based *BMDP PC* is the *BMDP DATA Manager*. Some users of older versions of *BMDP* report that the manipulation of data sets is not in the vocabulary of *BMDP*. But the addition of the *DATA Manager* permits data sets to be ordered, reconstructed, redistributed, and otherwise completely massaged to meet the needs of the user. Using a data set created first on *LOTUS 1-2-3*, this writer was able to input it into the *DATA Manager*, reconstruct it, run an analysis, read the output file with the *DATA Manager*, and export it back to the *LOTUS 1-2-3* spreadsheet from whence it originated. This type of data manipulation was unavailable in the older version of *BMDP*.

In summary, *BMDP PC* is, as advertised, a microcomputer version of the mainframe program: The operation is very similar to the mainframe, the commands are exactly the same, the output nearly the same, and the limitations are nearly the same. The addition of the *DATA Manager* plugs a hole in the package's ability to manipulate and change data sets. The separate modules of the package may be purchased for $60 each, and the whole main package costs $1800.

The support from BMDP is readily available for registered users. In addition to the usual telephone support, registered users are automatically placed on the BMDP communications mailing list and receive regular newsletters and up-to-date information. Licensed users are advised well in advance of changes, enhancements, and additions to the BMDP line.

Dyna-Stat
Dynamic Microsystems, Inc.
13003 Buccaneer Rd.
Silver Spring, MD 20904

Cost: List $195

System: Apple II series (IBM version due soon)

Memory: 128K

Version: Pro

Format: Menu

Functions: Descriptive, correlations, ANCOVA, ANOVA, cross tables, multiple analysis of variance, plots and charts, nonparametric, regression, multidimensional scales

Comments: Dyna-Stat is one of the few programs available for the Apple II series that can be called a full-system package. The program is entirely menu-driven with a straightforward selection of options. The breadth of statistical tests is broad enough to make the program quite useful as either a research or instructional tool. Despite the limited memory of an Apple II system, *Dyna-Stat* permits over 100 variables with up to 16,000 observations (these figures are based on the use of expanded memory boards or the use of a 640K IBM system). The process of manipulating data is handled through a built-in editor that, while slow, does handle a variety of operations, including merges and sorts.

Minitab

Minitab, Inc.
3081 Enterprise Dr.
State College, PA 16801

Cost: Academic $250; non-academic $500

System: IBM PC and compatibles (MS-DOS)

Memory: 512K

Version: 7.1

Format: Menu or Command

Functions: Descriptive, correlations, ANOVA, exploratory analysis, time series, cross tables, regression, distributions, plots and charts, nonparametric, regression

Comments: Minitab is another program that has migrated from a mainframe environment to a microcomputer setting, with the intention of providing a workable data-analysis system in the least constrictive context. *Minitab* was initially developed with academic settings (meaning colleges and universities) as its primary users. The advent of the microcomputer has made the movement of *Minitab* to noncollege surroundings a reality. *Minitab* operates using rectangular datasets of rows and columns for observations and data types. The resemblance between this format and that of spreadsheets has not been lost on the makers of *Minitab,* who have introduced an interface between *Minitab* and *LOTUS 1-2-3* called *Minitab/1 2 3* (release 2.1, cost $30). The interface permits the importation of *LOTUS 1-2-3* and *Symphony* spreadsheets directly into *Minitab* datasets. Furthermore, the translation is bidirectional, meaning that the output dataset of *Minitab* can be imported into *LOTUS 1-2-3.*

Minitab handles the direct assessment of data; the various features permit all but the most complex analyses, in particular multivariate (beyond regression). Perhaps because of its business-based origins, *Minitab* performs time-series analyses, handles exploratory data analysis, and is one of the few that will permit analyses in distributions other than a normal distribution (e.g., Poisson and binomial).

Through a licensing process, Minitab, Inc., will discount multiple purchases by institutions. The user of *Minitab* will find a series of support manuals and third-party publications available, plus a number of statistics texts that use *Minitab.*

PC SAS

SAS Institute, Inc.
Box 8000, SAS Circle
Cary, NC 27511

Cost: Licensing fee $2,700

System: IBM PC and compatibles (MS-DOS)

Memory: 512K (hard drive required)

Version: 6.03

Format: Menu or command

Functions: Descriptive, correlations, ANOVA, ANCOVA, factor analysis, cross tables, multiple analysis of variance, plots and charts, nonparametric regression

Comments: PC SAS is very close to the mainframe version of *SAS*. The primary difference is the window-like environment, familiar to those who have operated *SAS* in an interactive VMS environment. The multiple windows are used for the program, the log and the output for the analysis. Additional windows for the online *Help* features, menus, etc. are also available any time for the user.

The total *SAS* system includes:

1. Base *SAS:* for general data input and manipulation; includes most descriptive analyses procedures
2. STATS: for inferential analyses of data
3. GRAPH: for high-resolution and color graphics
4. AF: for full-screen, interactive applications
5. FSP: interactive, menu-driven applications for data entry, editing, and text processing
6. IML: interactive matrix applications

When fully loaded, the current version of *SAS* takes nearly 20 megabytes of storage. *SAS* requires a hard drive to operate properly. While *SAS* recommends a minimum of 332K free memory, the system operates more efficiently with 512K. *SAS* will use EMS (expanded memory), but the user should consult the *SAS Installation* manual about the correct activation requirements. Although a math co-processor is not required by *SAS*, it is highly recommended.

One of the major strengths of *SAS* for mainframe computers has been the extent of statistical analyses available. The microcomputer user of *SAS* will find the same depth and breadth of available procedures. While the processing time of *SAS* is somewhat slower on a microcomputer than on a mainframe, it is still very impressive. The ability of mainframe *SAS* to handle monster datasets and to work with these datasets and their subsets is also a characteristic of *PC SAS*.

SAS is particularly appropriate for the user who has programming skill and a rich understanding of the underlying theoretical models of statistical analysis. The manuals accompanying *SAS* are legion, yet necessary to grasp the full power of the system. User support has always been a hallmark of *SAS,* and continues to be so. Registered users will find consultants available from SAS Institute Inc. for almost any problem or inquiry, from installation to sophisticated statistical applications.

One final comment related to *SAS* is the current practice of the SAS Institute Inc. regarding the acquisition of *SAS* version 6.034. SAS will not sell a copy of PC SAS; instead you may only purchase a license for its use. This license is good for one year, at which time the program will cease operating unless a special code is input into the system.

The BASE module comes with a limited built-in editor for program creation and will read and write external ASCII files. The edition of the AF and FSP modules greatly enhances *SAS*'s ability to manipulate data and datasets. Furthermore, the effective application of these modules permits the skilled user to develop menu-driven applications for the less-skilled user. Version 6.03 does not directly import or export LOTUS, dBase, or DIF files.

A very useful addition to the BASE and STATS modules of *PC SAS* is the GRAPH package. While it is the largest *SAS* module (over 7 megabytes) the graphics produced by the module cover a full range, from plots to regional maps of the United States. The graphics produced by this package are crisp and as elaborate as needed by the user. One note of caution: the full implementation of the GRAPH module may require some time to master because of the complexity of the available options.

SPSS/PC+
SPSS Inc.
444 North Michigan Ave.
Chicago, IL 60611

Cost: List $795; Advanced Statistics package $295

System: IBM PC and compatibles (MS-DOS)

Memory: 384K (hard drive required)

Version: 4.0

Format: Command

Functions: Descriptive, correlations, ANOVA, ANCOVA, factor analysis, cross tables, multiple analysis of variance, plots and charts, nonparametric, regression

Comments: A knowledge of the parent system will help the user of *SPSS/PC+*. The program is large, taking nearly 10 megabytes of space on a hard disk, but it is complete when the advanced statistical package is added. The manuals for *SPSS/PC+* include some of the principles of statistical analysis.

SPSS/PC+ does import ASCII and *LOTUS 1-2-3* files and follows precisely the syntax of its mainframe brother. There are interactive and batch modes. The command structure is the same for the microcomputer version as it is for the mainframe version, and the general format appears to mirror the format and structure introduced with *SPSS-X*.

The newest release of *SPSS/PC+* has the ability to interface with *Microsoft Chart* (additional cost of $395), a graphics-presentation program. The integration of *SPSS/PC+* with *Microsoft Chart* permits the user to generate graphs, charts, and plots far more elaborate than those permitted within the *SPSS/PC+* system.

SPSS/PC+ is particularly helpful to the user who expects to perform a high number of statistical activities or share the program with another user. As with many of the microcomputer packages derived from mainframe packages, *SPSS/PC+* offers available support for registered users. The response time to user inquiries is relative to the complexity of the problem. The user may also be placed on the SPSS mailing list and receive regular newsletters about updates, changes, and system enhancements.

Statgraphics
STSC, Inc.
2115 East Jefferson St.
Rockville, MD 20852

Cost: List $795

System: IBM PC and compatibles (MS-DOS)

Memory: 512K

Version: 2.0

Format: Menu

Functions: Descriptive, correlations, limited analysis of variance, limited multivariate analysis, cross tables, time series, plots and charts, nonparametric, regression, factor analysis

Comments: Statgraphics is more than just a statistical package. It contains a limited word processor, a database manager, and a spreadsheet, as well as a complex graphics-generation capability. The whole program is menu-driven, which is an aid in developing the desired output. The limitation of the system is in the number of variables and the number of cases that can be in the dataset (although STSC Inc. indicates that a forthcoming version will increase these limits).

The real strength of *Statgraphics* is in its graphics generation. The program produces high resolution and color graphics in plots, charts, maps, etc., in two or three dimensions. The program has the capacity to produce 50 distinct presentations, when you include the addition of multiple fonts, icons, and freehand drawing, the actual number of top quality graphics is nearly limitless.

Statgraphics will import and export *LOTUS 1-2-3, dbase,* and ASCII files. The ease of movement of file types is a strong feature of the program.

In summary, *Statgraphics* is a full-featured package, but currently limited by the number of cases and variables. It produces excellent graphics based upon the statistical output of the program.

SYSTAT
Systat, Inc.
2902 Central St.
Evanston, IL 60201

Cost: List $595; Sygraph, $595; Together, $895

System: IBM PC and compatibles (MS-DOS); Macintosh

Memory: 256K (512K version available for large datasets)

Version: 5.0

Format: Command

Functions: Descriptive, correlations, ANOVA, ANCOVA, factor analysis, cross tables, multiple analysis of variance, plots and charts, nonparametric, regression, multidimensional scales

Comments: The recently released version 5.0 of *SYSTAT* represents a fully evolved and integrated statistical-analysis system. Analytic procedures are available from descriptive statistics to multidimensional scaling. The package comes with two editors, one for editing *SYSTAT* files and the other for editing ASCII data and command (CMD) files. The latter editor, called FEDIT, works from within the statistical modules.

The newest addition to *SYSTAT* is *Sygraph,* a full-featured graphics program. *Sygraph* includes high-resolution and color graphics for the following: (1) bar graphs, (2) box plots, (3) plots, (4) density plots, (5) pie graphs, and (6) probability plots. In addition, the user is able to utilize special map-generating, drawing, icon, multiple-font, and text-inclusion features, as well as two- and three-dimensional graphs. *Sygraph* is available as a set with *SYSTAT* 5.0 or may be purchased separately. Use of *Sygraph* requires at least 640K of memory for PC-compatible systems.

SYSTAT 5.0 comes on eight disks and *Sygraph* comes on six disks. While *SYSTAT* may be set up to run on dual floppies, a hard drive is strongly recommended. When fully installed, *Systat* and *Sygraph* use slightly less than three megabytes of memory. If your system is equipped with a math co-processor, *SYSTAT* will automatically utilize it to aid its calculations and graphics generation.

Transferring data to and from *SYSTAT* is accomplished through several import operations for *LOTUS 1-2-3, dBase,* and DIF files. Through this function, data is automatically converted into *SYSTAT* files. Exporting files to either of these file types is also readily accomplished through the *Export* procedure.

SYSTAT 5.0 continues the MACRO procedure that simplifies the construction of batch programs, called COMMAND (CMD). It enables the user to create complex strings of analyses and integrate programming commands into the operations. For example, *SYSTAT* command files may be used to generate scoring algorithms for a test, score the test, and produce a descriptive analysis as well as an item analysis of the test. The resulting data file may be retained and exported to another package (e.g., data base or word processor) for reporting and integration with additional information.

The manuals accompanying *SYSTAT* and *Sygraph* were designed for use by the statistically skilled user. The manuals explain *how* to use a particular procedure or operation, not *why.* Users unsure of the proper use of a procedure should consult the references provided by the manuals.

SYSTAT takes great pride in attempting to answer questions and queries from its registered users. While the response time for questions varies according to the complexity of the question, *SYSTAT* reports that there has always been a response. *SYSTAT* also will provide on-site licensing and training programs for users.

Overall, the *SYSTAT* package is very complete, requiring little recourse to other programs. *SYSTAT* is also available for Macintosh, CP/M, UNIX, and VAX systems. The complete *SYSTAT* family includes logistical regression

(LOGIT), probit analysis (PROBIT), regression analysis of censored data (TOBOT), two-stage least squares (2SLS), classical and latent trait analyses of tests (TESTAT), and experimental design tools (DESIGN).

The following programs were not available or not provided for review. Some limited information is offered to aid the researcher in identifying additional packages on the market. For further details, the reader is advised to seek other published reviews concerning these programs.

1,2,3 Forecast!
1,2,3 Forecast!
P. O. Box 9309
Salem, OR 97309

Cost: List $89.95

System: IBM PC and compatibles (MS-DOS)

Memory: 512K

Functions: Descriptive, correlations, plots and charts, ANOVA, regression

Comments: *1,2,3 Forecast* is a *LOTUS 1-2-3* template program that uses the *LOTUS 1-2-3* spreadsheet to compute limited statistical analyses.

ABstat
Anderson-Bell Corporation
South Pine Dr., Suite 411
Parker, CO 80134

Cost: List $395

System: IBM PC and compatibles (MS-DOS)

Memory: 256K

Format: Command

Functions: Descriptive, correlations, ANCOVA, ANOVA, cross tables, plots and charts, nonparametric, regression

Comments: *ABstat* is another of the "older" statistical programs that has survived. A command-driven, multifaceted program, *ABstat* performs a number of statistical operations. It has the ability to read and write *LOTUS 1-2-3* and *dBase III Plus* files.

ADDaSTAT
Abacus Scientific Software
4521 Victory Dr.
Irvine, CA 92715

Cost: $80

System: IBM PC and compatibles (MS-DOS)

Memory: 384K

Format: Menu or Command

Functions: Descriptive, correlations, ANCOVA, factor analysis, cross tables, t-tests, Chi-square, plots and charts, nonparametric, regression

Comments: AddaSTAT is another *LOTUS 1-2-3* template program that calculates limited inferential statistics. It is a quick and simple format for users familiar with *LOTUS 1-2-3*.

CRUNCH

Crunch Software Corporation
5335 College Ave., Suite 27
Oakland, CA 94618

Cost: Regular $495; student $99

System: IBM PC and compatibles (MS-DOS)

Memory: 256K

Format: Menu, Macros, Batch

Functions: Descriptive, correlations, t-tests, regression, cross-tabulation, principal component, factor analysis, ANOVA, nonparametric, ANCOVA

Comments: Crunch is a new program introduced in spring 1988. The system permits direct importation of ASCII, DIF, *dBase,* and *R-base* files. The number of observations is limited to memory but the maximum number of variables is 252. *Crunch* permits various file manipulations, including merges, sorts, transposing observations, and variables as well as subgrouping.

CSS—Complete Statistical System

StatSoft
2832 East 10th St., Suite 4
Tulsa, OK 74104

Cost: List $495

System: IBM PC and compatibles (MS-DOS)

Memory: 384K

Format: Menu

Functions: Descriptive, correlations, ANOVA, ANCOVA, factor analysis, cross tables, multiple analysis of variance, plots and charts, nonparametric, regression, time-series analysis

Comments: One of the new breed of statistical programs, *CSS* appears to blend the best of statistical analysis with an array of graphic capabilities. *CSS* imports files from *LOTUS 1-2-3, dBase III Plus,* ASCII sources, and reportedly can do screen captures from noncompatible systems. The program is primarily written in assembler code, which should add speed to the processing time of calculations.

Microstat
Ecosoft, Inc.
6413 North College Ave.
Indianapolis, IN 46220

Cost: List $375

System: IBM PC and compatibles (MS-DOS)

Memory: 64K

Format: Menu

Functions: Descriptive, correlations, ANOVA, cross tables, time series, plots and charts, nonparametric, regression

Comments: Microstat was originally designed for use in CP/M systems, for which it is still available. It offers a menu-driven format for entering statistical operations, and will read *LOTUS 1-2-3, dBase II,* and ASCII files.

NWA Statpak
Northwest Analytical, Inc.
520 N.W. Davis St.
Portland, OR 97209

Cost: List $495

System: IBM PC and compatibles (MS-DOS)

Memory: 512K

Format: Menu or Command

Functions: Descriptive, correlations, ANOVA, cross tables, probability calculations, plots and charts, nonparametric, regression

Comments: NWA Statpak will read *LOTUS 1-2-3* and ASCII files.

PC Statistician
Human Systems Dynamics
9010 Reseda Rd., Suite 222
Northridge, CA 91324

Cost: List $300

System: IBM PC and compatibles (MS-DOS)

Memory: 256K

Functions: Descriptive, correlations, ANOVA, cross tables, plots and charts, nonparametric, regression

Comments: PC Statistician also includes a database manager to aid in the construction of datasets. The program will accept *LOTUS 1-2-3,* DIF, and ASCII files.

Stat I—A Statistical Toolbox
Sugar Mill Software Corporation
1180 Kika Place
Kailua, HI 96734

Cost: List $130

System: IBM PC and compatibles (MS-DOS)

Memory: 256K

Format: Menu

Functions: Descriptive, correlations, ANOVA, ANCOVA, cross tables, multiple analysis of variance, plots and charts, nonparametric, regression

Comments: This program includes a full-screen editor and will read ASCII and *LOTUS 1-2-3* files.

STATA
Computing Resource Center
10801 National Blvd., 3rd Floor
Los Angeles, CA 90064

Cost: List $395 (graphic module $195)

System: IBM PC and compatibles (MS-DOS)

Memory: 256K

Format: Menu or Command

Functions: Descriptive, correlations, cross tables, plots and charts, regression

Comments: *STATA* and its companion *STATA/Graphics* produce univariate statistics, as well as a limited number of inferential statistics. A *STATA/Graphics* module is available.

StatPac Gold
Walonick Associates
6500 Nicollet Ave., S.
Minneapolis, MN 55423

Cost: List $595

System: IBM PC and compatibles (MS-DOS)

Memory: 192K

Format: Menu or Command

Functions: Descriptive, correlations, ANOVA, ANCOVA, factor analysis, cross tables, multiple analysis of variance, plots and charts, nonparametric, regression

Comments: *StatPac* is one of the oldest statistical packages on the market. It is a command-oriented program. The package has recently added a multiple-regression modeling program called *Goodness-of-Fit,* which sells for $595.

StatPlan III
The Futures Group
76 Eastern Blvd.
Glastonbury, CT 06033

Cost: List $179 (8087 version $189)

System: IBM PC and compatibles (MS-DOS)

Memory: 256K

Format: Menu

Functions: Descriptive, correlations, ANOVA, ANCOVA, cross tables, plots and charts, regression

Comments: StatPlan III will read files from ASCII sources as well as *LOTUS 1-2-3* and *dBase III Plus*.

Statpro
Penton Software, Inc.
420 Lexington Ave., Suite 2846
New York, NY 10017

Cost: List $795

System: IBM PC and compatibles (MS-DOS)

Memory: 256K

Format: Menu

Functions: Descriptive, correlations, ANOVA, ANCOVA, factor analysis, cross tables, multiple analysis of variance, plots and charts, nonparametric, regression, time series

Comments: Statpro integrates graphic and database management facilities in a comprehensive package. The graphic output are in multiple colors, and the program reads *LOTUS 1-2-3* and ASCII files.

Table of Critical Features for Statistical Programs

FEATURE	PROGRAM NAME							
	SYSTAT	PC-SAS	BMDP-PC	SPSS-PC	Minitab	Stat-graphics	Dyna-Stat	1,2,3 Forecast
Descriptive	X	X	X	X	X	X	X	X
Correlations	X	X	X	X	X	X	X	X
ANOVA	X	X	X	X	X	X	X	X
ANCOVA	X	X	X	X	X		X	
MANOVA	X	X	X	X	X	X	X	
Cross Tabulations	X	X	X	X	X	X	X	
Plots	X	X	X	X	X	X	X	X
Charts	X	X	X	X	X	X	X	X
Nonparametric	X	X	X	X	X	X	X	
Regression	X	X	X	X	X	X	X	X
Multidimensional Scales	X						X	
Factor Analysis	X	X	X	X		X		
Time Series					X	X		
Exploratory Analysis	X				X			
Reads ASCII	X	X	X	X	X	X		X
Reads LOTUS Files		X			X	X		X
Memory Required	512K	512K	640K	384K	512K	512K	128K	512K
System(s)	IBM/MACINTOSH	IBM	IBM	IBM	IBM	IBM	Apple	IBM

PROGRAM NAME											
ADDa-STAT	StatPac Gold	AB-stat	STATA	Stat-Plan III	PC Stat-istician	CSS	Micro-stat	NWA Statpac	Stapro	CRUNCH	STAT 1A
X	X	X	X	X	X	X	X	X	X	X	X
X	X	X	X	X	X	X	X	X	X	X	X
X	X	X		X	X	X	X	X	X	X	X
X	X	X				X			X		X
	X					X			X		X
X	X	X	X	X	X	X	X	X	X	X	X
X	X	X	X	X	X	X	X	X	X	X	X
X	X	X	X	X	X	X	X	X	X	X	X
Chi Sq Poisson	X	X			X	X	X	X	X		X
X	Limited	X	X	X	X	X	X	X	X	X	X
							X		X		
						X	X		X		
X	X	X	X	X	X	X	X	X	X		X
X		X		X	X	X	X	X	X		X
384K	192K	256K	256K	256K	256K	384K	64K	512K	256K	256K	256K
IBM	IBM	IBM	IBM	IBM	IBM	IBM	IBM	IBM	IBM	IBM	IBM

SECTION C

Software Applications in Statistics

CHAPTER

1

Introduction

Purpose

The purpose of chapters 2 through 5 in Section C is to provide a simplified description of the program commands needed to perform the statistical procedures in Section A, using four commonly available statistical software packages; namely, *SAS* (Chapter 2), *SYSTAT* (Chapter 3), *SPSS-X* (Chapter 4), and *Minitab* (Chapter 5). With the information contained in this section, the reader will be able to carry out the majority of the statistical analyses needed to examine data from descriptive, correlational, and experimental studies. Also presented are samples of the output produced by the various program commands.

Intended Audience

This section, like the entire volume, has been designed to serve two primary audiences. First, there are students who early in their training need a user-friendly text to help them operate the various computer programs needed to implement the statistical procedures they are studying in research-oriented courses. Getting started with computer applications is often a difficult and frustrating task. Section C has been designed to alleviate this frustration by addressing in straightforward terms those issues that are of most importance to the student. An effort has been made to make this both interesting and enjoyable.

Researchers previously skilled with one or more software packages constitute the second intended audience. Computer software, like hardware, changes so frequently that it is sometimes difficult for professionals to keep current of the latest versions of even their favorite programs. Knowledge of software syntax fades fast when it is not regularly used. Section C has been designed to function as four mini-manuals that will enable the experienced, but rusty, researcher to review and locate easily the proper program code for performing the desired statistical procedure.

Section Overview

The material presented here is not designed to substitute for the more comprehensive documentation that normally accompanies the four software programs that constitute Section C. Instead, this section provides abridged versions of the procedures necessary to implement these programs, which keep in mind the needs of the two targeted audiences.

In the chapters on *SAS, SYSTAT, SPSS-X,* and *Minitab*, the following general divisions have been used to organize the material. Occasionally other division titles will also be found due to special circumstances in the individual chapters. The statistical procedures under a heading in Section C are the same as those under the heading in Section A.

PROGRAM BASICS

This initial section in each software chapter provides a brief overview of the program, describing its basic features and principal coding conventions.

DATA ENTRY

Describes one or more ways to format, configure, and input data into the program. When appropriate, tricks or shortcuts for inputting data are offered.

DATA MANIPULATION

Described under this heading are procedures for reformatting, transforming, recoding, or otherwise changing the data after data entry and before statistical analyses are performed. The use of logical operators (e.g., *if . . . then*), arithmetic operations, and data partitioning are among the manipulations that are explained.

DESCRIPTIVE STATISTICS

Procedures for obtaining frequency distributions, indices of normality (skewness and kurtosis), measures of central tendency, and variability are described and demonstrated. Also described are graphic techniques, such as creating bar graphs and histograms. This material encompasses statistical procedures **A4-1** to **A4-13**.

MEASURES OF RELATIONSHIPS

In this category are examined procedures for obtaining measures of relationships for different types of variables in the two-variable case. Included are statistical procedures **A4-14** to **A4-30**, as applicable. Part and partial correlation (**A4-26** and **A4-27**) are referenced here; however, in some chapters they are also treated under the regression category.

REGRESSION ANALYSIS

This section discusses basic procedures for simple and multiple regression (**A4-31** to **A4-39**). Part and partial correlation, residual analyses, and various options for drawing bivariate plots demonstrating different aspects of the regression analysis are shown.

COMPARISONS: ONE SAMPLE

Most one-sample statistical tests (**A4-40** to **A4-47**) are performed more easily by using hand-held calculators and the output from some basic descriptive analyses. Nevertheless, program statements for performing one-sample tests are described.

COMPARISONS: TWO SAMPLES (PARAMETRIC)

This section covers statistical procedures **A4-48** to **A4-56**, as applicable. Some of these tests are not easily performed by machine. For example, the F-test for independent-sample variances might be more easily executed by having the software calculate the variances within groups and then performing the F-ratio by hand.

COMPARISONS: TWO SAMPLES (NONPARAMETRIC)

The nonparametric two-samples comparisons (**A4-57** to **A4-65**) are described and illustrated under this heading.

COMPARISONS: TWO OR MORE SAMPLES (NONPARAMETRIC)

This section describes and demonstrates the nonparametric tests presented in **A4-66** to **A4-69**.

COMPARISONS: TWO OR MORE SAMPLES—ANALYSIS OF VARIANCE (ANOVA)

This section contains several subsections; namely, factorial ANOVA (**A4-71** to **A4-78**), repeated-measures/split-plot ANOVA (**A4-79** to **A4-83**), nested/hierarchical ANOVA (**A4-84** to **A4-86**), general ANOVA designs (**A4-87** and **A4-88**), and related designs (**A4-89** and **A4-90**).

MULTIPLE-COMPARISON PROCEDURES

Multiple-comparison methods (**A4-91** to **A4-97**) are treated as a separate section instead of being placed under each applicable ANOVA subsection. The multiple-comparison procedures are quite general, and they are illustrated with applications to a variety of ANOVA situations.

Delimitations

Decisions regarding what material to include and what to exclude from Section C were often based on the authors' experiences and personal preferences. The selection of *SAS, SYSTAT, SPSS-X,* and *Minitab* does not constitute an endorsement of these programs over others; instead, it indicates an opinion that these programs have found wide acceptance by researchers in education and that the two targeted audiences would be familiar with these packages.

The coverage of statistical topics within the four programs also reflects judgmental decisions. Not all statistical procedures have been included, nor are those that are addressed treated with the same detail or depth within each of the programs. These decisions were dependent, in part, on the degree of completeness and clarity with which the different software packages address the individual statistical procedures.

Using Section C and the Index

Within the four chapters of Section C described above are programming statements (referred to as "code" or "coding") that have been printed in capital letters. Additionally, examples of programming codes have often been set off from the text by placing them between horizontal parallel lines with a shaded background. Output that results from a programming statement has been placed within a box. Item entry codes for the statistical procedures are those used in Section A. For example, **A4-2** refers to Section A, Chapter 4, item 2 (the mean).

Many of the computer procedures are illustrated using data from what are referred to as "Data Set 1" up through "Data Set 7". These data sets are described in detail in Chapter 2 (*SAS*), and referenced in Chapters 2-5.

The index at the back of this book will help the reader use all three sections, especially the interworking of Sections A and C. A double-entry system has been used. Researchers who want to utilize a particular statistical procedure can find the index entry for that procedure. Following the entry will be a page reference to the basic description in Section A and to the appropriate programming pages in *SAS, SYSTAT, SPSS-X,* and *Minitab*. Conversely, the researcher who wants to determine how a particular software program executes a specific statistical procedure can also look under the index entry for that program and find reference to the page that addresses the procedure. Also listed under the software names are page references to the principal codes.

CHAPTER

2

Statistical Analysis System (SAS)

Introduction

SAS* (Statistical Analysis System) is a powerful and popular multipurpose program with a wide range of data-entry, manipulation, format, and statistical-analysis options. Applications of *SAS*, version 5, for a mainframe computer will be illustrated in this chapter. The microcomputer version, *SAS/PC*, provides many of the same procedures with very similar coding. The complete documentation for *SAS* is contained in three manuals published by SAS Institute, Inc. These are *Introductory Guide, Version 5 Edition; SAS User's Guide: Basics, Version 5 Edition;* and *SAS User's Guide: Statistics, Version 5 Edition.* The manuals are comprehensive, clear, and provide very useful information about the statistical procedures performed by *SAS*. The information in this chapter is also consistent with *SAS* version 6 which supersedes version 5.

Text Format

Examples of *SAS* commands and output will be shown throughout this chapter. The *SAS* codes will be printed in CAPITAL letters when referenced in a paragraph and will be printed in a shaded area between parallel lines when SAS statements appear on separate lines that illustrate the code. In practice, the *SAS* code can appear in any column position on a screen line; however, examples in this chapter will be written as if the left-most *SAS* code is in the first column of the input screen. Examples of *SAS* output will be displayed in boxes to set them apart from the text.

---

*SAS and *SAS/PC* are registered trademarks of SAS Institute, Inc., Cary, NC.

Some SAS Basics

A standard form for a *SAS* job on a mainframe would have the features shown in the sample job below. System JCL (Job Control Language) comes first. The *SAS* program is initiated, desired options (e.g., LS=72) are listed, and the DATA-set name is given and signals the beginning of the data step. The INPUT statement names the variables and indicates how they are to be read. All *SAS* names, used for data sets or variables, must begin with a letter, cannot have embedded blanks, and cannot be longer than eight characters.

Data manipulation follows as part of the data step and is often used to create new variables. Data are supplied and formatted as specified in the INPUT statement.

JCL   System-required Job Control Language, or JCL, usually provides an account number, password, and other system-control information

// EXEC SAS   {Initiates the execution of SAS}
OPTIONS LS=72;   {Sets the line size to 72 characters, makes reading output on a CRT much easier}
DATA EXAMPLE;   {Provides a data-set name}
INPUT TEST1 1-2 TEST2 4-5;   {Names two input variables and identifies column locations for data entry}
DIFF=TEST2-TEST1;   {Defines a new variable, DIFF, as the difference between TEST2 and TEST1}
CARDS;   {Signals the beginning of data}
17 20   {TEST1 and TEST2 values for the first observation}
15 26   {TEST1 and TEST2 values for the second observation}
13 29   {TEST1 and TEST2 values for the third observation}
PROC MEANS;
VAR TEST1 TEST2 DIFF;
TITLE 'EXAMPLE OF SAS JOB';

JCL   System JCL to signal end of job.

Central to *SAS* statistical analyses are the procedure, or PROC, statements. For example, the *SAS* codes PROC MEANS and PROC CORR produce means and correlations, respectively, for all variables to which the procedures are applied. In general, *SAS* PROCs are of the form:

PROC (procedure name) (options);

where PROC indicates the beginning of a *SAS* procedure, the "procedure name" indicates the specific procedure being employed, and "options" indicates particular options related to the *SAS* procedure. All *SAS* statements, including *SAS* PROCs, end with a semicolon.

SAS PROCs are applied to all appropriate variables unless a variable list is included. The variable list usually begins with the term VAR and lists the names of the variables to which the procedure is to be applied. For PROC FREQ, used to obtain frequency distributions, the variables list begins with the word TABLES. Details about data entry, data manipulation, and *SAS* PROCs are provided in the sections that follow.

Some Helpful SAS Features

Comment Statements. Statements beginning with an asterisk (*) in *SAS* are comment statements used to document a program. These statements are notes used to explain the program flow and operation. Comment statements are not executed, do not affect the program, and must end with a semicolon.

PROC PRINT. The PRINT procedure prints all information that has been read or created by *SAS*. PROC PRINT is useful for verifying the accuracy of the data entry for relatively small data sets. With large data sets, a variable list on a VAR statement should be used.

Variable Statement. The application of many *SAS* procedures can be restricted to a specific set of variables named on a variable, or VAR, statement.

TITLE Cards. TITLE cards at the end of *SAS* procedures are used to label output for future reference. TITLE cards are of the form: TITLE 'title goes here, inside single quotes';.

Groups of Variables. Groups or strings of variables can be referred to by specifying the first and last variable, separated by two hyphens. For example, variables A, B, X, and Y, can be listed on a VAR statement as VAR A--Y;. The variables must have been read or created by *SAS* in the order A, B, X, and Y.

DO Statements. DO statements initiate a repetitive sequence of executable steps. The completion of a DO sequence is signaled by an END statement.

BY Statements. BY statements specify the execution of *SAS* commands for the different levels of the variable in the BY statement. The *SAS* code

```
PROC SORT; BY GENDER; PROC PRINT; BY GENDER;
```

would print data separately for subjects in the levels of GENDER. Data must first be sorted according to the BY variable.

OUTPUT Option. An OUTPUT option can be used with many *SAS* procedures. The OUTPUT option creates a new data set which includes variables created by the procedure.

Data Entry (Input)

SAS provides a variety of input formats to meet the needs of users with different types of data sets. *SAS*'s input formats all begin with the word INPUT. Some *SAS* input formats specify the fields or columns in which data are located. Data can be anywhere within the specified fields. Four commonly used input formats are illustrated and explained below. A fifth input format is described in the section of this chapter titled "Repeated Measures/Split-Plot Analysis of Variance," on page 211.

FORMAT 1:

The following is an example of Format 1.

```
INPUT STUDENT 1-2 GROUP 4 IQ 6-8 PRETEST 10-11 POSTTEST
13-14 TRSP 16;
CARDS;
 1 1 112 21 35 2
 2 1 109 17 33 2
 3 1 105 19 31 1
 4 1 118 25 36 2
 5 1  95 18 25 1
 6 2 111 19 24 1
 7 2 107 20 29 2
 8 2 102 11 24 1
 9 2  98 17 27 1
10 2 105 13 21 1
```

As mentioned previously, the "I" in INPUT should be thought of as being in column 1 of the input screen. Format 1, like all input procedures, begins with the word INPUT. The variables are then given *SAS* names. In this hypothetical example, henceforth referred to as Data Set 1, the variables are defined as follows:

$$
\begin{aligned}
\text{STUDENT} &= \text{Student ID number (two-digit maximum in this example)} \\
\text{GROUP} &= \text{Two groups, coded "1" or "2"} \\
\text{IQ} &= \text{Score on a standardized IQ test} \\
\text{PRETEST} &= \text{Score on a class pretest} \\
\text{POSTTEST} &= \text{Score on a class posttest} \\
\text{TRSP} &= \text{Teacher rating of student potential (1 or 2)}
\end{aligned}
$$

The fields or columns containing the data for each variable are specified after the variable name. For example, data for the variable STUDENT are contained in columns 1 and 2. If data for a variable are written with letters, rather than numbers, there must be a dollar sign ($) after the variable name. For example,

```
INPUT GENDER $ 3;
```

would be needed if data for the variable *GENDER* were coded F and M.

FORMAT 2:

Format 2 is very similar to the first format except that the fields or columns are not specified. For instance,

> INPUT STUDENT GROUP IQ PRETEST POSTTEST TRSP;

This format can be used as long as there is at least one space between the data for each variable. This format is quite flexible and convenient but must be used with care. *SAS* expects to read data for six variables for each case or record because six variables are listed in the input statement. For each record, data are read and assigned to each variable in the order specified in the input statement. If the value of GROUP for STUDENT=1 were omitted, GROUP would be assigned a value of 112 (the IQ score), IQ would have a value of 21 (the PRETEST score), and so on. Thus the order of actual input is very important! To avoid difficulties, a period (.) should be entered for missing data.

FORMAT 3:

A third *SAS* input format is especially useful and efficient when variable names are repetitious. Such situations occur when several measures are taken over different occasions or when other types of multiple measures are used. Consider a situation in which there are four scores for each subject. Each subject is designated by a three-digit ID number in columns 1-2-3. Input Formats 1, 2, and 3 for this example are as follows:

Format 1:

> INPUT ID 1-3 SCORE1 5-6 SCORE2 8-9 SCORE3 11-12 SCORE4 14-15;

Format 2:

> INPUT ID SCORE1 SCORE2 SCORE3 SCORE4;

Format 3:

> INPUT ID (SCORE1-SCORE4) (@4 4*3.);

Input Format 3 contains three new features that were not seen in the first two input formats. First, some variable names are not explicitly listed; for example, SCORE2 and SCORE3. This can be done whenever the variable names have a repetitious literal portion and a unique and sequential numeric ending. The expression SCORE1-SCORE4 is read by *SAS* as SCORE1 SCORE2 SCORE3 SCORE4. Second, the symbol @, which is called a "pointer" in *SAS*, is used. The expression @4 instructs *SAS* to go to column 4 to begin reading SCORE1. The @ symbol can be used to direct the input statement to read any designated column. The third element introduced in Format 3 is the expression 4*3. This expression indicates that there are four fields and that each field contains three columns.

Format 3 would be used as follows:

```
INPUT ID 1-3 (SCORE1-SCORE4) (@4 4*3.);
CARDS;
001 21 25 30 35
002 17 22 28 33
003 19 24 27 31
004 25 28 33 36
005 18 21 23 25
```

FORMAT 4:

With Format 4, data are entered continually across the line of input. The input format for this approach is shown below, using the same data illustrated in Format 3.

```
INPUT ID (SCORE1-SCORE4) (4*3.) @@;
001 21 25 30 35   002 17 22 28 33   003 19 24 27 31
004 25 28 33 36   005 18 21 23 25
```

The double @@ symbol at the end of the input line instructs the program to use the same input specifications across lines of input data. In this example, the variable ID is read first. Next, the variables SCORE1, SCORE2, SCORE3, and SCORE4 are read using the code (4*3.). The code (4*3.) indicates four fields, each of which is three columns in length. These three columns include a blank for convenience in inputting and checking data entry and a two-column field for the scores. After the fourth field in the (4*3.) sequence is read, the input sequence is repeated beginning with the variable ID. Missing data must be physically represented in the input, normally using a period (.), or the data will be misread.

Data Manipulation (Transformation)

Data that have been inputted can be manipulated or transformed in many different ways with a variety of easy-to-use expressions or commands. Arithmetic, algebraic, trigonometric, statistical, and logical manipulations are easily managed. Two new variables can be created in Data Set 1 with the following simple arithmetic manipulations:

```
GAIN = POSTTEST - PRETEST;
TOTAL = PRETEST + POSTTEST;
```

The variable GAIN is the difference between each student's performance on POSTTEST and PRETEST. The variable TOTAL is the combined number of points each student earns on the PRETEST and POSTTEST. GAIN and TOTAL will be assigned missing values if either PRETEST or POSTTEST is missing.

SAS includes an expression or operator for calculating the sum of a set of variables. It takes the form

```
VARIABLE NAME = SUM (OF list of variables to be added);
```

Using this notation, the variable TOTAL, the sum of the PRETEST and POSTTEST, can be written as

```
TOTAL = SUM (OF PRETEST POSTTEST);
```

This code for summing a set of variables is especially convenient when variables have been named using the serial notation of the form SCORE1-SCORE4. The sum of SCORE1, SCORE2, SCORE3, and SCORE4 can be written:

```
TOTAL = SUM (OF SCORE1-SCORE4);
```

Test scores defined as the number of items answered correctly can be changed to percentages by simply defining a new variable as the test score divided by the total number of items on the test and multiplying the result by 100. For example, the percentage of items each student answers correctly on PRETEST and POSTTEST can be named as PREPCT and POSTPCT, respectively. These variables are defined as a student's test score divided by 40 (the number of items on the test), using the following *SAS* code:

```
PREPCT = (PRETEST/40) * 100;
POSTPCT= (POSTTEST/40) * 100;
```

The *SAS* symbols for the four arithmetic operations are the plus sign (+) for addition, the minus sign (−) for subtraction, an asterisk (*) for multiplication, and a forward slash (/) for division. Exponents are indicated with two asterisks (**). Thus the PRETEST scores could be squared and called PRESQRED with either of the following notations:

```
PRESQRED = PRETEST*PRETEST;
PRESQRED = PRETEST**2;
```

Four basic logical operators used in conjunction with IF statements can be used to manipulate data in *SAS*. The logical operators are:

> LT = less than
> LE = less than or equal to
> GT = greater than
> GE = greater than or equal to

These logical operations are used in expressions that take the form:

> IF IQ GE 107 THEN IQGROUP=2; ELSE IQGROUP=1;

This line of *SAS* code would define a new variable named IQGROUP. A student would be assigned to IQGROUP=2 if his/her IQ score were greater than or equal to (GE) 107; otherwise (ELSE), if the IQ score were less than 107, the assignment would be to IQGROUP=1.

The power and flexibility of data manipulation in SAS has only been briefly described in this section. Readers who anticipate working with *SAS* can become more familiar with these data-manipulation capabilities by a careful examination of the *SAS* manuals.

Descriptive Statistics (A4-1 to A4-13)

A variety of *SAS* procedures can be used to obtain basic descriptive statistics. The most common procedures used for descriptive statistics are PROC MEANS, PROC UNIVARIATE, and PROC FREQ. PROC CHART can be used to obtain graphic representations of the frequency distribution for the values of each variable. PROC PLOT can be used to show the relationship between pairs of variables.

PROC MEANS routinely produces the number of observations, N, used in calculating the descriptive statistics for each variable, the mean (**A4-2**), standard deviation (**A4-8**), minimum and maximum values for the variable, standard error of the mean (standard deviation divided by the square root of the number of observations), the sum of the values of the variable, the variance (**A4-7**), and the coefficient of variation (standard deviation divided by the mean, multiplied by 100). PROC MEANS can also produce other information, including the skewness (**A4-12**) and kurtosis (**A4-13**).

The following *SAS* coding will produce basic descriptive statistics, including the skewness and kurtosis:

> PROC MEANS MAXDEC=3 SKEWNESS KURTOSIS;
>   VAR PRETEST POSTTEST;
>   TITLE 'MEANS PROCEDURE, DATA SET 1';

The option MAXDEC=3 instructs the program to print all statistics with three decimal places. PROC MEANS is applied only to the PRETEST and POSTTEST, since they are listed on the VAR statement. For Data Set 1, the PRETEST mean and standard deviation are 18.000 and 6.812, respectively, while the mean and standard deviation for POSTTEST are 28.500 and 5.126.

PROC UNIVARIATE routinely produces the mean (**A4-2**), median (**A4-3**), mode (**A4-4**), range (**A4-5**), variance (**A4-7**), standard deviation (**A4-8**), skewness (**A4-12**) and kurtosis (**A4-13**). Additional information is provided by PROC UNIVARIATE, such as various percentiles and descriptive data plots. An example of PROC

UNIVARIATE, applied to POSTTEST from Data Set 1, uses the following coding:

```
PROC UNIVARIATE;
  VAR POSTTEST;
  TITLE 'UNIVARIATE PROCEDURE, DATA SET 1';
```

A frequency distribution (**A4-11**) can be obtained by using the *SAS* procedure FREQ. For example, the frequency distribution for the variable TRSP in Data Set 1 is obtained with the following coding:

```
PROC FREQ;
  TABLES TRSP;
  TITLE 'PROC FREQ, VARIABLE TRSP, DATA SET 1';
```

Output from PROC FREQ includes the raw frequency and percentage of observations in each category of the variable, as well as the cumulative raw frequency and cumulative percentage in each category. The use of the FREQ procedure with variables that take on many different values, for instance, test scores and physical measurements, provides useful information. Inspection of the frequency distribution indicates the approximate value of the median (**A4-3**), mode (**A4-4**), various percentile positions, and the general shape of the distribution (**A4-12** and **A4-13**).

The FREQ procedure can be used to show the joint distribution of two variables in the form of a contingency table. This is done as follows:

```
PROC FREQ;
  TABLES GROUP*TRSP;
  TITLE 'PROC FREQ, TWO VARIABLES, DATA SET 1';
```

This coding produces raw frequencies and percentages for cells, rows, and columns. *SAS* produces a variety of statistics for contingency tables and these will be described later.

The *SAS* procedures MEANS, UNIVARIATE, and FREQ provide very useful descriptive information. Much of this information can be shown graphically using a number of *SAS* procedures. The CHART procedure is easy to use and generates bar charts in both the vertical and horizontal formats. For example, the *SAS* lines

```
PROC CHART;
  HBAR TRSP;
  TITLE 'BAR CHART, VARIABLE TRSP, DATA SET 1';
```

produce a horizontally displayed bar chart. The horizontal format includes data about the raw and proportional frequency distribution (**A4-11**) for the variable. A vertically displayed bar chart is produced with the expression VBAR TRSP instead of HBAR TRSP.

It is sometimes useful to control the number of categories plotted in the chart for continuous variables that can assume a large number of values. For example, in a large data set, there may be 60 or more values for a variable like IQ. Such data can be grouped into a smaller number of categories, or levels, using the LEVELS option. For example, the coding

```
PROC CHART;
  VBAR IQ / LEVELS=5;
  TITLE ' BAR CHART, FIVE LEVELS OF IQ';
```

would produce a vertical bar chart grouped into five categories, or levels, regardless of how many values were present for the variable IQ.

PROC PLOT is used to show the joint distribution of two continuous variables. The coding

```
PROC PLOT;
  PLOT POSTTEST*PRETEST;
  TITLE 'SCATTERPLOT OF POSTTEST ON PRETEST';
```

would produce a scatterplot of POSTTEST (Y-axis) onto PRETEST (X-axis).

Measures of Relationships (A4-14 to A4-30)

Different measures of association, or relationship, can be obtained using a number of *SAS* procedures. The measure of association chosen depends, of course, on the nature of the research situation and the type of data involved. Correlations can be obtained by using the CORRELATION procedure with the abbreviation PROC CORR. This procedure is applied to all numeric variables in the data set unless a VAR list or other form of data editing is employed. A particularly useful option with PROC CORR is the use of a VAR list and a WITH statement. This produces the correlation of the variable(s) in the VAR list with the variable(s) listed in the WITH statement. The use of this *SAS* PROC will be illustrated using data from Ryan (1977)* with 23 children (observation) on five variables. This data set will be referred to as Data Set 2. The variables are:

AGEMONTH = children's age in months
CSYN = children's score on a measure of syntactic (grammar) ability
MSYN = mothers' prediction of children's score on the syntax test
CVOCAB = children's score on a vocabulary test
MVOCAB = mothers' prediction of children's vocabulary score

---

*J. P. Ryan. *Expansions in the Speech of Mothers to Their Young Children.* Doctoral dissertation, University of Chicago, 1977.

The data are entered as follows:

```
DATA MOMSKIDS;
INPUT AGEMONTH CSYN MSYN CVOCAB MVOCAB @@;
CARDS;
61  9   8 68 56  48  7 10 41 47  72 10 10 66 68  49  5  4 46 48
39  8   6 19 33  90  9 10 64 79  43  9 10 53 54  84 10 10 64 75
72 10  10 79 65  29  7  3 27 39  48  7  8 49 63  64  7  6 59 61
41  5   7 42 47  76  9 10 65 69  42  7  7 38 49  46  5 10 44 51
63 10   8 60 63  72  8 10 59 89  70  9 10 60 76  37  5  9 37 57
70  8  10 57 53  40  9 10 52 58  45  6  6 15 20
PROC CORR;
TITLE 'CORRELATIONS AMONG VARIABLES, DATA SET 2';
```

PROC CORR will produce basic descriptive statistics, including the mean (**A4-2**) and standard deviation (**A4-8**), as well as the Pearson product-moment correlation coefficient (**A4-16**). Using Data Set 2, the correlations produced by PROC CORR are as follows:

	AGEMONTH	CSYN	MSYN	CVOCAB	MVOCAB
AGEMONTH	1.0000	0.63054	0.54107	0.78177	0.75206
	0.000	0.0013	0.0077	0.0001	0.0001
CSYN	0.63054	1.00000	0.48652	0.65710	0.53032
	0.0013	0.0000	0.0186	0.0007	0.0092
MSYN	0.54107	0.48652	1.00000	0.60028	0.62201
	0.0077	0.0186	0.0000	0.0025	0.0016
CVOCAB	0.78177	0.65710	0.60028	1.00000	0.81209
	0.0001	0.0007	0.0025	0.0000	0.0001
MVOCAB	0.75206	0.53032	0.62201	0.81209	1.00000
	0.0001	0.0092	0.0016	0.0001	0.0000

The correlation between AGEMONTH and CSYN is .63054. The significance level for this correlation, testing the hypothesis that the correlation differs from zero (**A4-44**), is listed directly below the correlation. This is a two-tailed test and the significance level must be divided by two to obtain the significance for a one-tailed test. The correlation between AGEMONTH and CSYN is significant at .0013.

The correlation of AGEMONTH with the other variables, but not the intercorrelations among the other variables, could be produced with the code

```
PROC CORR;
   VAR AGEMONTH;
   WITH CSYN--MVOCAB;
```

The Spearman rank order correlation (Rho) (**A4-17**) and Kendall's Tau (**A4-18**) can be calculated with PROC CORR. These correlation coefficients are obtained with the following *SAS* code:

```
PROC CORR SPEARMAN KENDALL;
```

The point biserial correlation (**A4-19**) and biserial correlation (**A4-20**) are special cases of the Pearson product-moment correlation, and thus are produced by PROC CORR when data for the variables have the appropriate form. The point biserial correlation is obtained when one variable is a discrete dichotomous and the other is continuous. The biserial correlation is obtained when one variable is dichotomous but has an underlying continuous normal distribution and the other variable is continuous.

The Phi-coefficient (**A4-21**) and contingency coefficient (**A4-23**) are produced by the FREQ procedure with the specification of the CHISQ option. The Phi-coefficient and contingency coefficient for the variables GROUP and TRSP in Data Set 1 are produced by the following *SAS* code:

```
PROC FREQ;
    TABLES GROUP*TRSP / CHISQ;
    TITLE 'PHI AND CONTINGENCY COEFFICIENTS, DATA SET 1';
```

Part (**A4-26**), partial (**A4-27**), and multiple correlation (**A4-28**) will be described within the context of regression analysis (**A4-31** to **A4-39**) in the next section.

SAS provides a comprehensive approach to factor analysis (**A4-30**) with a wide range of options for PROC FACTOR. A variety of factor analyses, component analyses, and rotations can be carried out with this procedure. A useful description and illustration of PROC FACTOR would be too lengthy for this chapter; consequently, the interested reader is referred to *SAS User's Guide: Statistics, Version 5*, Chapter 17.

Regression Analysis (A4-31 to A4-39)

Two *SAS* procedures can be used to perform regression analysis. These are the regression and general linear model procedures, typically referred to as PROC REG and PROC GLM. These procedures are very similar and can be used interchangeably for most applications. The following *SAS* code generates the simple linear regression (**A4-32**) of CSYN onto AGEMONTH. This examines the prediction of children's scores on the syntax measure from their age in months.

```
PROC GLM;
    MODEL CSYN = AGEMONTH /P;
```

PROC REG could have been used instead of PROC GLM. The model statement indicates that the dependent measure CSYN is predicted by the independent variable AGEMONTH. The option /P requests the listing of the observed and predicted values for the dependent variable, and the residuals from the regression (e.g., observed minus predicted values), in raw score form. A portion of the output produced by PROC GLM is as follows:

DEPENDENT VARIABLE: CYSN

SOURCE	DF	SUM OF SQUARES	MEAN SQUARE	F-VALUE	R-SQUARE
MODEL	1	26.206	26.206	13.86	0.397
ERROR	21	39.706	1.890	ROOT MSE 1.375	
CORRECTED TOTAL	22	65.913			

SOURCE	DF	TYPE I SS
AGEMONTH	1	26.206

PARAMETER	ESTIMATE
INTERCEPT	4.120896
AGEMONTH	0.064734

Estimates of the regression coefficient (**A4-33**) and Y-intercept (**A4-34**) are listed under the headings PARAMETER and ESTIMATE. The regression coefficient is 0.064734 and the Y-intercept is 4.120896. Thus the regression equation is

$$CSYN = .064734(AGEMONTH) + 4.12.$$

The standard error of estimate (**A4-35**) is the standard deviation of the errors in the prediction. These errors are the difference between the observed and predicted values of the dependent variable. The variance of these errors is the mean square due to error, referred to as MSE. In this example, MSE=1.89. The square root of the variance is the standard deviation of the errors or the standard error. The standard error of estimate (**A4-35**) is listed under the heading ROOT MSE. The ROOT MSE is the square root (ROOT) of the mean square due to error (MSE). The R-squared (**A4-37**) is 0.397 and is listed on the right side of the first line of numerical output under the heading R-SQUARE.

Part correlation (**A4-26**) and partial correlation (**A4-27**) can be calculated with PROC GLM or PROC REG by using the output option. Consider the following lines of code:

```
PROC GLM;
   MODEL CSYN = AGEMONTH;
   OUTPUT OUT=B R=RESCSYN;
PROC GLM;
   MODEL CVOCAB = AGEMONTH;
   OUTPUT OUT=C R=RESCVCAB;
PROC CORR;
   VAR AGEMONTH CSYN CVOCAB RESCSYN RESCVCAB;
```

The output produced by PROC CORR, showing part and partial correlations, is shown below:

	AGEMONTH	CSYN	CVOCAB	RESCSYN	RESCVCAB
AGEMONTH	1.0000	0.63054	0.78117	0.00000	0.00000
	0.000	0.0013	0.0001	0.0000	0.0000
CSYN	0.63540	1.00000	0.65710	0.77615	0.26326
	0.0013	0.0000	0.0007	0.0001	0.2249
CVOCAB	0.78117	0.65710	1.00000	0.21151	0.62357
	0.0001	0.0007	0.0000	0.3326	0.0015
RECSYN	0.00000	0.77615	0.21151	1.00000	0.33919
	0.0000	0.0001	0.3326	0.0000	0.1133
RESCVCAB	0.00000	0.26326	0.62357	0.33919	1.00000
	0.0000	0.2249	0.0015	0.1133	0.0000

The correlation between CSYN and CVOCAB is .657, P = .0007. The correlation between CVOCAB and RESCSYN is a part correlation (**A4-26**) and is equal to .212 (not significant, n.s.). This is the correlation between children's vocabulary scores and scores on the syntax test, with the influence of age having been removed from the syntax scores (RESCSYN). The correlation between CSYN and RESCVCAB is .263 (n.s.). This is the part correlation between children's scores on the syntax measure (CSYN) and scores on the vocabulary test, with the influence of age having been removed from the vocabulary scores (RESCVCAB). The partial correlation (**A4-27**) between children's vocabulary and syntax scores is reflected in the correlation between RESCVCAB and RESCSYN (.339, n.s.). This is the correlation between children's vocabulary and syntax scores, removing from both variables the influence that the children's age has on the correlation. (Note: The significance levels reported by *SAS* will not be completely correct because a degree of freedom was used in each regression analysis, which *SAS* does not account for in calculating the significance levels in PROC CORR.)

Multiple regression (**A4-36**), which includes curvilinear regression (**A4-38**) as a special case, can be performed with PROC GLM, PROC REG, PROC RSQUARE, and PROC STEPWISE. Multiple regression with PROC GLM would use the following code:

```
PROC GLM;
   MODEL CSYN = AGEMONTH CVOCAB MSYN MVOCAB;
   TITLE 'MULTIPLE REGRESSION, DATA SET 2';
```

The regression coefficients (**A4-33**) using this model for AGEMONTH, CVOCAB, MSYN, and MVOCAB are, respectively, .033, .049, .116, and −.020. The Y-intercept (**A4-34**) is 3.547. The standard error of estimate (**A4-35**) is 1.37 (labeled in *SAS* as the "Root Mean Square"). The R-squared (**A4-37**) is .484. The multiple correlation (**A4-28**) is the square root of the R-squared, or .696.

Curvilinear regression can be performed by using nonlinear transformations of the independent variable(s). For example, the following coding would be used if the relationship between CSYN and children's age were thought to be curvilinear:

```
PROC GLM;
   MODEL CSYN = AGEMONTH AGEMONTH*AGEMONTH;
   TITLE 'CURVILINEAR (QUADRATIC) MULTIPLE REGRESSION';
```

This model would predict CSYN from two independent variables; namely, AGEMONTH and the quadratic version of AGEMONTH, (AGEMONTH*AGEMONTH).

The *SAS* RSQUARE procedure is useful for exploring the relationships among a dependent variable and a set of independent variables in the context of multiple regression. The following code using PROC RSQUARE is similar to the PROC GLM example:

```
PROC RSQUARE;
   MODEL CSYN = AGEMONTH CVOCAB MSYN MVOCAB;
```

This procedure will list the R-squared, in ascending order, for predicting CSYN from each of the independent variables separately, for each set of two independent variables, for each set of three independent variables, and for all four independent variables.

Stepwise multiple regression (**A4-39**) is performed with the STEPWISE procedure. Using the same example as before, this procedure is written

```
PROC STEPWISE;
   MODEL CSYN = AGEMONTH CVOCAB MSYN MVOCAB;
   TITLE 'STEPWISE REGRESSION EXAMPLE';
```

The STEPWISE procedure enters the independent variables into the regression equation in a series of steps. The R-squared and other relevant statistics are reported for each step. Options used in the model statement control the way variables are entered into or are removed from the equation at each step. The options for variable selection are:

FORWARD — Variables are added to the model, one at a step, in order of significance and only if they meet a specified significance level. Once added, a variable remains in the model (*SAS* code F).

BACKWARD — All variables are included in the model and then variables are eliminated, one at a step, based on which variable makes the smallest contribution to the model (*SAS* code B).

STEPWISE — Operates like the FORWARD procedure. Variables are added to the model, one at a step, in order of significance and only if they meet a specified significance level. Variables entered in earlier steps are removed if they no longer contribute significantly (*SAS* code STEPWISE).

MAXIMUM R-SQUARED IMPROVEMENT — Attempts to find the best one-variable model, the best two-variable model, and so forth, finding the combination of variables at each step that produces the maximum improvement in R-squared (*SAS* code MAXR).

MINIMUM R-SQUARED IMPROVEMENT — Attempts to find the best one-variable model, the best two-variable model, and so forth, finding the combination of variables at each step that produces the minimum improvement in R-squared. Often identifies the same model as MAXR for a given number of variables in the model (*SAS* code MINR).

The previous PROC STEPWISE example, illustrating the use of all selection options, is written as follows:

```
PROC STEPWISE;
MODEL CSYN = AGEMONTH CVOCAB MSYN MVOCAB / F B STEPWISE
       MAXR MINR ;
TITLE 'STEPWISE REGRESSION EXAMPLE';
```

Comparisons: One Sample (A4-40 to A4-47)

Hypotheses related to one sample involve the comparison of a mean (**A4-41**), variance (**A4-42**), proportion (**A4-43**), or correlation (**A4-44** and **A4-45**) to some value, or the comparison of a set of frequencies to some expected frequency values (**A4-46** and **A4-47**). Two general approaches can be used to apply SAS to test such hypotheses. First, *SAS* can be used to calculate the appropriate sample statistic and then the actual test statistic (z, Chi-square, t, or F) for testing the hypothesis (**A4-41** to **A4-47**). These can also be calculated with a hand calculator. This is often a practical and sensible approach when a small number of statistical tests are involved. PROC MEANS or PROC UNIVARIATE can be used to obtain sample means and variances. Sample proportions and frequencies can be produced with PROC FREQ. PROC CORR will calculate sample correlation coefficients and test the hypothesis that the correlations differ from zero (0) (**A4-44**). *SAS* performs a two-tailed test.

A second approach involves using *SAS* to calculate the test statistics for hypotheses related to one sample. This is advisable if there are a large number of statistical tests. Hypotheses comparing the mean to some value (A) or the variance to some value (A) can be tested with the MEANS procedure combined with an OUTPUT statement and some variable manipulations in the new data set created by the OUTPUT statement. An example of this coding is as follows:

```
PROC MEANS;
  VAR AGEMONTH;
  OUTPUT OUT=B N=N MEAN=MEANAGE VAR=VARAGE STDERR=SEAGE;
DATA C; SET B;
* T-TEST COMPARING THE MEAN AGEMONTH TO SOME VALUE, A;
  T=(MEANAGE-A)/SEAGE;
* ;
* CHISQUARE TEST COMPARE THE VARIANCE OF AGEMONTH TO A;
  CHISQ=((N-1)*VARAGE)/A;
  PROC PRINT; VAR T CHISQ:
```

The lines that begin with an asterisk (*) are comment lines used to document the program. Comment lines are not executed by *SAS*. The OUTPUT statement with PROC MEANS produces a new data set, DATA = B. DATA set B has four variables and one observation or record. The four variables are N, which is the number of observations on the basis of which the mean was calculated; MEANAGE is the mean of the variable AGEMONTH; VARAGE is the variance of AGEMONTH; and SEAGE is the standard error of the mean for AGEMONTH. The *SAS* code T=(MEANAGE-A)/SEAGE;, calculates the t-statistic described in **A4-41** for comparing the mean (MEANAGE) to some value (A). The *SAS* code CHISQ=((N-1)*VARAGE)/A, calculates the Chi-square statistic described in **A4-42** for comparing the variance (VARAGE) to some value.

Hypotheses in which a sample proportion is compared to some population value (**A4-43**) are tested with a z-statistic. This z-statistic can be calculated in the *SAS* data step. If PS represents the sample proportion, PP represents the population proportion, and N is the sample size, the z-statistic is given by the following *SAS* coding:

```
Z = (PS - PP)/(SQRT((PS*(1-PS))/N))
```

Sample correlations are frequently examined to determine if they are significantly different from zero (**A4-44**). This test is automatically performed by the CORR procedure and the exact significance level is reported for a two-tailed test. The significance level for a one-tailed test is obtained by dividing the significance level produced by *SAS* by two. Hypotheses comparing sample correlations to some other hypothesized value (**A4-45**) may also be of interest. Such comparisons require the Fisher Z-transformation of both the sample correlation and the hypothesized correlation. This transformation has the following form:

```
ZR = .5*(log((1+r)/(1-r)));
```

If the sample correlation is noted as SCORR, the hypothesized value is noted as HCORR, and N is the sample size, the following *SAS* coding will produce the z-statistic needed to test the hypothesis that the two correlations are equal. This coding would appear in the data step:

```
ZRSCORR = .5*(LOG((1+SCORR)/(1-SCORR)));
ZRHCORR = .5*(LOG((1+HCORR)/(1-HCORR)));
Z = (ZRSCORR-ZRHCORR)/SQRT(1/(N-3));
```

It is useful to note that the *SAS* coding for the one-sample hypotheses about a proportion (**A4-43**) or a correlation compared to some value (**A4-45**) basically employs *SAS* as a calculator. As mentioned before, these calculations may be more efficiently performed on a hand calculator. The Chi-square goodness-of-fit test (**A4-46**) and the Kolmogorov-Smirnov one-sample test (**A4-47**) can also be calculated by *SAS*. However, in most cases, it is simpler to use PROC FREQ to obtain the necessary frequencies or proportions and then perform the test calculation with a hand calculator.

Comparisons: Two Samples (Parametric) (A4-48 to A4-56)

Comparisons involving two samples include both independent and dependent sample tests. Hypotheses for independent and dependent samples, respectively, examine questions about means (**A4-49** and **A4-50**), variances (**A4-51** and **A4-52**), proportions (**A4-53** and **A4-54**), and correlations (**A4-55** and **A4-56**). As in the one-sample case, two general approaches can be used in applying *SAS* to test these hypotheses. First, *SAS* can be used to calculate the appropriate statistics for each of the two samples. The actual test statistic (z, Chi-square, t, or F) for examining differences between the sample statistics (**A4-49** to **A4-56**) can then be determined easily with a hand calculator. This is often a practical and sensible approach when a small number of statistical tests are involved, especially if the tests relate to comparisons of variances, proportions, and correlations. A second approach involves using *SAS* to calculate the actual test statistics for hypotheses related to two samples. This approach is advisable if there is a large number of statistical tests, and it is especially appropriate for comparisons of means.

SAS is well suited for testing hypotheses about the means for two independent or dependent samples. The comparison of independent sample means (**A4-49**) is performed with the TTEST procedure. The use of this procedure can be illustrated with Data Set 1 using the following SAS code:

```
PROC TTEST;
   CLASS GROUP;
   VAR POSTTEST;
   TITLE 'INDEPENDENT SAMPLE T-TEST, DATA SET 1';
```

The CLASS statement defines the independent variable with two levels and the VAR statement defines the dependent variable(s). Multiple dependent variables can be listed on the VAR statement. The following is a portion of the output produced by the coding PROC TTEST:

VARIABLE POSTTEST

GROUP	N	MEAN	STD DEV	VARIANCES	t	DF	PROB
1	5	32.0	4.358	Unequal	2.93	7.2	.021
2	5	25.0	3.082	Equal	2.93	8.0	.018

HO: Variances are Equal, F = 2.00, with 4 and 4 DF, Prob=.51

For each group, this output includes the number of subjects, mean, standard deviation, and the t-statistic, which tests the hypothesis that the means are statistically equivalent. The validity of this statistical test assumes that the variances for the two groups are equivalent (homogeneity of variances). PROC TTEST produces two different significance levels for the t-statistic, based on the variances being UNEQUAL or EQUAL, by adjusting the degrees of freedom (DF).

PROC TTEST actually carries out the statistical test examining the assumption that the variances for the two groups are equal (**A4-51**). The F-statistic, degrees of freedom, and significance level for the comparison of the independent sample variances are produced by PROC TTEST. In this example, F = 2.00, P = .51, indicating that the assumption of homogeneity of variance cannot be rejected.

The comparison of means for two dependent samples involves a data manipulation to create a new variable and the use of the MEANS procedure with appropriate options. The variable GAIN was created in Data Set 1 with the code

```
GAIN = POSTTEST - PRETEST;
```

The comparison of the PRETEST mean to the POSTTEST mean is equivalent to comparing the mean of GAIN to zero. This is performed by PROC MEANS as follows:

```
PROC MEANS MEAN STDERR T PRT;
   VAR GAIN;
   TITLE 'DEPENDENT SAMPLE T-TEST, DATA SET 1';
```

This coding generates the following output for PROC MEANS, showing the dependent-sample t-test:

VARIABLE	MEAN	STD ERROR OF MEAN	t	PROB.
GAIN	10.50	1.067	9.84	0.0001

For the variable(s) in the VAR statement, in this case GAIN, this procedure produces the mean, standard error (STDERR), the t-statistic related to the hypothesis that the mean is zero, and the significance level for the t-statistic and appropriate degrees of freedom.

The comparison of the variances for two independent samples (**A4-51**) is performed by the TTEST procedure when the test of the assumption of homogeneity of variance is performed. The dependent-sample t-test comparing two correlated variances (**A4-52**) is calculated using the two variances, the correlation between the two variables, and the number of subjects involved. In most cases, it is relatively simple to calculate this t-statistic with a hand calculator using the variances, correlation, and sample size obtained from PROC MEANS and PROC CORR.

Procedures for comparing proportions for two independent or dependent samples are described in **A4-53** and **A4-54,** respectively. In most cases, these comparisons can be performed efficiently with a hand calculator using information from PROC FREQ. With Data Set 1, for example, the coding

```
PROC FREQ;
   TABLES TRSP*GROUP;
```

would produce a 2 × 2 contingency table. The information in this table could be used to compare the proportion of Group 1 students at level 1 of TRSP with students in Group 2 at the same level. This would be a comparison of proportions for independent samples (**A4-53**). In addition, the comparison of the proportion of students in Group 1 who are in levels 1 and 2 of TRSP also could be calculated directly from information contained in the 2 × 2 table. This would be a comparison of proportions for dependent samples (**A4-54**). (Note: This example with Data Set 1 is used for illustrative purposes only. The small sample size of Data Set 1 would make the procedures in the illustration statistically inappropriate.)

The comparison of independent-sample correlations requires the calculation of the correlation between the two variables separately for each of two groups. The use of a BY statement in conjunction with PROC CORR performs these calculations quite easily. The required *SAS* coding is shown below for Data Set 1:

```
PROC SORT;
  BY GROUP;
PROC CORR;
  VAR PRETEST POSTTEST;
  BY GROUP;
  TITLE 'PRETEST-POSTTEST CORRELATIONS, BY GROUP, DATA
SET 1';
```

Data must first be sorted into the levels of the variable that defines the comparison categories or groups. This is accomplished with PROC SORT; BY GROUP;. The CORR procedure is then executed for each level of the variable in the BY statement. This produces the standard output for PROC CORR separately for the subjects in each level of GROUP. The correlation between PRETEST and POSTTEST for Group 1 and Group 2 can then be compared as described in **A4-55**.

The correlation coefficients needed for the comparison of two-sample correlation coefficients for dependent samples **(A4-56)** are generally produced with the CORR procedure. A typical application of this test can be illustrated with Data Set 1. A researcher might be interested in knowing whether the correlation between PRETEST and IQ is significantly different from the correlation between POSTTEST and IQ. The correlations needed to calculate the t-statistic **(A4-56)** for testing whether the difference between these correlations is significant are produced by the following *SAS* coding:

```
PROC CORR;
  VAR PRETEST POSTTEST IQ;
  TITLE 'CORRELATIONS AMONG PRETEST, POSTTEST, AND IQ';
```

Significance Levels for z-, Chi-square, t-, and F-statistics

Comparisons involving one sample and two samples, and many other statistical procedures, frequently result in the calculation of a z-, Chi-square, t-, or F-statistic. *SAS* can produce the significance level associated with each of these statistics through the use of functions designed for that purpose. These functions take the following forms:

$$PZ = 1 - PROBNORM(z);$$

The proportion of the z-distribution above the given value of z. This is a one-tailed test. The two-tailed test can be calculated as PZ = (1 − PROBNORM(ABS(z)))*2; ABS takes the absolute value of z.

206 SOFTWARE APPLICATIONS IN STATISTICS

$$PCHI = 1 - PROBCHI(C, DFC);$$

The proportion of the Chi-square distribution that lies above the value C for DFC degrees of freedom. DFC is the number of degrees of freedom in the Chi-square.

$$PT = 1 - PROBT(t, DFT);$$

The proportion of the t-distribution that lies above the value t with DFT degrees of freedom. DFT is the degrees of freedom to the t-statistic. This is a one-tailed test. The two-tailed test can be calculated as

$$PT = (1 - PROBT(ABS(t),DFT))*2;$$

ABS takes the absolute value of t.

$$PF = 1 - PROBF(F, DFN, DFD);$$

The proportion of the F distribution that lies above the value F with DFN degrees of freedom in the numerator and DFD degrees of freedom in the denominator.

Comparisons: Two Samples (Nonparametric) (A4-57 to A4-65)

The *SAS* NPAR1WAY procedure is used to perform the two-sample median test **(A4-58)** and Wilcoxon rank sum test **(A4-59)**. The Wilcoxon is statistically equivalent to the Mann-Whitney U-Test **(A4-60)**. *SAS* instructions for using PROC NPAR1WAY, using Data Set 1, are as follows:

```
PROC NPAR1WAY WILCOXON MEDIAN;
   CLASS GROUP;
   VAR POSTTEST;
   TITLE 'WILCOXON, (MANN-WHITNEY), AND MEDIAN TESTS';
```

The CLASS statement defines the independent variable, levels of which constitute the groups being compared. The VAR statement specifies the dependent variable(s).

The Chi-square test of independence **(A4-61)**, more commonly considered the Chi-square test of homogeneity, can be carried out as an option in the FREQ procedure. Again using Data Set 1 to illustrate, the appropriate coding is written as follows:

```
PROC FREQ;
    TABLES GROUP*TRSP / EXPECTED CHISQ;
    TITLE 'CHI-SQUARE TEST OF INDEPENDENCE (HOMOGENEITY)';
```

The Wilcoxon matched-pairs–signed-rank test **(A4-65)** can be performed in *SAS*. This procedure is carried out by PROC UNIVARIATE applied to a new variable created by taking the difference between the two matched variables.

Comparisons: Two or More Samples (Nonparametric) (A4-66 to A4-69)

The sign test for K-independent samples (or K-sample median test) **(A4-67)** and the Kruskal-Wallis rank test for K-independent samples **(A4-68)** are performed with PROC NPAR1WAY. The example of NPAR1WAY in the previous section serves as an example. If the independent variable in the CLASS statement has three or more levels, the WILCOXON option in PROC NPAR1WAY produces the Kruskal-Wallis test **(A4-68)**, and the MEDIAN option produces the K-sample median test **(A4-67)**.

The Friedman two-way analysis of variance by ranks **(A4-69)** can be performed in *SAS*. This procedure requires the use of a data set that is created by PROC RANK and data manipulation that is beyond the scope of this chapter.

Comparisons: Two or More Samples—Analysis of Variance (ANOVA) (A4-70 to A4-90)

Analysis of variance, used to compare the means of two or more samples, is a powerful and general technique used to test hypotheses generated from a number of basic research designs. Factorial **(A4-71 to A4-78)**, repeated measures/split-plot **(A4-79 to A4-83)**, and nested or hierarchical **(A4-84 to A4-86)** analyses of variance are described in Section A, Chapter 4. The use of *SAS* to perform these analyses will be demonstrated only for the most commonly occurring models. Repetitious examples will be omitted.

Two *SAS* procedures are commonly used for analysis of variance. These are the ANOVA and the GLM (general linear model) procedures. The ANOVA procedure is used when there is an equal number of subjects in each cell of the design. The GLM procedure can also be used in the equal-cell-size case; however, it is also appropriate when cell sizes are not equal. Many researchers routinely use the GLM procedure, since it is more broadly applicable.

Factorial Analysis of Variance (A4-71 to A4-78)

Factorial ANOVA procedures will be illustrated with Data Set 3, which involves 18 subjects in a two-factor design. The factors are (A) treatment, with three levels (E1, E2, and C) and (B) gender, with two levels (F and M). The dependent measure is a test score. These data are as follows:

	Group	
E1	E2	C
1E F 110	2E F 112	3C F 95
1E F 95	2E F 106	3C F 86
1E F 98	2E F 99	3C F 91
1E M 79	2E M 84	3C M 68
1E M 73	2E M 92	3C M 79
1E M 72	2E M 86	3C M 72

These 18 records are read in *SAS* as follows:

```
INPUT GROUP $ GENDER $ SCORE;
CARDS:
E1 F 110
E1 F  95
 .  .   .
 .  .   .
C M  72
```

A one-way ANOVA **(A4-71)**, with GROUP as the independent variable, is performed with PROC ANOVA with the following *SAS* coding:

```
PROC ANOVA;
   CLASSES GROUP;
   MODEL SCORE = GROUP;
   MEANS GROUP;
   TITLE 'ONE-WAY ANOVA, BY GROUP, THREE LEVELS';
```

The CLASSES statement defines the independent variable(s), the levels of which form classification categories. The MODEL statement takes the general linear model form, with the dependent variable on the left of the equal sign and the independent variable(s) and effects on the right. The MEANS statement instructs the program to produce the means of the dependent variable for each level of the independent variable. If the cell sizes are unequal, PROC GLM must be used. In such cases, the statement MEANS GROUP; must be replaced with LSMEANS GROUP; to obtain the means adjusted for cell sizes. The portion of the output produced by the PROC ANOVA coding listed above includes the following information for this one-way ANOVA:

DEPENDENT VARIABLE: SCORE					
SOURCE	DF	SUM OF SQUARES	MEAN SQUARE	F-VALUE	PROB
MODEL	2	652.44	326.22	2.05	0.163
ERROR	15	2389.16	159.27	ROOT MSE	
				12.62	
CORRECTED TOTAL	17	3041.61			
SOURCE	DF	ANOVA SS	F-VALUE	PROB	
GROUP	2	652.44	2.05	0.16	

A two-way ANOVA (**A4-72**), treatment (3) × gender (2), is the appropriate analysis for the information in Data Set 3. The two-way ANOVA, using PROC GLM, is coded as follows:

```
PROC GLM;
   CLASSES GENDER GROUP;
   MODEL SCORE = GENDER GROUP GENDER*GROUP;
   MEANS GROUP GENDER GROUP*GENDER;
   TITLE 'TWO-WAY FIXED EFFECTS ANOVA, DATA SET 3';
```

Both the MODEL and the MEANS statements can be written in simplified form using the vertical line symbol (|) as follows:

```
PROC GLM;
   CLASSES GENDER GROUP;
   MODEL SCORE = GENDER | GROUP;
   MEANS GENDER | GROUP;
   TITLE 'TWO-WAY FIXED EFFECTS ANOVA, DATA SET 3';
```

The vertical symbol instructs the program to generate the full model with all interaction effects.

The ANOVA and GLM procedures both perform the analysis for a fixed-effects model (**A4-72**). Other models can be examined using the TEST statement. This statement is of the form

```
TEST H-Effect1 Effect2 etc. E=Error;
```

Effect1, Effect2, etc. refer to any main effect or any interaction effects in the model for which the researcher wishes to test hypotheses. "Error" refers to the error terms (denominators in the F-statistic) used to test the previously listed effects.

Effects using the same error term are listed on a single TEST statement. Effects using different error terms must be listed on separate TEST statements with the appropriate error term.

The TEST statement is central to many analyses and can be illustrated with Data Set 3 under the assumption that GROUP is a random effect and GENDER is a fixed effect. These assumptions make Data Set 3 an example of a two-way factorial mixed model **(A4-74)**. The correct error term for the fixed factor (GENDER) is the interaction term. The correct error term for the interaction effect and the random effect (GROUP) is the within-cell variance (MSW). The correct analysis for this case is coded as

```
PROC GLM;
   CLASSES GENDER GROUP
   MODEL SCORE = GENDER | GROUP;
   MEANS GENDER | GROUP;
   TEST H-GENDER E-GENDER*GROUP;
   TITLE 'TWO-WAY MIXED EFFECTS ANOVA, DATA SET 3';
```

SAS will first perform the default analysis; namely, a two-way fixed-effects analysis. This will result in the correct hypothesis test for GROUP and the interaction effect using the mean square within. The TEST statement will then result in the mean square for GENDER being divided by the mean square due to interaction, which is the correct test for the fixed effect in a two-way mixed model **(A4-74)**. A portion of the output from this analysis, using PROC GLM for a two-way fixed-effect ANOVA, is shown below:

DEPENDENT VARIABLE: SCORE					
SOURCE	DF	SUM OF SQUARES	MEAN SQUARE	F-VALUE	PROB
MODEL	5	2664.94	532.98	16.98	0.0001
ERROR	12	376.66	31.38	ROOT MSE	
				5.60	
CORRECTED TOTAL	17	3041.61			
SOURCE	DF	TYPE I SS	F-VALUE	PROB	
GENDER	1	1942.72	61.89	0.0001	
GROUP	2	652.44	10.39	0.0024	
GENDER*GROUP	2	69.77	1.11	0.3607	
TEST OF HYPOTHESES USING MS FOR GENDER*GROUP AS ERROR TERM					
SOURCE	DF	TYPE III SS	F-VALUE	PROB	
GENDER	1	1942.7	55.68	0.0175	

The same basic coding used for the two-way ANOVAs applies to three-way ANOVA models. Consider a hypothetical case with independent variables A, B, and C, with A fixed, and B and C being random **(A4-76)**. The *SAS* coding to analyze data from this design would be as follows:

```
PROC GLM;
    CLASSES A B C;
    MODEL Y = A B C A*B A*C B*C A*B*C;
    MEANS A | B | C;
    TEST H=B C          E=B*C;
    TEST H=A*B A*C      E=A*B*C;
```

The default procedure will result in the data first being analyzed with the fixed-effects model, **(A4-75)** using the mean square within (MSW) as the error term. This is the correct error term for B*C and A*B*C. The TEST statement will then result in the effects due to B, C, and the A*B and A*C interactions being tested against their respective error terms. The A effect in this model is treated by dividing the mean square for A by the sum of the mean squares for A*B and A*C, minus the A*B*C mean square. This computation must be done by hand. The spacing on the TEST statement does not affect the execution of the program.

Repeated-Measures/Split-Plot Analysis of Variance (A4-79 to A4-83)

The one-way repeated measures **(A4-79)** design generally involves a single group of subjects measured on two or more occasions. The most straightforward way to analyze such data in *SAS* is to conceptualize the design as a two-way mixed model **(A4-74)** with "subjects" as the random factor and "occasions" as the fixed factor. Differences across occasions, generally the hypothesis of interest, are tested with a TEST statement by dividing the mean square due to occasions by the mean square due to interactions.

A major task in repeated-measures/split-plot analyses usually involves reading or formatting the data. Consider Data Set 1 with 10 subjects and 2 occasions (pretest and posttest). For the purpose of illustration, imagine that 2 additional measures were taken between the pretest and the posttest. Thus, there are 10 subjects measured 4 times. This revised data set will be referred to as Data Set 4.

These data could be entered as 40 records, with 4 records for each person. Each single record would list the subject's ID (1 through 10), the occasion (1, 2, 3, or 4), and the score (pretest for occasion 1, the second test for occasion 2, the third test for occasion 3, and the posttest for occasion 4). This approach reflects the data structure in a simple fashion but is cumbersome and inefficient with large data sets.

A more realistic procedure for reading the data involves the *SAS* DO statement. The DO statement instructs *SAS* to carry out a repetitious procedure a specified number of times. DO statements can be embedded, one within the other. The *SAS* coding using the DO statements for the repeated measures data is as follows:

```
DATA REPEATED;
  DO SUBJECT=1 TO 10;
    DO OCCASION=1 TO 4;
      INPUT SCORE @;
      OUPUT;
    END;
  END;
CARDS;

21 25 30 35   17 22 28 33   19 24 27 31   25 28 33 36
18 21 23 25   19 21 23 24   20 23 25 29   11 15 19 24
17 20 25 27   13 15 19 21
PROC ANOVA;
  CLASSES SUBJECT OCCASION;
  MODEL SCORE = SUBJECT OCCASION SUBJECT*OCCASION;
  TEST H=OCCASION E=SUBJECT*OCCASION;
  TITLE ' ONE-WAY REPEATED MEASURES';
```

The DO statements operate on the INPUT statement, which is embedded inside the two DO statements. The coding works as follows. The first DO statement sets SUBJECT=1. The second DO statement sets OCCASION=1. The INPUT statement reads the first value for the variable SCORE, and the OUTPUT statement writes a record in which SUBJECT=1, OCCASION=1, SCORE=21. The first END statement encountered returns control to the closest Do statement and changes OCCASION=1 to OCCASION=2. The INPUT statement then reads the next value for the variable SCORE, and the OUTPUT statement writes a record in which SUBJECT=1, OCCASION=2, and SCORE=25. The first END statement read relates to the DO OCCASION=1 TO 4. After this DO statement has been executed the specified number of times (in this instance, 4), the program goes to the next line.

On the next line is the second END statement, which relates to the DO SUBJECTS=1 To 10 statement. After OCCASION=4 for SUBJECT=1 is reached, the second END statement is read and program control is returned to the first DO statement, which sets SUBJECT=2. The next line sets OCCASION=1, the INPUT STATEMENT reads the fifth value for the variable SCORE, and the OUTPUT statement writes a record in which SUBJECT=2, OCCASION=1, and SCORE=17. The program continues to loop through the DO statements until SUBJECT=10 and OCCASION=4. At this point all the data have been read, the second END statement has been executed, and the PROC ANOVA has been performed. This analysis is a two-way mixed-model analysis, virtually identical in content and output to the preceding example. The fixed effect, OCCASION, is tested against the interaction mean square. The effect due to SUBJECT and the SUBJECT*OCCASION effect are generally not of particular interest with this design.

The two-way split-plot ANOVA **(A4-80)** involves two factors; namely, one between-subjects factor, such as treatment groups, and one within-subjects factor, such as occasions. The analysis of split-plot designs of this sort can be illustrated with Data Set 4B (Data Set 4 restructured) which contains GROUP, a between-subjects factor with two levels, and OCCASION, a within-subjects factor with four levels. The data for this analysis can be read with three levels of DO statements as follows:

```
DATA SPLITPLT;
DO GROUP=1 To 2;
   DO SUBJECT=1 TO 5;
      DO OCCASION=1 TO 4;
         INPUT SCORE @;
         OUTPUT;
      END;
   END;
END;
CARDS:
21 25 30 35   17 22 28 33   19 24 27 31   25 28 33 36
18 21 23 25   19 21 23 24   20 23 25 29   11 15 19 24
17 20 25 27   13 15 19 21
PROC ANOVA;
CLASSES SUBJECT GROUP OCCASION;
   MODEL SCORE=GROUP SUBJECT (GROUP) OCCASION GROUP*OCCASION
              OCCASION*SUBJECTS (GROUP);
   MEANS GROUP | OCCASION;
   TEST H=GROUP          E=SUBJECTS (GROUP);
   TEST H=OCCASION GROUP*OCCASION
     E=OCCASION*SUBJECT (GROUP);
   TITLE 'TWO-WAY SPLIT PLOT, ONE-BETWEEN,ONE-WITHIN FACTOR';
```

The functioning of the DO statements is easier to follow if they are read from the inside DO to the outside DO. The DO statements input the variable SCORE for each of four occasions (the innermost DO), for each of five subjects (the next DO), and defines group membership (GROUP = 1 or GROUP = 2), the outermost DO. The expression SUBJECTS(GROUP) is read "subjects within groups," even though it might appear to be GROUP within SUBJECTS. (GROUP corresponds to Factor A, and OCCASION corresponds to Factor B in the description given in **A4-80.**) Sample output for this PROC ANOVA for a two-way split-plot ANOVA is shown below:

DEPENDENT VARIABLE: SCORE

TEST OF HYPOTHESES USING ANOVA FOR SUBJECT (GROUP) AS ERROR TERM

SOURCE	DF	ANOVA SS	F-VALUE	PROB
GROUP	1	308.025	7.30	0.027

(*continued on next page*)

TEST OF HYPOTHESES USING ANOVA FOR SUBJECT (GROUP) AS ERROR TERM

SOURCE	DF	ANOVA SS	F-VALUE	PROB
OCCASION	3	623.47	98.38	0.0001
GROUP*OCCASION	3	12.07	1.91	0.1557

It is useful to note that SAS first analyzes the model as a fixed-effects design. The estimates of the sums of squares for each effect in the fixed-effect analysis are correct, but the F-tests are meaningless since the wrong error terms are used and there are no degrees of freedom within cells.

The two-way split-plot can be read and analyzed more efficiently with the use of the REPEATED statement in PROC GLM or PROC ANOVA. With this approach, data are read and analyzed with the following code:

```
DATA SPLITPLT;
INPUT GROUP PRETEST TEST2 TEST3 POSTTEST;
CARDS;
1 21 25 30 35
1 17 22 28 33
1 19 24 27 31
1 25 28 33 36
1 18 21 23 25
2 19 21 23 24
2 20 23 25 29
2 11 15 19 24
2 17 20 25 27
2 13 15 19 21
PROC GLM;
CLASSES GROUP
MODEL PRETEST TEST2 TEST3 POSTTEST = GROUP / NOUNI;
REPEATED OCCASION 4 / NOM;
MEANS GROUP
TITLE 'TWO-WAY SPLIT PLOT, ONE-BETWEEN, ONE-WITHIN FACTOR';
```

This coding produces the same results as the preceding coding, which used the embedded DO statements. The REPEATED statement leads to the use of a multivariate approach in the data analysis. The four measures are included as dependent variables on the left side of the equal sign in the model statement. The options NOUNI, after the slash in the model statement, stands for "no univariate statistics." This instructs the program not to perform separate univariate ANOVAs for the dependent variables. The REPEATED statement specifies the name used to refer to the within subjects factor (OCCASION) and may not be a name used elsewhere. The number of levels (four) of the within-subjects factor are specified (4) after the factor name. The number of levels specified for the within-subjects factor must equal the number of variables listed to the left of the equal sign in the model statement. The NOM option suppresses the printing of the multivariate results.

The REPEATED statement is quite helpful in more-complex split-plot designs. The analysis for the three-way split-plot, using one between and two within-subjects factors (**A4-81**), will be illustrated with Data Set 5. This hypothetical example deals with training subjects to recall information. Subjects' recall is measured on three occasions; namely, before, during, and after a training program. Subjects on each occasion are given four trials at recalling key pieces of information after being given the stimulus material. The dependent measure is the number of key pieces of information recalled. Males and females are involved in the study and are to be compared. GENDER is the between-subjects factor. OCCASION and TRIALS are within-subjects factors. GENDER, OCCASION, and TRIALS correspond to Factors A, B, and C in **A4-81**. The data from the study are read and analyzed as follows:

```
DATA SP1B2W;
*SP1B2W STANDS FOR SPLIT PLOT, 1 BETWEEN, 2 WITHIN;
INPUT GENDER $ RECALL1-RECALL12;
OCCAS1 = MEAN (OF RECALL1-RECALL4);
OCCAS2 = MEAN (OF RECALL5-RECALL8);
OCCAS3 = MEAN (OF RECALL9-RECALL12);
TRIAL1 = MEAN (OF RECALL1 RECALL5 RECALL9);
TRIAL2 = MEAN (OF RECALL2 RECALL6 RECALL10);
TRIAL3 = MEAN (OF RECALL3 RECALL7 RECALL11);
TRIAL4 = MEAN (OF RECALL4 RECALL8 RECALL12);
RECALL = MEAN (OF RECALL1-RECALL12);
CARDS;
M   1 3 6 8   3 5 6 7   4 5 8 8
M   2 4 4 6   2 4 6 6   3 6 6 8
M   3 3 4 5   1 2 5 7   2 2 6 7
F   4 5 6 7   4 4 7 8   4 5 7 9
F   1 2 4 5   2 4 5 7   3 6 6 8
F   1 3 3 4   2 4 4 5   2 4 5 7
PROC SORT; BY GENDER;
PROC MEANS; BY GENDER; VAR RECALL;
PROC MEANS;
VAR OCCAS1 OCCAS2 OCCAS3 TRIAL1 TRIAL2 TRIAL3 TRIAL4;
PROC GLM;
    CLASS GENDER;
    MODEL RECALL1-RECALL12 = GENDER/NOUNI;
    REPEATED OCCASION 3, TRIALS 4/NOM;
MEANS GENDER;
TITLE1 '3-WAY SPLIT PLOT, GENDER (2) BETWEEN';
TITLE2 'OCCASION (3) AND TRIALS (4) WITHIN';
```

The MEANS statement within PROC GLM and PROC ANOVA cannot be applied to within-subjects variables read with the REPEATED statement; therefore, the mean responses for each level of OCCASION and each level of TRIALS are created in the data step with the MEAN(OF) function.

Each subject in this design is actually measured 12 times, RECALL1-RECALL 12. The REPEATED statement describes the structure of these 12 measurements. There are three levels of OCCASION (3), and four levels of TRIALS (4) within each level of OCCASION. The pertinent output from this analysis, using PROC GLM to show a three-way split-plot with one-between and two-within factors, is illustrated below:

TESTS OF HYPOTHESES FOR BETWEEN-SUBJECTS EFFECTS				
SOURCE	DF	TYPE III SS	F-VALUE	PROB
GENDER	1	0.01	0.000	0.97
ERROR	4	41.22		
TESTS OF HYPOTHESES FOR WITHIN-SUBJECTS EFFECTS				
SOURCE	DF	TYPE III SS	F-VALUE	PROB
OCCASION	2	28.69	18.28	0.001
OCCASION*GENDER	2	00.86	0.55	0.598
ERROR	8	6.27		
TRIALS	3	189.37	82.14	0.0001
TRIALS*GENDER	3	1.81	0.79	0.522
ERROR	12	9.22		
OCCASION*TRIALS	6	2.08	0.55	0.768
OCCASION*TRIALS *GENDER	6	1.47	0.39	0.881
ERROR	24	15.27		

The analysis indicates significant main effects due to OCCASION and TRIALS but no other significant effects. The means for OCCASION levels 1, 2, and 3 are respectively 3.91, 4.58, and 5.45. The means for levels 1, 2, 3, and 4 for TRIALS are respectively 2.44, 3.94, 5.44, and 6.77. Inspection of the means indicates that subjects' recall increases both across OCCASION and across TRIALS. Additional analysis, using multiple-comparison procedures **(A4-91 to A4-97)**, are needed to further clarify the nature of these differences.

The three-way split-plot, with two between-subjects factors and one within-subjects factor **(A4-82)**, will be illustrated with a hypothetical study involving learning strategies of subjects to solve rapidly three-digit subtraction problems. Data in this example will be referred to as Data Set 6. Three different methods are used to teach the subjects and both male and female subjects are taught under all three methods. METHOD, with three levels, and GENDER, with two levels, are between-subjects factors. Subjects are measured before, during, and after the instruction. The three measurements are the within-subjects factor. The dependent measure is the number of problems correctly solved in three minutes. GENDER, METHOD, and OCCASION correspond to Factors A, B, and C in **A4-82**. The following coding will read the data and perform the appropriate analysis:

```
DATA SP2B1W;
INPUT METHOD GENDER $ OCCAS1-OCCAS3 @ @;
CARDS;
1 M 11 23 37     1 M 13 22 39     1 M  9 24 41
1 F  9 25 41     1 F  8 21 37     1 F 11 22 37
2 M  7 18 35     2 M 13 25 46     2 M  8 21 37
2 F 11 21 41     2 F  7 24 36     2 F  7 21 38
3 M  9 23 38     3 M  9 20 39     3 M 12 21 38
3 F 12 24 41     3 F  9 21 37     3 F  7 20 38
PROC GLM;
    CLASSES METHOD GENDER;
    MODEL OCCAS1-OCCAS3 = METHOD GENDER METHOD*GENDER / NOUNI;
    REPEATED OCCASION 3 / NOM;
    MEANS METHOD | GENDER;
    TITLE1   '3-WAY SPLIT PLOT, METHOD (3) AND GENDER (2), BETWEEN';
    TITLE2   '                       OCCASION (3) WITHIN';
```

The INPUT statement uses the symbol @@. This symbol allows data to be read across the line. The spacing of the data does not affect the execution of the program but makes it easier to read and check the accuracy of the information. The output describing the results of this analysis is quite similar in format to the previous example.

NESTED/HIERARCHICAL ANOVA (A4-84 TO A4-86)

Nested, or hierarchical, analyses (**A4-84** to **A4-86**) are employed for designs in which only some of the levels of one factor appear under only one of the levels of another factor. For example, a two-way nested design may have instructional method as Factor 1, with teachers (Factor 2) *nested* under Factor 1. The nested relationship between factors 1 and 2 means that some of the teachers use one method of instruction while a different set of teachers use another method. This would not be a factorial design unless teachers were *crossed* with method, that is, all teachers taught all methods.

Nested designs can be analyzed with ANOVA, GLM, or the NESTED procedures. By way of illustration, consider five teachers using Method 1, five different teachers using Method 2, and each teacher teaching three students. The analysis of these data, using both PROC ANOVA and PROC NESTED, are shown below using Data Set 7.

```
DATA NESTED;
INPUT METHOD TEACHER SCORE @ @;
CARDS;

1 1 25     1 1 30     1 1 29     1 2 27     1 2 28     1 2 24
1 3 26     1 3 25     1 3 24     1 4 27     1 4 30     1 4 24
1 5 18     1 5 24     1 5 23     2 6 10     2 6 8      2 6 12
2 7 9      2 7 9      2 7 10     2 8 7      2 8 6      2 8 9
2 9 7      2 9 10     2 9 9      2 10 9     2 10 10    2 10 11

PROC ANOVA;
    CLASSES METHOD TEACHER;
    MODEL SCORE = METHOD TEACHER (METHOD);
    MEANS METHOD TEACHER TEACHER (METHOD);
    TEST H = METHOD E = TEACHER (METHOD);
    TITLE '2-WAY NESTED ANOVA, TEACHER NESTED IN METHOD';
PROC SORT;
    BY METHOD TEACHER;
PROC NESTED;
    CLASS METHOD TEACHER;
    VAR SCORE;
    TITLE 'TWO-WAY NESTED ANOVA, TEACHER NESTED IN METHOD';
```

The correct error term for testing the effect due to METHOD is TEACHER(METHOD), which is read "teacher nested in method." The TEST statement in the ANOVA procedure supplies the proper instructions for carrying out this test.

The NESTED procedure requires that the data be sorted by the two factors, METHOD and TEACHER, listed in the CLASS statement. The VAR statement lists dependent variable(s).

General ANOVA Designs (A4-87 to A4-88)

Randomized-blocks designs (A4-87) are a general class of designs that can be employed in a variety of analysis of variance (ANOVA) procedures that have already been described. The blocking variable is simply one of the factors used in the ANOVA procedure. Latin square designs (A4-88) can be analyzed in *SAS* but it is beyond the scope of this book to explain the procedure because of the manipulations of the data structure required.

Related Designs (A4-89)

Analysis of covariance (ANCOVA) (A4-89) improves on the statistical precision of analysis of variance by including a covariate(s) to reduce the proportion of variation attributed to error. Data Set 1 can be used as an illustration. A simple one-way ANOVA can be performed with POSTTEST as the dependent variable and GROUP as the independent variable. The data could also be analyzed as a one-way ANCOVA by including PRETEST as a covariate. A common procedure for analyzing such data would involve testing whether the groups were significantly different on the covariate (PRETEST), on the dependent variable (POSTTEST), and whether the assumption of homogeneity of regression has been met. The coding for these procedures is as follows:

```
PROC GLM;
  CLASSES GROUP;
  MODEL PRETEST POSTTEST = GROUP;
  MEANS GROUP;
  TITLE 'ONE-WAY ANOVA, DEP. VARS. PRETEST AND POSTTEST';
PROC GLM;
  CLASSES GROUP;
  MODEL POSTTEST = GROUP PRETEST PRETEST*GROUP;
  TITLE 'CHECK ASSUMPTION OF HOMOGENEITY OF REGRESSION';
PROC GLM;
  CLASSES GROUP;
  MODEL POSTTEST = PRETEST GROUP;
  TITLE 'ONE-WAY ANCOVA USING PRETEST AS COVARIATE';
```

The first GLM procedure performs a one-way ANOVA, testing whether the groups are significantly different on the PRETEST and POSTTEST. The second GLM procedure tests the assumption that the regression lines of POSTTEST onto PRETEST are parallel for the different levels of GROUP. If the PRETEST*GROUP effect is significant, the assumption is rejected. The final GLM procedure performs the one-way ANCOVA where POSTTEST is the dependent variable, PRETEST is the covariate, and GROUP is the independent variable. A portion of the results of the sample output for PROC GLM showing one-way ANCOVA are illustrated below:

DEPENDENT VARIABLE: POSTTEST

SOURCE	DF	SUM OF SQUARES	MEAN SQUARE	F-VALUE	PROB
MODEL	2	167.39	83.69	8.48	0.0135
ERROR	7	69.11	9.87	ROOT MSE 3.14	
CORRECTED TOTAL	9	236.50			

SOURCE	DF	TYPE III SS	F-VALUE	PROB
PRETEST	1	44.89	4.55	.0704
GROUP	1	33.33	3.38	.1088

Multiple-Comparison Procedures (A4-91 to A4-97)

Multiple-comparison procedures are performed by *SAS* with the PROC GLM or PROC ANOVA procedures. Contrasts are specified as an option in the MEANS statement for factorial effects, in the REPEATED statement for within-subjects effects, or explicitly with the use of a CONTRAST statement. The one-way ANOVA for Data Set 3 is used below to illustrate the coding needed to perform all pairwise contrasts using orthogonal contrasts **(A4-92)**, orthogonal polynomial contrasts **(A4-93)**, Scheffé **(A4-94)**, Tukey **(A4-95)**, Newman-Keuls (coded as SNK) **(A4-96)**, and Duncan **(A4-97)** multiple-comparison procedures, and the contrast of Groups 1 and 2 compared to Group 3.

```
PROC GLM;
    CLASSES GROUP;
    MODEL SCORE = GROUP;
    MEANS GROUP / SCHEFFE TUKEY SNK DUNCAN;
    CONTRAST '1 & 2 VS.   3' GROUP  1  1 -2;
    CONTRAST 'LINEAR       ' GROUP -1  0  1;
    CONTRAST 'QUADRATIC  ' GROUP  1 -2  1;
    TITLE1  '   ONE-WAY ANOVA';
    TITLE2  '   EXAMPLES OF CONTRASTS';
```

The SCHEFFE, TUKEY, SNK (Student Neuman Keuls), and DUNCAN procedures are specified as options in the MEANS statement. Particular contrasts are specified with the CONTRAST statement. A label within the single quotation marks following the word CONTRAST, for instance, '1 & 2 VS. 3,' is used in the output. This label is followed by the name of the factor or independent variable in question, in this case, GROUP, and the coefficients of contrast that are of interest; 1, 1, and −2. Sample output for PROC GLM related to the analyses of the various contrasts appears as follows:

CONTRAST	DF	SS	F-VALUE	PROB
1 & 2 vs 3	1	427.11	2.68	0.122
LINEAR	1	108.00	0.68	0.423
QUADRATIC	1	544.44	3.42	0.084

A similar approach is used when these same data are analyzed as a two-way ANOVA (gender × group). The following coding is used in this case:

```
PROC GLM;
   CLASSES GENDER GROUP;
   MODEL SCORE = GENDER | GROUP;
   MEANS GENDER | GROUP;
   CONTRAST 'GROUP                       ' GENDER 1 -1;
   CONTRAST 'GROUP 1 & 2 VS. 3' GROUP 1 1 -2;
   CONTRAST 'GENDER*GROUP, 2 VS 3' GENDER*GROUP 0 1 -1 0 -1 1;
   TITLE 'CONTRASTS WITH TWO-WAY ANOVA';
```

The contrasts of levels 1 and 2 for gender, GENDER 1 −1, is trivial since there are only two levels of gender. The contrasts for GROUP, 1 1 −2, examines main effects only, collapsing across levels of gender. The 1 1 −2 contrast compares treatment groups 1 and 2, combined, with Treatment Group 3. The interaction contrast coefficients are related to the way the cells of the design are read. The six cells, GENDER (2) × GROUP (3), are read beginning with the first level of the first independent variable in the model statement, GENDER, and reading across all levels of the second independent variable in the model statement, GROUP. Thus the contrast coefficients, 0 1 −1 0 −1 1, compare GENDER=1, GROUP=2 and 3, and GENDER=2, GROUP=2 and 3, or cells 12, 13, 22, and 23.

Contrasts related to within-subjects factors of split-plot designs are specified in the REPEATED statement. The analysis of the two-way split-plot (A4-80) will be used to illustrate this procedure.

```
PROC GLM;
   CLASSES GROUP;
   MODEL PRETEST TEST2 TEST3 POSTTEST = GROUP / NOUNI;
   REPEATED OCCASION 4 CONTRAST (1) / SUMMARY NOM;
   MEANS GROUP;
   TITLE 'TWO-WAY SPLIT PLOT, ONE-BETWEEN,ONE-WITHIN
   FACTOR';
```

The word CONTRAST in the REPEATED statement indicates that contrasts are to be performed on the within-subjects factor. The number in the parenthesis, (1), indicates the reference level of the within-subjects factor against which the other levels are to be tested, in this case, the first level (1). In this example, TEST2, TEST3, and POSTTEST will be contrasted with the PRETEST, the first (1) level of the factor. The last level of the within-subjects factor is used if no reference level is specified. The SUMMARY option after the slash (/) results in the printing of the univariate analyses for each of the one-degree-of-freedom-tests specified by the CONTRAST option.

Polynomial contrasts are often of particular interest when a within-subjects factor is examined. Such contrasts are specified with the following code:

```
REPEATED OCCASION 4 POLYNOMIAL / SUMMARY NOM;
```

In many research settings the spacing between the levels of the within-subjects factor is not merely ordinal, as in levels 1, 2, 3, and 4, but has some particular structure. For example, if the within-subjects factor relates to a measure of children's language proficiency at 24, 30, 36, and 48 months of age, this spacing can be incorporated into the contrast as

> REPEATED OCCASION 4 (24 30 36 48) POLYNOMIAL / SUMMARY NOM;

In this case, the polynomial terms will be tested with the differences between levels actually reflecting the spacing in months.

CHAPTER 3

SYSTAT

Introduction

The *SYSTAT** package is one of the most popular statistical packages available for microcomputers today. The package is available in version 5.0 for DOS-based systems and version 4.0 for the Macintosh. The package is actually a set of modules that operate within the framework of the *SYSTAT* system. The primary package consists of the following modules:

CLUSTER	Provides cluster analyses for rectangular or symmetric data matrices.
CORR	Performs general bivariate and multivariate correlations, including Pearson, Spearman, and Kendall's Tau.
DATA	Data manipulation and conversion.
EDIT	A full-screen editor for data entry and direct data manipulation.
FACTOR	Produces a variety of principal-component and principal-axis factor analyses.
GRAPH	Produces a variety of nongraphic, or character-based graphs, including plots, histograms, bar, box plots, and stem-leaf.
MACRO	A programming tool designed to permit the user to produce batch-type programming, thus permitting the user to create and analyze data in a single pass rather than moving from module to module within the *SYSTAT* system.
MDS	Provides multidimensional scaling for matrices up to five dimensions.
MGLH	Produces a variety of inferential tests based upon the general linear model. Statistical tests include ANOVA, ANCOVA, regression, MANOVA, discriminant analysis, and canonical analysis.
NONLIN	Produces a series of nonlinear tests using general algebraic

---

*SYSTAT is a registered trademark of *SYSTAT*, Inc., Evanston, IL.

	models, employs two minimization strategies: Quasi-Newton or Simplex.
NPAR	Produces a full spectrum of nonparametric statistical tests for groups and pairs of variables. Tests include the sign test, Wilcoxon, Friedman two-way, Kruskal-Wallis one-way, and Kolomogorov-Smirnov.
SERIES	Provides analysis of time-series data under a set of models, including exponential, ARIMA, Adjseason, Smooth, and Fourier.
STATISTICS	Calculates almost all primary descriptive statistics, as well as some bivariate inferential statistics, including independent and dependent t-tests and Duncan, Tukey, and Newman-Keul's pairwise *post hoc* tests.
SYGRAPH	A high-resolution graphics module that permits the user to create and print some of the most outstanding graphics available.
TABLES	Produces two-way and multi-way tables to create appropriate cross-tabulations, produces a series of statistical tests, including Phi coefficient, Cramer's V, Chi-square, McNemar's test, Gamma, and Kendall's Tau.

The documentation for the *SYSTAT* system is contained in two manuals: *SYSTAT: The System for Statistics* and Sygraph: *The System for Graphics*. Both of these manuals accompany the full SYSTAT package. They are extremely well written and documented, but are not intended to teach statistics, just SYSTAT.

Text Format

Examples of *SYSTAT* commands and output will be shown throughout this chapter. The *SYSTAT* code will be printed in CAPITAL letters when referenced in a paragraph, and it will be printed in a shaded area between parallel lines when *SYSTAT* statements appear on separate lines that illustrate the code. In practice, the *SYSTAT* code can appear in any column position on a screen line; however, examples in this chapter will be written as if the left-most *SYSTAT* code is in the first column of the input screen. Examples of *SYSTAT* output will be displayed in boxes to set them apart from the text.

Some SYSTAT Basics

USING SYSTAT'S FULL-SCREEN EDITOR

As a microcomputer-based package, *SYSTAT* requires that the data be entered into a format that can be interpreted readily by the modules within the system. The easiest method for entering data into SYSTAT is to use the full-screen editor that accompanies the package. The steps for activating SYSTAT are:

1. Change to the directory containing SYSTAT
2. Type SYSTAT and press <R>*
3. At the main SYSTAT menu type 1 and press <R>.
4. You should now see the EDIT screen.

The EDIT screen consists of two windows: the data window and the command window. The data window is where data is inputted and edited like the data input of a spreadsheet, but do not mistake SYSTAT's DATA module for a spreadsheet. It is not.

The unnumbered top lines are for variable labels: up to eight characters plus a dollar sign ($) to indicate an alphanumeric variable. For example, two acceptable labels would be TIMEWORK and TIMEWORD$. Both contain eight characters; TIMEWORK, however, is a label for a numeric variable; whereas, TIMEWORD$ is a label for an alphanumeric (nonnumeric) variable. These should be typed by preceding the label with double quotation marks (' ') or a single quotation mark ('). All nonnumeric data as well should be typed in by preceding it with the single quotation mark (').

The data in the following table represent the results of 1989 statewide testing of reading skills conducted by the State of Illinois for the school districts surrounding the city of Chicago. The data were inputted into the editor first by listing the

---

*The <R> symbol means to press the RETURN key

Case	POP	AREA$	GR3	GR6	GR
1	1067.000	W	347.000	295.000	30
2	396.000	N	344.000	327.000	34
3	348.000	N	340.000	331.000	29
4	800.000	W	328.000	302.000	31
5	1322.000	SW	324.000	284.000	27
6	476.000	N	323.000	316.000	32
7	875.000	N	317.000	329.000	31
8	1315.000	N	316.000	322.000	31
9	1212.000	N	314.000	306.000	30
10	625.000	N	312.000	288.000	29
11	4017.000	N	312.000	288.000	30
12	1578.000	S	307.000	314.000	29
13	1176.000	N	304.000	293.000	29
14	315.000	N	303.000	261.000	30
15	2239.000	SW	303.000	269.000	27

variable names, POP for district population, AREA$ for area designation (S=South, N=North, W=West, and SW=Southwest), GR3 for third-grade average (mean), GR6 for sixth-grade mean, GR8 for eighth-grade mean, LOINCOM for percentage of students classified as low income, and COST for the district cost per pupil. This data set was named COOK1.

When inputting the labels or the data simply press the <R> (RETURN) key and the pointer (a solid block) will move to the next cell to the right. The DOWN arrow and the UP arrow on the cursor-control pad will also work with input and will move the pointer down one cell or up one cell; however, none of the other cursor arrows will work. Thus, using either the <R> key or the DOWN or UP cursor-control arrows will permit movement through the data window as data is being entered.

Once the data are completely entered you may exit the data window by pressing the <ESC> (ESCAPE) key. This puts the user in the command window. Now the data may be saved by typing: SAVE [filename]. The [filename] may be any eight-character name, excluding the following characters: !, £, –, $, %, ^, &, *, -, __, +. To exit the EDIT module, type QUIT.

ENTERING AND MODIFYING DATA

Before we exit the EDIT module we need to explore some of the additional features of it. Data cannot be directly manipulated within any of the modules other than the EDIT and DATA modules. In the EDIT module press the <ESC> key and move into the Command window. At the command line enter any of the data manipulations that you wish to perform that directly relate to the data currently residing in the Data window.

For example, assume that we want to create a new variable for our COOK1 data set. This new variable is to be the average of the GR3, GR6, and GR8 scores and will be called GRANDX. Typically, we might first sum the three variables and then compute a mean value for the data; however, using one of *SYSTAT*'s built-in functions, we are able to write a simple statement such as:

```
LET GRANDX = AVG(GR3,GR6,GR8)
```

This creates the new variable label and calculates the average value of the three earlier variables. *SYSTAT* permits six types of arguments or functions; namely, one-variable numeric, distribution, multivariable, two-variable numeric, relational, and logical. Using either the EDIT or DATA modules the user may create the necessary variables for analysis.

The logical and relational transformations are ones often employed in data analysis to divide numeric data into two or more alphanumeric categories. For example, suppose we want to group our districts in the COOK1 data set by their cost per pupil values (COST). At the command line the user should type two transformation commands:

```
IF COST <=5000 THEN LET NEWCOST$ = 'LOW'  <R>
IF COST >=5001 THEN LET NEWCOST$ = 'HIGH' <R>
```

Several items need to be noted at this point. First, the logical operant of <= means less than or equal to, while >= means greater than or equal to. Additional operations include = (equal), <> (does not equal).

The process can be as elaborate as combining relational factors of several variables to create another variable. Suppose we categorize districts by a combination of their percentage of low income students (LOINCOM) and the NEWCOST variable. Such a transformation would take four lines of activity:

```
IF LOINCOM <=7.5 AND NEWCOST = 'LOW' THEN LET
GRP = 1
IF LOINCOM <=7.5 AND NEWCOST = 'HIGH' THEN LET
GRP = 2
IF LOINCOM >=7.51 AND NEWCOST = 'LOW' THEN LET
GRP = 3
IF LOINCOM >=7.51 AND NEWCOST = 'HIGH' THEN LET
GRP = 4
```

The result of such a transformation would be a new variable called GRP, which would contain the new variable's categories.

The user should now save the new data set with a new name or with the old name. If the old name (COOK1) is chosen, *SYSTAT* will ask if you realize that you have requested to save a file to an existing file. Pressing Y for "yes" will automatically tell *SYSTAT* to overwrite the old data set with the new data-set information. The user may now exit the EDIT module by typing QUIT or switch to another data set by typing SWITCHTO [module name]. If the user utilizes the latter procedure, the system will immediately switch to the named module with the current data set as the active data set for that module.

USING THE DATA MODULE

The second method for inputting data into a *SYSTAT* data set is to use the DATA module contained within the *SYSTAT* package. The DATA module permits the creation of data sets directly from keyboard input, from existing ASCII files, and from either *LOTUS 1-2-3* or *dBASE III* files using a special feature called IMPORT. As do many other statistical packages, *SYSTAT* permits the user to utilize either free-format or fixed-format data sets.

"Fixed-format" data sets resemble a set of columns evenly aligned and corresponding to a particular column number that you must tell *SYSTAT* in order to input the data set. For example, the following data set, called Data Set 1 in the *SAS* chapter, (and stored as DASET1) could have a fixed format:

```
                           1         2         3
                  1234567890123456789012345678 90

             1  1  112 21 35 2
             2  1  109 17 33 2
             3  1  105 19 31 1
             4  1  118 25 36 2
             5  1   95 18 25 1
             6  2  111 19 24 1
             7  2  107 20 29 2
             8  2  102 11 24 1
             9  2   98 17 27 1
            10  2  105 13 21 1
```

The first two lines represent column markers and are used only for illustration. The format for the variables are Student 1 2, Group 4, IQ 6 8, Pretest 10 11, Posttest 13 14, and TRSP 16. To input this data set into *SYSTAT* using a fixed format, the user first must save the file in ASCII format. Any editor or word-processing program that saves files in a nondocument or unformatted form may be used for this; however the file must be saved with a .DAT extension. The second step is to enter the DATA module from the main *SYSTAT* menu.

Once in the DATA module the user types the following:

```
GET DASET1
SAVE DASET1
INPUT (STUDENT, GROUP, IQ, PRETEST, POSTTEST, TRSP),
(#2,>,#1,>,#3,>,#2,>,#2,>,#1)
RUN
```

The final result of the process is the creation of a *SYSTAT* data set called DASET1.SYS. While the process is rather direct, several points within the process need to be discussed in more depth.

1. The first line of the command tells *SYSTAT* to find the ASCII file called DASET1.DAT. The user does not need to tell *SYSTAT* the .DAT extension. *SYSTAT* assumes that the file was saved with this extension. Had the file been saved with any other extension, the user would have to identify the full name of the data set (e.g., DASET1.PRN).
2. While it is convenient to save and use a file from the same directory or disk that contains the *SYSTAT* program, this is not always practical or possible, particularly if the user is operating from a floppy-disk-only system. If the file resides on another drive or directory, then the user must provide a full path name for *SYSTAT* (e.g., GET A:\DASET1 or GET D:\DATA\DASET1).
3. The second line tells *SYSTAT* to save the file that will be created with the name DASET1. *SYSTAT* will automatically add the .SYS extension to the file name. As noted earlier, to save the file to another disk or drive that is different from the *SYSTAT* disk or drive, the user must provide a full path name (e.g., SAVE A:\DASET1 or SAVE D:\DATA\DASET1).

4. The INPUT line consists of two parts, separated by enclosing each part within parentheses.
5. The first part of the INPUT line consists of the names of the variables to be read into the *SYSTAT* data set. Please note that none of the above variables were alphanumeric. If one or more of the variables had been alphanumeric the INPUT line would have had to be modified by placing a dollar sign ($) after the variable name.
6. The second part of the INPUT line tells *SYSTAT* the formatting to be used in the creation of the data set. For example, the "greater than" (>) symbol tells *SYSTAT* to move the pointer one column to the right. The numbers preceded by a pound sign (#) tell *SYSTAT* how many places to read for each variable. If the variables had been alphanumeric the pound sign would have been replaced by a dollar sign ($)
7. The first part of the INPUT line ends in a comma (,) to tell *SYSTAT* that the line of information continues down on the next line of input. This is a strategy that *SYSTAT* uses throughout its programs to indicate continued information.
8. The RUN statement on the last line tells *SYSTAT* to perform the data transformation and to produce a *SYSTAT* data set.

INPUT TECHNIQUES

In the INPUT format described above, the "greater than" symbol (>) was used to tell *SYSTAT* to move the column pointer one column to the right. The following symbols and their meanings may be used when writing INPUT format statements:

<	Move one column to the left
>	Move one column to the right
^n	Move to column n
#n	Read a numeric variable in the next n columns
$n	Read an alphanumeric variable in the next n columns
n*r	Repeat r n times, where r is any of the above commands
\	Leaves the pointer on the current line for the next case
/	Moves the pointer to the first column of the next line
%n	Moves the pointer to the first column of the n line

As an example of some of the above, let's take the same data set (DASET1.DAT) and change the spacing.

```
                     1         2         3
            1234567890123456789012345678 90

        1   1   112   21   35        2
        2   1   109   17   33        2
        3   1   105   19   31        1
        4   1   118   25   36        2
        5   1    95   18   25        1
        6   2   111   19   24        1
        7   2   107   20   29        2
        8   2   102   11   24        1
        9   2    98   17   27        1
       10   2   105   13   21        1
```

The INPUT statement would then be changed to the following:

> INPUT (STUDENT, GROUP, IQ, PRETEST, POSTTEST, TRSP),
> (#2,^6,#1,^10,#3,^16,#2,^20,#2,^30,#1)

The second line now reads: (1) start on column 1 and read a numeric variable in two places, (2) move to column 6 and read a one-place numeric variable, (3) move to column 10 and read a three-place numeric variable, (4) move to column 16 and read a two-place numeric variable, (5) move to column 20 and read a two-place numeric variable, and (6) move to column 30 and read a one-place numeric variable.

Since the data in the above data set (DASET1) is written in spaced columns it could have been inputted in a free-format style. This style could have been employed whether we had the evenly spaced set in the first example or the widely spaced set in the second example. The resulting input format would have been:

> INPUT STUDENT, GROUP, IQ, PRETEST, POSTTEST, TRSP

As can be readily seen, when data sets are created with correct spacing the input is much more simple in free format than in fixed format. Sometimes, however, the data cannot be spaced so nicely and the fixed-format strategy is preferred. For example, suppose the following data set were created representing the scores and ID of students in a class:

```
                     1         2         3
            123456789012345678901234567890

            AFO1          293021294655
            AMO2          303224296973
            BMO3          293424299093
            BMO4          353833349594
            AFO5          283223285364
            BM25          222618258885
```

The above data set actually represents nine different variables, but the input format is blocked into two sets with the first identifying the student's class (A or B), gender (M or F), and student number. The second set contains each student's scores on four quizzes and two unit tests. The appropriate INPUT statement would be:

> INPUT (CLASS$, SEX$, ID, QUIZ(1-4), TEST(1-2)), ($1,>,$1,>,#2,^
> 15,4*#2,>,2*#2)

This input format illustrates various types of strategies for dealing with fixed-format data sets that lack uniform spacing. The INPUT lines perform several operations:

1. The variables CLASS and SEX are identified as alphanumeric (or character) variables.
2. The variable QUIZ is identified as a subscripted variable with four elements, and this creates four variables with the labels QUIZ(1), QUIZ(2), QUIZ(3), and QUIZ(4).
3. The variable TEST is also created as a subscripted variable, with two elements that have the labels TEST(1) and TEST(2).
4. The format line tells SYSTAT to (1) read the first variable in column one as a one-place character variable, (2) move to the next column, (3) read the next variable as a one-place character variable, (4) move to the next column and read a two-place numeric variable, (5) move to column 15 and read a set of four two-place numeric variables, and (6) move to the next column and read a set of two two-place numeric variables.

Another type of formatting is used when the data are constructed to repeat several cases across a single line or record. Such a data set was introduced earlier in the *SAS* chapter as Data Set 2. The data set consists of the following variables: (1) AGEMONTH, (2) CSYN, (3) MSYN, (4) CVOCAB, and (4) MVOCAB. The data have been stored in a format where the scores for up to four subjects are repeated in sets across a single line (or record). The data set looks like this:

61	9	8	68	56	48	7	10	41	47	72	10	10	66	68	49	5	4	46	48
39	8	6	19	33	90	9	10	64	79	43	9	10	53	54	84	10	10	64	75
72	10	10	79	65	29	7	3	27	39	48	7	8	49	63	64	7	6	59	61
41	5	7	42	47	76	9	10	65	69	42	7	7	38	49	46	5	10	44	51
63	10	8	60	63	72	8	10	59	89	70	9	10	60	76	37	5	9	37	57
70	8	10	57	53	40	9	10	52	58	45	6	6	15	20					

Since the data are evenly spaced in columns, the user can refrain from using a fixed-format input strategy and may employ free format. However, since the data for several subjects is repeated across a single line the input statement must be constructed to tell *SYSTAT* to repeat itself within the sets. The simplest strategy would be the following:

```
GET DASET2
SAVE DASET2
INPUT AGEMONTH, CSYN, MSYN, CVOCAB, MVOCAB \
RUN
```

The input line uses straightforward free-format style but ends with a backslash (\) symbol. This symbol tells *SYSTAT* to hold the pointer because another set of student data begins on the same line. *SYSTAT* follows this strategy until it runs out of data and moves down to the next line of data, where it repeats the process.

OTHER INPUT STRATEGIES

The previous strategies illustrate the primary methods of creating *SYSTAT* data sets, but there are two additional strategies that may be employed by the user. The first involves a technique of direct input from the DATA module command line. This strategy would be similar to the following lines:

```
SAVE SAMPLE
INPUT VAR1, VAR2, NAME$
23 24 TOM
25 26 BILL
27 28 SUE
29 30 JANE
etc.
RUN
```

The result would be a data set named SAMPLE.SYS. This strategy is recommended only when the data set is extremely small; otherwise, it is recommended that the user employ the *SYSTAT* full-screen editor in the EDIT module.

The final strategy for inputting data into *SYSTAT* files would be to use data that have been created within the framework of another program and to import that data directly into *SYSTAT*. If the data exist in either a *LOTUS 1-2-3* or *dBASE III* file, the IMPORT facility of *SYSTAT* may be used. The exact format of the IMPORT operation is specific to the individual microcomputer environment (DOS or Macintosh) and the user should consult the *SYSTAT* manual for directions.

Another technique is to employ a translation package designed to import data directly from one file type to *SYSTAT* and vise versa. Two packages exist that perform this translation without difficulty. SYSTAT, Inc. produces a package called *Sytran* that translates 28 different file types into *SYSTAT* files (and back again). The other package is called *DBMS/COPY,* from Conceptual Software, Inc.* *DBMS/COPY* translates data files between more than 40 different programs, including *SYSTAT*. Interested users should contact the publishers for information and details concerning these programs.

---

*Conceptual Software, Inc., P.O. Box 56627, Houston, TX 77256, (713) 667-4222.

WORKING WITH SYSTAT DATA SETS

The previous sections have dealt with the techniques of data input and data-set creation within the modules of *SYSTAT*. One of *SYSTAT*'s strengths is its flexibility in manipulating data to meet the various needs of the user. For example, several of the modules produce a variety of output that may be directly saved into a *SYSTAT* data file for later analysis. While this is a real asset, the importance of the process becomes more evident when the user is able to combine (merge) the newly created data set with the existing data set to extend the depth of analysis. The process of merging data sets may also be accomplished for data sets created through the more traditional methods described in the previous sections.

Whether the data sets to be merged were created by traditional input methods or through output *SYSTAT* files, the process is exactly the same. In order to merge a data set the user must be in the DATA module. Let's assume we have the two data sets below, identified as A and B.

SETA1			SETB1		
AF1	21	29	AF1	29	30
AM2	24	29	AM2	30	32
AM3	24	29	AM3	29	34
AM4	33	34	AM4	35	38
AF5	23	28	AF5	28	32
AF6	25	29	AF6	26	32
AF7	22	29	AF7	23	31
AF8	22	27	AF8	25	29
AM9	30	32	AM9	32	36
AM10	21	29	AM10	28	32

Each of these data sets represent parallel information for the same subjects. The first data set contains pretest results for a group of subjects using the variables ID$, PROBLEM1, and ATTITUD1. The second data set contains the posttest results for the same subjects with the variables ID$, PROBLEM2, and ATTITUD2. The task for the user is to create one data set containing all the unique information, but only a single ID$ variable. Before combining the two data sets the user must ensure that the critical, or merge, variable (ID$) is sorted exactly the same for both data sets. One method for doing this is to use the sort capability of DATA. The following lines will accomplish this goal:

```
USE SETA
SAVE SETA1
SORT ID$
RUN
```

The result is a new sorted data set called SETA1. The SORT command within DATA is usually quite adequate, but *SYSTAT* also has a special module called SSORT for DOS and MacIntosh systems, which not only performs high speed and/or multiple sorts but also permits the organization of the data in ascending or descending order.

Once the data sets have been sorted correctly, the following lines in DATA will produce a merged data set:

```
USE SETA1 SETB1 / ID$
SAVE BIGSET
RUN
```

The resulting data set, BIGSET, contains the variables ID$, PROBLEM1, ATTITUD1, PROBLEM2, and ATTITUD2. It is ready for further analysis.

AF1	29.000	30.000	21.000	29.000
AM2	30.000	32.000	24.000	29.000
AM3	29.000	34.000	24.000	29.000
AM4	35.000	38.000	33.000	34.000
AF5	28.000	32.000	23.000	28.000
AF6	26.000	32.000	25.000	29.000
AF7	23.000	31.000	22.000	29.000
AF8	25.000	29.000	22.000	27.000
AM9	32.000	36.000	30.000	32.000
AM10	28.000	32.000	21.000	29.000

Descriptive Statistics (A4-1 to A4-13)

The descriptive statistics explained earlier in this book may be obtained, for the most part, with the use of the STATISTICS, TABLES, or GRAPH modules. *SYSTAT* utilizes the STATISTICS module as the primary source of descriptive statistics. From this module the user may obtain the following: number of cases, minimum, maximum, range (**A4-5**), mean (**A4-2**), variance (**A4-7**), standard deviation (**A4-8**), standard error of measure, skewness (**A4-12**), kurtosis (**A4-13**), sum, and coefficient of variation.

The general procedure for obtaining the descriptive statistics is to enter the STATISTICS module from the main menu. Once in the module, enter the name of the data set to be utilized with the USE command and type the STATISTICS command. The following is an example using the DASET1 data set created earlier:

```
USE DASET1
STATISTICS [PRETEST POSTTEST] / [ALL]
```

The names of the variables to be studied follow the STATISTICS command and the option ALL following the forward slash (/) directs *SYSTAT* to produce all of the available statistics. The option could have been any of the above statistics in any combination. The results of the above command set would produce the following output with the DASET1 data set:

	PRETEST	POSTTEST
N OF CASES	10	10
MINIMUM	11.000	21.000
MAXIMUM	25.000	36.000
RANGE	14.000	15.000
MEAN	18.000	28.500
VARIANCE	15.556	26.278
STANDARD DEV	3.944	5.126
STD. ERROR	1.247	1.621
SKEWNESS (G1)	−0.172	0.133
KURTOSIS(G2)	−0.180	−1.295
SUM	180.000	285.000
C.V.	0.219	0.180

If the user simply types the key word STATISTICS and identifies the variables without specifying the options, *SYSTAT* will produce values for number of cases, minimum, maximum, range, and standard deviation.

It is unfortunate, but *SYSTAT* is unable to produce directly a statistic for MODE (**A4-4**). The easiest method to obtain a modal estimate would be to use the BAR command within GRAPH to obtain a horizontal bar graph that will indicate a frequency distribution for the data set. The example below is once again from the DASET1 data set on the PRETEST variable:

```
USE DASET1
BAR PRETEST
```

BAR GRAPH OF VARIABLE PRETEST, N = 10

VALUE	COUNT	PERCENT
11.000	1	10.00
13.000	1	10.00
17.000	2	20.00
18.000	1	10.00
19.000	2	20.00
20.000	1	10.00
21.000	1	10.00
25.000	1	10.00

To obtain a value for median (**A4-3**), employ the GRAPH module and the key word STEMLEAF. While this will produce a stem-and-leaf plot, part of the output will be a median value for the data set as well. The output below illustrates this process.

```
USE DASET1
STEMLEAF PRETEST
```

```
STEM AND LEAF PLOT OF VARIABLE:   PRETEST, N = 10
MINIMUM IS:              11.000
LOWER HINGE IS:            17.000
MEDIAN IS:               18.500
UPPER HINGE IS:            20.000
MAXIMUM IS:              25.000
                     1    1
                ***OUTSIDE VALUES***
                     1    3
                     1
                     1  H  77
                     1  M  899
                     2  H  01
                ***OUTSIDE VALUES***
                     2    5
```

Additional commands within the GRAPH module may also be used to obtain more complete visual pictures of the shape of distributions. For the DASET 1 data set, the following commands will produce the histogram shown at the top of page 237:

```
USE DASET1
HISTOGRAM PRETEST / [BARS=3]
```

The final strategy for obtaining descriptive statistics is the creation of frequency distributions through the TABLES module. The TABLES module produces a variety of frequency distributions as well as nonparametric statistical tests and categorical correlations (more on these later). In general, the command:

```
TABULATE [variables] [/] [PERCENT, ROWPCT, COLPCT, SORT]
```

will produce a full cross-tabulation for the variables. The options following the forward slash (/) may be used to provide percentage breakdowns of the cells. The SORT option should always be used with TABULATE since *SYSTAT* does not automatically sort data within the TABLES module, and the SORT option will produce output sorted either alphabetically or numerically. A sample of output using the DASET1 data set follows at the top of page 238.

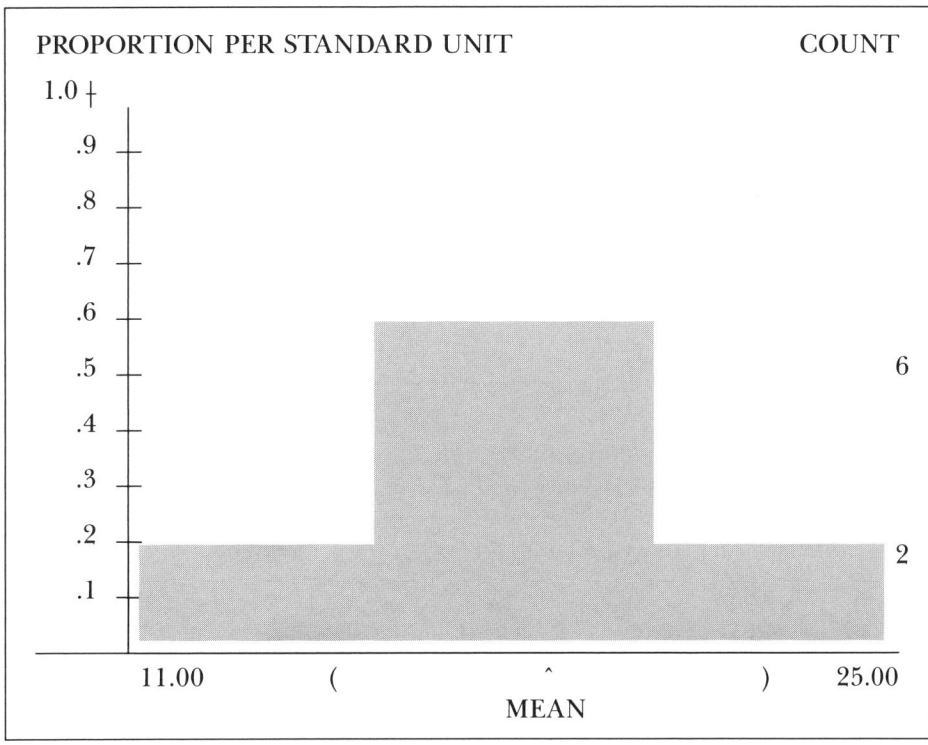

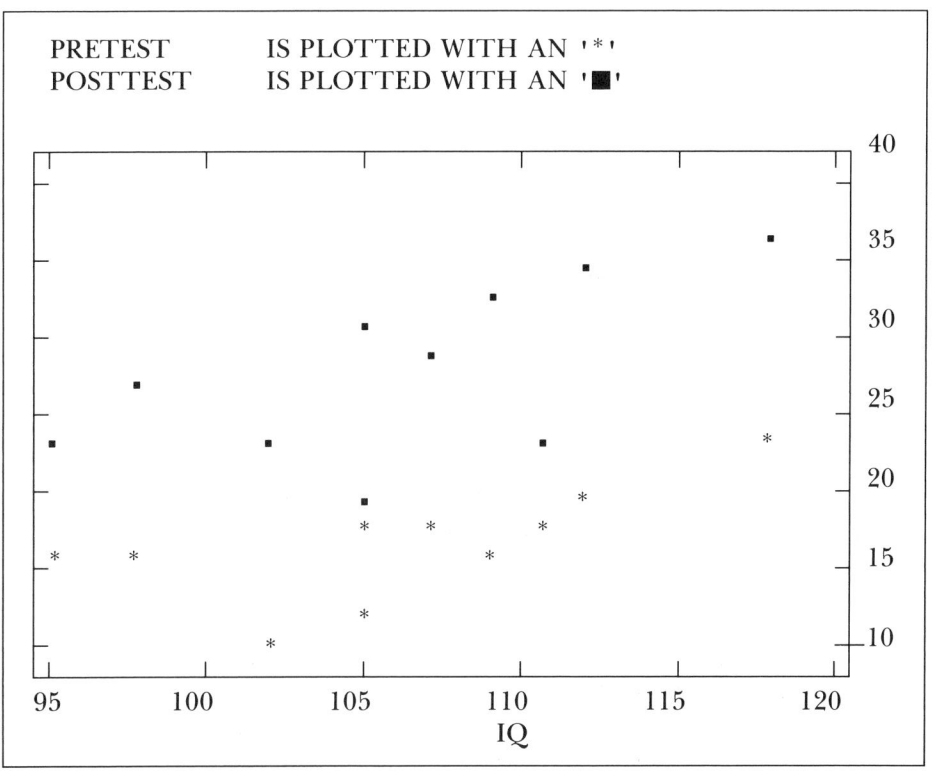

TABLE OF	TRSP	(ROWS) BY	GROUP	(COLUMNS)
PERCENTS OF TOTAL OF THIS (SUB)TABLE				
	1.000	2.000	TOTAL	N
1.000	20.00	40.00	60.00	6.00
2.000	30.00	10.00	40.00	4.00
TOTAL	50.00	50.00	100.00	
N	5	5	10	

Further descriptive information concerning variables can be obtained through the GRAPH module and the PLOT command to obtain a picture of a bivariate, or joint, distribution. The following commands will produce a plot of DASET1's PRETEST and POSTTEST data:

```
GRAPH
USE DASET1
PLOT PRETEST*POSTTEST / SYMBOL='*'
```

The SYMBOL='*' option is used to illustrate *SYSTAT*'s ability to provide a series of symbols that represent the data points. Additional analyses combining overlays of variables would be possible utilizing this option. For example, suppose we are interested in plotting Pretest and Posttest against IQ. The command would be:

```
GRAPH
USE DASET1
PLOT PRETEST, POSTTEST* IQ / SYMBOL='*'
```

The result of the above set of commands is a scatterplot of the PRETEST and POSTTEST variables as illustrated on the bottom of page 237.

Measures of Relationships (A4-14 to A4-30)

The various measures of association may be obtained through a variety of SYSTAT modules. The most common measures of association for continuous or ranked variables, such as Pearson product-moment correlation (**A4-16**), Spearman's Rho (**A4-17**), and Kendall's Tau (**A4-18**) may be obtained directly with the CORR module. Moreover, several of the specialized correlations, such as covariance (**A4-15**) and point biserial (**A4-19**) may be obtained with the same module. The user who wishes to calculate measures of association for categorical data should employ a special feature of the TABLES module. Correlations that may be obtained with this module include Phi coefficient (**A4-21**), contingency coefficient (**A4-23**), Kendall's Tau (**A4-18**), Goodman-Kruskal Gamma, and Cramer V. To obtain the part (**A4-26**), partial (**A4-27**), and multiple R (**A4-28**) correlations, the MGLH module and the CORR module must be used with a little finessing of the data. These will be described in the next section of this chapter on regression analysis.

Pearson Product-Moment Correlation (A4-16)

The general process for obtaining a correlation coefficient for continuous data is to employ the CORR module and to specify the type of correlation with a command statement. For example, the following will produce a correlation matrix for the DASET2 data set:

```
CORR
USE DASET2
PEARSON / PROB
```

The result is a set of correlation matrices, with the top set containing the correlation coefficients for each of the variables, and the second set (activated by using the / PROB option) containing the probability estimates (based upon a t-test of significance).

Below is an example of the output of this command set:

PEARSON CORRELATION MATRIX

	AGEMONTH	CSYN	MSYN	CVOCAB	MVOCAB
AGEMONTH	1.000				
CSYN	0.631	1.000			
MSYN	0.541	0.487	1.000		
CVOCAB	0.782	0.657	0.600	1.000	
MVOCAB	0.752	0.530	0.622	0.812	1.000

BARTLETT CHI-SQUARE STATISTIC: 64.933 DF= 10 PROB= .000

MATRIX OF PROBABILITIES

	AGEMONTH	CSYN	MSYN	CVOCAB	MVOCAB
AGEMONTH	0.000				
CSYN	0.001	0.000			
MSYN	0.008	0.019	0.000		
CVOCAB	0.000	0.001	0.002	0.000	
MVOCAB	0.000	0.009	0.002	0.000	.000

NUMBER OF OBSERVATIONS: 23

Spearman's Rho — Rank Correlation (A4-17)

Spearman's Rho is a special form of the Pearson product-moment correlation and is used when the data are ordinal and in ranked order. The limitations of the *SYSTAT* CORR module for the SPEARMAN command is that the data must first be ranked and then sorted. This first task is accomplished via the DATA module. The commands below are employed by applying the DASET2 data set to the variables of CVOCAB and MVOCAB. The reader should understand that these are interval data treated as ordinal for illustrative purposes. The following commands would create the necessary data set for our Spearman analysis:

```
USE DASET2
SAVE DASET2B
LET RANK1 = CVOCAB
LET RANK2 = MVOCAB
RUN
SORT RANK1, RANK2
RUN
[CASELIST]
[RUN]
```

The last two optional commands are used to get a listing of the results of our manipulation. This listing is illustrated below for the first six cases:

		RANK 1	RANK 2
CASE	1	1.000	1.000
CASE	2	2.000	2.000
CASE	3	3.000	3.000
CASE	4	4.000	12.000
CASE	5	5.000	7.000
CASE	6	6.000	4.500

The command for the correlation, after switching to the CORR module, is straightforward:

```
USE DASET2B
SPEARMAN RANK1 RANK2
```

The output for this is a simple correlation matrix; however, unlike the PEARSON command, SPEARMAN does not have a / PROB option to print out the resultant probability tests; therefore, a table of correlations and probabilities would have to be used.

OTHER CONTINUOUS OR RANKED COEFFICIENTS

Kendall's Tau coefficient (**A4-18**) may also be obtained by using the CORR module. The command procedure is very similar to the SPEARMAN command.

```
USE DASET2B
TAU RANK1 RANK2
```

As with the SPEARMAN command, the / PROB option has no effect, and the results would have to be checked for significance in an appropriate table.

The point biserial correlation (**A4-19**) is considered a special form of the Pearson product-moment correlation where one variable is continuous and the other is a true dichotomous variable. Such a correlation is often employed in item analyses for tests where the correlation describes the impact of each item on the overall test score. The example below is drawn from a test where answers were scored as zero (0) for wrong and one (1) for right. Using the CORR module, the intercorrelation matrix for the first five items (out of a total of 50 items) and the total test score was calculated. The correlations for the TOTAL variable (indicated in italics) are point biserials:

```
PEARSON CORRELATION MATRIX

              Y(1)         Y(2)         Y(3)         Y(4)         Y(5)

Y(1)          1.000
Y(2)         -0.070        1.000
Y(3)          0.234       -0.000        1.000
Y(4)          0.037        0.322        0.395        1.000
Y(5)          0.281       -0.289        0.277        0.062        1.000
TOTAL         0.127        0.415        0.600        0.688        0.425

   MATRIX OF PROBABILITIES

              Y(1)         Y(2)         Y(3)         Y(4)         Y(5)

Y(1)          0.000
Y(2)          0.705        0.000
Y(3)          0.198        1.000        0.000
Y(4)          0.840        0.073        0.025        0.000
Y(5)          0.119        0.109        0.124        0.736        0.000
TOTAL         0.489        0.018        0.000        0.000        0.015

   NUMBER OF OPERATIONS:    32
```

In *SAS*, the PROC CORR can employ the WITH option to print only a single correlation figure. *SYSTAT* does not possess this feature; however, this is a minor inconvenience since the result is exactly the same. The command lines used to produce the previous output were:

```
GET TEST1
PEARSON Y(1-5), TOTAL / PROB
```

As noted earlier, the biserial correlation (**A4-20**) is another special form of the Pearson product-moment correlation with the understanding that while one variable is continuous and the other is dichotomous, the dichotomous variable is assumed to have an underlying normal distribution. This is a statistic that is also employed in test-item analysis. The difference between the point biserial and the biserial is the theoretical model of the shape of the distribution of the dichotomous variable: that is, the former is not assumed to be normal, whereas the latter is assumed to have a normal distribution.

Correlations for Categorical Data

Measures of association for categorical data may be obtained with the TABLES module. The available measures include the Phi-coefficient (**A4-21**) and the contingency coefficient (**A4-23**). While the processes to obtain either of these estimations are exactly the same, the TABULATE command *must* be preceded by a special *SYSTAT* command: PRINT = LONG. When the TABULATE command is preceded by this command, *SYSTAT* prints a full table of statistical tests associated with contingency tables, including nonparametric tests (more on these tests later) and categorical measures of association. The following commands would produce a set of measures for the cross tabulation of TRSP and GROUP from the DASET2 data set:

```
TABLES
GET DASET2
PRINT = LONG
TABULATE GROUP*TRSP / SORT
```

The result of these commands would look like this:

```
TABLE OF          GROUP     (ROWS)     BY     TRSP     (COLUMNS)
FREQUENCIES
                  1.000     2.000             TOTAL
        1.000       2         3               5
        2.000       4         1               5
TOTAL               6         4              10


TEST STATISTIC                             VALUE       DF        PROB
PEARSON CHI-SQUARE                         1.667        1        0.197
LIKELIHOOD RATIO CHI-SQUARE                1.726        1        0.189
MCNEMAR SYMMETRY CHI-SQUARE                0.143        1        0.705
YATES CORRECTED CHI-SQUARE                 0.417        1        0.519
FISHER EXACT TEST (TWO-TAIL)                                     0.524
COEFFICIENT                                VALUE     ASYMPTOTIC STD ERROR
    PHI                                   -0.408
    CONTINGENCY                            0.378
    GOODMAN-KRUSKAL GAMMA                 -0.714         0.353
    KENDALL TAU-B                         -0.408         0.285
    STUART TAU-C                          -0.400         0.283
    YULE Q                                -0.714         0.353
    YULE Y                                -0.420         0.297
    COHEN KAPPA                           -0.400         0.284
    SPEARMAN RHO                          -0.408         0.285
    SOMERS D       (COLUMN DEPENDENT)     -0.400         0.283
    LAMBDA         (COLUMN DEPENDENT)      0.250         0.484
    UNCERTAINTY    (COLUMN DEPENDENT)      0.128         0.185
```

244   SOFTWARE APPLICATIONS IN STATISTICS

As can be readily seen, the output from the TABULATE command is very complete. The user is advised to select the appropriate measure of association according to the nature of the categorical variables and the statistical model employed (Section A, Chapter 2 in this book describes these decision elements).

Regression Analysis (A4-31 to A4-39)

SYSTAT performs the majority of its regression analyses through the MGLH module. The MGLH module employs a general linear model to produce statistical analyses, including regression analysis, analysis of variance, and multiple analysis of variance. The general format is to enter the module by typing MGLH at the menu screen, select the data set with the USE command, and follow this with the appropriate set of commands to produce the desired output.

To produce a simple linear regression **(A4-32)** using the DASET2 data set, the format would be:

```
MGLH
USE DASET2
MODEL CSYN = CONSTANT + AGEMONTH
SAVE ESTDS3 / MODEL
ESTIMATE
```

The above command lines each tell *SYSTAT* to perform a particular operation. The first critical command, after establishing the data set to be used, is the MODEL command. This command is based upon the mathematical model that *SYSTAT* will employ as it analyzes the data. The command identifies the dependent variable and the independent variable as well as telling *SYSTAT* to include a constant within the model. This expression parallels the traditional statistical model ($y = \alpha + \beta x_1$).

The command SAVE ESTDS3 / MODEL is used to create an output data set (in *SYSTAT* format) containing the following elements: the values of the model variables (CSYN and AGEMONTH), the predicted values, residual values, leverage values for each observation, studentized residuals, Cook's D values, and the standard error of predicted values (the latter three are created only when there is one dependent variable in the MODEL statement). The SAVE command is essential to obtain predicted values and the use of the / MODEL option creates a data set with all the essential elements for later analyses and graphing of the model.

The final command, ESTIMATE, is the command that tells *SYSTAT* to perform the analysis. It *must* be the last command because *SYSTAT* immediately begins to process after the <R> key is pressed following this command.

The output below is the output obtained from this series of commands (the actual spacing has been slightly modified to fit the margins of this book):

DEP VAR: CSYN   N: 23   MULTIPLE R: .631
SQUARED MULTIPLE R: .398
ADJUSTED SQUARED MULTIPLE R: .369
STANDARD ERROR OF ESTIMATE: 1.357

VARIABLE	COEFFIC	STD ERR	STD COEF	TOL	T	P (2 TAIL)
CONSTANT	4.121	1.025	0.000	.	4.022	0.001
AGEMONTH	0.065	0.017	0.631	1.0	3.723	0.001

ANALYSIS OF VARIANCE

SOURCE	SUM-OF-SQUARES	DF	MEAN-SQUARE	F-RATIO	P
REGRESSION	26.206	1	26.206	13.860	0.001
RESIDUAL	39.707	21	1.891		

DURBIN-WATSON D STATISTIC   2.646
FIRST ORDER AUTOCORRELATION   -.348

RESIDUALS HAVE BEEN SAVED

The estimate of the regression coefficient (**A4-33**) and the Y-intercept (**A4-34**) are listed under the column COEFFICIENT in the first table in the output. The regression coefficient is .065 and the Y-intercept is 4.121, consequently the regression equation is: CSYN = 4.121 + AGEMONTM (.065). The standard error of estimate (**A4-35**), with a value of 1.375, is directly stated in the above printout and, once again, is interpreted as being the square root of the mean square due to error. The R-squared (**A4-37**) value of .398 is listed beside the heading SQUARED MULTIPLE R.

The output data file, called ESTDS3.SYS, contains the predicted values, residuals, studentized residuals, and so forth. The following is a sample listing of the first five cases of this new data set:

ESTIMATE	RESIDUAL	LEVERAGE	COOK	STUDENT	SEPRED	CSYN	AGEMONTH
8.070	0.930	0.047	0.012	0.684	0.297	9.000	61.000
7.228	−0.228	0.055	0.001	−0.167	0.323	7.000	48.000
8.782	1.218	0.082	0.038	0.921	0.393	10.000	72.000
7.293	−2.293	0.053	0.082	−1.803	0.315	5.000	49.000
6.646	1.354	0.093	0.055	1.036	0.419	8.000	39.000

The values listed under the ESTIMATE column reflect the predicted values, the LEVERAGE values reflect the influence of each observation on the size of the mean square error,* the COOK column provides the Cook's D value for each observation, and the column labeled SEPRED provides the standard error of prediction for each observation.

To obtain the part (**A4-26**) and partial (**A4-27**) correlations, another data set needs to be created. In our example we will use, once again, the DASET2 data set, with our interest centered on the part and partial correlations among CSYN, CVOCAB, and AGEMONTH. The easiest way to obtain these correlations is to use the MGLH module and the CORR module. The first step is to employ the MGLH module:

```
MGLH
GET DASET2
MODEL CSYN, CVOCAB = CONSTANT + AGEMONTH
SAVE ESTDS3 / MODEL
ESTIMATE
```

While the output of this command set will produce the partial correlation for our model, it is the output data set, ESTDS3.SYS, that is of primary interest since it contains the residuals for the CSYN = CONSTANT + AGEMONTH model, labeled RESIDUAL(1), and the residuals for the CVOCAB = CONSTANT + AGEMONTH model, labeled RESIDUAL(2), as well as the old values for CSYN, CVOCAB, and AGEMONTH.

---

*L. Wilkinson. *SYSTAT: The System for Statistics*. Evanston, IL: SYSTAT, Inc., 1988. p. 469.

We now switch to the CORR module and issue the following commands:

```
USE ESTDS3
PEARSON RESIDUAL(1), RESIDUAL(2), CSYN, CVOCAB,
AGEMONTH / PROB
```

The table below is the output of this series of commands and contains the correlation coefficients:

PEARSON CORRELATION MATRIX

	RESIDUAL(1)	RESIDUAL(2)	CSYN	CVOCAB	AGEMONTH
RESIDUAL(1)	1.000				
RESIDUAL(2)	0.339	1.000			
CSYN	0.776	0.263	1.000		
CVOCAB	0.212	0.624	0.657	1.000	
AGEMONTH	0.000	0.000	0.631	0.782	1.000

ERROR
CORRELATION MATRIX IS NOT POSITIVE DEFINITE.
INDIVIDUAL SIGNIFICANCE TESTS ARE SUSPECT.

MATRIX OF PROBABILITIES

	RESIDUAL(1)	RESIDUAL(2)	CSYN	CVOCAB	AGEMONTH
RESIDUAL(1)	0.000				
RESIDUAL(2)	0.113	0.000			
CSYN	0.000	0.225	0.000		
CVOCAB	0.333	0.001	0.001	0.000	
AGEMONTH	1.000	1.000	0.001	0.000	0.000

NUMBER OF OBSERVATIONS: 23

The correlation between CSYN and CVOCAB is .657, p = .001. The part correlation (**A4-26**) for CVOCAB and CSYN, which is the correlation between these two variables with the influence of AGEMONTH removed from the CSYN scores, is .212. This is the correlation in the table between RESIDUAL(1) and CVOCAB. The part correlation for CVOCAB and CSYN, with the influence of AGEMONTH removed from the CVOCAB scores, is .263. This is the correlation of RESIDUAL(2) and CSYN.

The partial correlation (**A4-27**) is the correlation between CSYN and CVOCAB with the influence of AGEMONTH removed from both variables. This correlation is the correlation between RESIDUAL(1) and RESIDUAL(2) and it is .339. The user is once again cautioned that the probability tests are incorrect because a degree of freedom was used in each analysis, which *SYSTAT* does not account for in its calculations. *SYSTAT* provides a warning to the user with the ERROR statement above. Actual significance levels should be verified through the appropriate table of significance values for correlations.

Multiple regression (**A4-36**) and its special case of curvilinear regression (**A4-38**) may be obtained with an extension of the MGLH procedure that was used to obtain the simple linear regression model. To obtain a multiple regression model of CSYN with the predictor variables of AGEMONTH, CVOCAB, MSYN, and MVOCAB, the user should use the following set of commands:

```
MGLH
USE DASET2
MODEL CSYN = CONSTANT + AGEMONTH + CVOCAB + MSYN
   + MVOCAB
SAVE ESTDS4 / MODEL
ESTIMATE
```

Once again, we are saving our output in a new data set (ESTDS4.SYS) for additional analysis later.

The output from this command set parallels the simple linear regression output with the addition of three new predictor variables in the printout. The multiple correlation (**A4-28**) among the variables is .696, and the squared multiple correlation (**A4-37**) is .484. The standard error of estimate is 1.375. The regression coefficients (**A4-33**) for each of the variables are: AGEMONTH .034, CVOCAB .050, MSYN .117, and MVOCAB −.020. The Y-intercept is the CONSTANT value, 3.548.

Suppose we have reason to believe that the relationship between AGEMONTH and CSYN is curvilinear. For example, we have created a scatterplot using the GRAPH module and the plot command PLOT CSYN*AGEMONTH. The plot indicates a nonlinear model. The next step would be to move to a quadratic model such as:

```
MODEL CSYN = AGEMONTH+AGEMONTH*AGEMONTH
```

The user should enter the DATA module and create a new variable in the data set that represents the value of AGEMONTH squared. The following commands would accomplish this:

```
DATA
GET DASET2
SAVE DASET2SQ
LET AGEMOSQ = AGEMONTH*AGEMONTH
```

This would produce the necessary variable for our new model. Then, switching to MGLH, issue the following commands:

```
USE DASET2SQ
MODEL CSYN = CONSTANT + AGEMONTH + AGEMOSQ
SAVE ESTDSSQ / MODEL
ESTIMATE
```

The output would represent two independent variables, labeled AGEMONTH and its quadratic form, AGEMOSQ.

The use of stepwise multiple regression (**A4-39**) programs in statistical packages has become quite common. While *SYSTAT* does support this particular procedure, it does so with certain limitations. The STEP command within the MGLH module employs a forward selection technique wherein it calculates the *best fit* model for the independent predictor variables. It does not, however, produce any statistical output regarding the model it selects. The user is advised to enter a new model statement to establish the parameters of the model. The commands to perform this operation are:

```
MGLH
USE COOK1
MODEL GRANDX = CONSTANT + POP + COST + LOINCOM
STEP
```

SYSTAT will then produce an analysis indicating which of the predictor(s) variables *best fits* the entry and exit characteristics (default values of $p = .15$). Using the COOK1 data set that was created earlier in this chapter, we would find that the model would include the COST and LOINCOM variables and would exclude the POP variable.

Using Command Programs in SYSTAT

Perhaps one of the primary limitations for acceptance of microcomputer statistical packages has been their dependence upon menu-based processing schemes. In other words, the user must operate within the framework of someone else's idea of an interface. While *SYSTAT* utilizes a menu-like framework for entry to the various modules, it does not prevent the experienced researcher from using a command-based analysis strategy. Even so, many experienced programmers developed their skills in a mainframe environment and prefer the idea of personally programming an analysis.

SYSTAT supports this so-called batch process in two ways. First, the user may construct what are called "command files" and submit the command files for execution within the *SYSTAT* system. A command file is an ASCII file created with either *SYSTAT*'s own ASCII editor, called FEDIT, or any other editor. The command file may have any file name the user desires; however, it must have as its extension name .CMD. The command file is then submitted at the main *SYSTAT* menu by typing: SUBMIT [filename].

The following is an example of a *SYSTAT* command file designed to (1) perform a stepwise selection of variables from the COOK1 data set, (2) produce a complete full-scale multiple regression analysis of the best-fit model, and (3) print a plot of the studentized residuals and the predicted values using the high-resolution graphics module SYGRAPH.

```
MGLH
PRINT = LONG
USE COOK1
MODEL GRANDX = CONSTANT+POP+COST+LOINCOM
STEP
MODEL GRANDX = CONSTANT+COST+LOINCOM
SAVE GRANDX1 / MODEL
ESTIMATE
[this line left blank on purpose]
HYPOTHESIS
ALL
TEST
[this line left blank on purpose]
SWITCHTO SYGRAPH
OUTPUT PRINTER
USE GRANDX1
PLOT STUDENT*ESTIMATE/SPIKE=0,XMAX=350,XMIN=150,
SYM=3,FIL=1,
TITLE 'PLOT OF STUDENT RESIDUALS BY PREDICTED
VALUES'
QUIT
```

The second technique for programming *SYSTAT* operations is to utilize the newest module of the *SYSTAT* system, MACRO. MACRO permits the user to write some rather extensive programming features and to employ these features in a variety of ways. The development of these extensive features is beyond the scope of this book, but the user is advised to explore this module as a vital part of *SYSTAT* operations.

Comparisons: One Sample (A4-40 to A4-47)

Hypotheses involving the comparison of a mean (**A4-41**), variance (**A4-42**), proportion (**A4-43**), or correlation (**A4-44**, **A4-45**) with some value, or the comparison of a set of frequencies with some expected frequency values (**A4-46**, **A4-47**), may be accomplished through the use of *SYSTAT*'s built-in functions. *SYSTAT* offers functions for eight different distributions: uniform, normal, t, F, Chi-square, exponential, Gamma, and Beta. Each of these distributions is offered as a cumulative function, an inverse function, and as a random-variate function.

Although many of these functions are employed within the various modules in the *SYSTAT* system, the user may choose to employ them individually to calculate specific tests for one-sample cases. The user is cautioned, once again, that to employ *SYSTAT* as a glorified hand calculator is somewhat akin to using an elephant gun to go rabbit hunting. You'll get the same results, but the process of getting there may be more work than the effort was worth. Because *SYSTAT* can readily be used to calculate the sample statistic, it is recommended that the actual tests of significance of this type be performed using a hand calculator on all data sets except extremely large ones.

If the choice of action is to use *SYSTAT* to calculate the test statistics, the strategy is very similar to the one described in the *SAS* chapter. The best strategy is to write a command file containing all the numerous commands and operations. The primary reason for this strategy is *SYSTAT*'s inability to store more than one output variable from its STATISTICS module. The process requires the creation of separate data sets containing single values for each of the critical statistics, merging these data sets, and, finally, entering the DATA module to create the test statistics and then listing the results. Such a command file would resemble the following:

```
STATISTICS
GET DASET2
SAVE NEWDS2A
STATISTICS AGEMONTH / N
NOTE 'CREATES OUTPUT DATA SET WITH N'
SAVE NEWDS2B
STATISTICS AGEMONTH / MEAN
NOTE 'CREATES OUTPUT DATA SET WITH MEAN'
SAVE NEWDS2C
STATISTICS AGEMONTH / SD
NOTE 'CREATES OUTPUT DATA SET WITH STANDARD
DEVIATION'
SAVE NEWDS2D
STATISTICS AGEMONTH / SEM
NOTE 'CREATES OUTPUT DATA SET WITH STANDARD
ERROR'

SWITCHTO DATA
USE NEWDS2A
LET AGEMONTH = NAGE
SAVE NEWDS2A
Y

USE NEWDS2B
LET AGEMONTH = MAGE
SAVE NEWDS2B
Y

USE NEWDS2D
LET AGEMONTH = SDAGE
SAVE NEWDS2D
Y

SAVE NEWDS2E
LET AGEMONTH = SEAGE
SAVE NEWDS2E
Y
```

(continued on next page)

```
USE NEWDS2A NEW2B
SAVE BIGDS1
RUN

USE NEWDS2C NEWDS2D
SAVE BIGDS2
RUN

USE NEWDS2E BIGDS2
SAVE BIGDS3
RUN

USE BIGDS1 BIGDS3
SAVE GRANDDS
RUN

USE GRANDDS
LET T = (MAGE-A)/SEAGE
LET SQ= ((N-1)*VARAGE)/A
SAVE GRANDDS
Y

CASELIST T SQ
RUN
```

As noted earlier, the task would be simpler if the researcher solely used the STATISTICS module to perform the descriptive statistics and completed the calculations by hand. Another strategy would be to use a Terminate-and-Stay-Resident (TSR) calculator that would permit the user to "pop up" the calculator, perform the calculations, and return to the *SYSTAT* module.

Another strategy, assuming that the user preferred not to go through the entire programming strategy outlined above, would be to create a new data set containing the descriptive information. Such a procedure would use STATISTICS to perform the descriptive analysis, move to the DATA module, input the variables of interest from the keyboard, and save the file. Once the file is saved the user may add the necessary command operations to create the tests.

A similar strategy, as described earlier for t-tests and Chi-square tests, *could* be employed if the user wanted to utilize the *SYSTAT* DATA module to calculate z-statistics (**A4-41**) or to compare two correlations (**A4-44, A4-45**) through the use of a Fisher's Z-transformation. The command lines would be entered into the DATA module, assuming the appropriate data set had previously been created through a command file or keyboard input. The command line in the DATA module for the Z-test would be:

```
Z = (PS - PP) / (SQR((PS*(1-PS))/N))
```

The command lines for a Fisher's Z-transformation test would be:

```
ZRSCORR = .5*(LOG((1+SCORR)/(1−SCORR)))
ZRPCORR = .5*(LOG((1+PCORR)/(1−PCORR)))
ZR = (ZRSCORR−ZRPCORR)/SQR(1/(N−3))
```

It should be noted yet again that the process of creating such a file with *SYSTAT* is more work than necessary. A good scientific hand calculator would perform the same set of tasks based upon the output of the CORR module.

Comparisons: Two Samples (Parametric) *(A4-48 to A4-56)*

As noted in the *SAS* chapter, comparisons involving two samples may be calculated either by using *SYSTAT* to produce the statistics of interest and then performing the tests by hand or by using *SYSTAT* to perform the actual tests. Which process to select is ultimately up to the user, but since *SYSTAT* is designed to perform most of the appropriate two-sample tests the easiest route would be to use it.

The t-test has two forms, a dependent t-test (**A4-49**) and an independent t-test (**A4-50**). The procedures are essentially the same for both within the *SYSTAT* system. The process is carried out in the STATISTICS module using the TTEST command. The critical difference for *SYSTAT* is that the user must identify two continuous variables for the independent t-test (**A4-50**) and a grouping variable for the dependent t-test (**A4-49**). The command lines for the dependent t-test would be:

```
STATS
USE DASET1
TTEST PRETEST,POSTTEST
```

The output for this would be:

```
PAIRED SAMPLES T-TEST ON   PRETEST VS POSTTEST     WITH    10 CASES

MEAN DIFFERENCE =         −10.500
SD DIFFERENCE =             3.375
T =       −9.839      DF = 9 PROB = .000
```

The commands for the independent t-test would be:

```
STATS
USE DASET1
TTEST POSTTEST*GROUP
```

The output would be:

```
INDEPENDENT SAMPLES T-TEST ON POSTTEST    GROUPED BY    GROUP
       GROUP          N          MEAN         SD
       1.000          5          32.000       4.359
       2.000          5          25.000       3.082
POOLED VARIANCES   T =2.932   DF =8   PROB = .019
```

Critical to these t-tests is the assumption of homogeneity of variance. Unlike *SAS*, which automatically provides t-test results for both unequal and equal variances, *SYSTAT* does not. It is up to the researcher to verify the variances through the use of formula **A4-51**, F-test for significant differences.

Another strategy to employ when testing an independent t-test is to use the PRINT=LONG command and a BY GROUP within the STATISTICS module. This will produce an independent t-test and the Bartlett test of homogeneity of variance as part of the output. Should the user find that the variances are different, then the use of Satterthwaite's correction formula* and an adjusted calculation for the standard error of difference** would be necessary to calculate the t-test. The process first requires that the data set has been sorted by the grouping variables. Assuming that the DATA module has been used to create a data set with sorted data, the following would produce the necessary statistics and test:

```
USE SORTDS1
BY GROUP
STATISTICS POSTTEST
```

A partial listing of the output would be the following:

```
SUMMARY STATISTICS FOR POSTTEST
BARTLETT TEST FOR HOMOGENEITY OF GROUP VARIANCES =      .471
APPROXIMATE F =              .418 DF =  1,     192 PROBABILITY = .519
OVERALL MEAN =       28.500 STANDARD DEVIATION =         5.216
POOLED WITHIN GROUPS STANDARD DEVIATION =       3.775
T STATISTIC =           2.932 PROBABILITY =   .019
```

Once again, the actual t-test is based upon the assumption that the variances are equal. Should the Bartlett test provide evidence to the contrary, then the Satterthwaite correction would have to be employed.

$$*df = \frac{(s_1^2/n_1 + s_2^2/n_2)}{\sqrt{[(s_1^2/n_1)^2 /(n_1-1)] + [(s_2^2/n_2)^2 /(n_2-1)]}}$$

$$**s_{x1-x2} = \sqrt{(s_1^2/n_1) + (s_2^2/n_2)}$$

Although the process described above does a test of homogeneity of variances for *independent* samples, no such test is carried out for the *dependent* samples within *SYSTAT*. Nevertheless, the test of two sample variances (**A4-52**) is readily calculated by hand using information derived from the STATISTICS and CORR modules.

The procedure for using *SYSTAT* to test for independent (**A4-53**) or dependent (**A4-54**) proportions is the same as described in the *SAS* chapter, with the exception that the TABLES module is used. The general format would be:

```
TABLES
USE DASET1
PRINT=LONG
TABULATE TRSP*GROUP /SORT
```

The formula for independent proportions (**A4-53**) may then be used, or the Chi-square statistic may be used if the sample size is not met. Although the same strategy may be employed for the dependent samples (**A4-54**) and then calculated by hand, using the above set of commands would be adequate since the procedure produces a McNemar statistic that is equivalent to **A4-54**.

The comparison of two correlations for independent groups (**A4-55**) can be accomplished by several steps within *SYSTAT*. The first step is to switch to the DATA module and create a data set sorted by the group or class variable. The second step is to switch to the CORR module and calculate a correlation coefficient for each of the two groups. The final step would be to either calculate **A4-55** by hand or use the procedure described earlier (in the section on one-sample comparisons) to create a data set within the DATA module using the descriptive data, and then use DATA commands to create the appropriate Fisher's Z-transformation values. Assuming the user intends to calculate the actual test (**A4-55**) by hand calculator, the *SYSTAT* steps would be:

```
DATA
USE DASET1
SAVE SORTDS1
SORT GROUP
SWITCHTO CORR
USE SORTDS1
BY GROUP
PEARSON POSTTEST
```

The information needed to perform the calculation for a t-test of two-sample dependent correlations (**A4-56**) can be directly obtained through the CORR module. The command set would be:

```
CORR
USE DASET1
PEARSON PRETEST, POSTTEST, IQ
```

The data must then be used in the formula found in **A4-56**. The user is reminded that the correct number of degrees of freedom is N-3, not N-2.

Significance Levels for z-, Chi-square, t-and F-statistics

SYSTAT employs tactics very similar to the strategies discussed in the *SAS* chapter to calculate significance levels for z-, Chi-square, t-, or F-distributions. To calculate the significance level associated with each of these distributions, the user should employ one of the following built-in *SYSTAT* functions within the DATA module:

LET PZ = 1 − ZCF(Z)
  This would calculate the proportion of the z-distribution above the given value of z. The two-tailed test would be:
  LET PZ = (1− ZCF(ABS(Z)))*2; where ABS is the absolute value of z.

LET PCHI = 1 − XCF(C,DFC)
  This would calculate the proportion of the Chi-square distribution that lies above the value C for DFC degrees of freedom.

LET PT = 1 − TCF(T,DFT)
  This would calculate the proportion of the t-distribution that lies above the value of t with DFT degrees of freedom. This is a one-tailed test. The two-tailed test would be:
  LET PT = (1 − TCF(ABS(T),DFT))*2; where ABS is the absolute value of t.

LET PF = 1 − FCF(F,DFN,DFD)
  This would calculate the proportion of the F-distribution that lies above the value of F with DFN degrees of freedom in the numerator and DFD degrees of freedom in the denominator.

Comparisons: Two Samples (Nonparametric) (A4-57 to A4-65)

SYSTAT's NPAR module offers two comparison tests for related samples, the sign test for dependent samples (**A4-64**) and the Wilcoxon matched-pairs (**A4-65**). For the independent two-sample tests, NPAR offers the Kolmogorov-Smirnov (**A4-62**) and the Mann-Whitney U-Test (**A4-60**), which is produced as part of the Kruskal-Wallis test (**A4-68**) when the grouping variable has only two categories. The process for implementing these tests is the same: identify the test, the independent variable, and, when appropriate, the grouping variable.

For example, in the DASET1 data set, suppose we wanted to test the pretest scores to see if the two scores for the two sample groups were drawn from the same population distribution. We would use the Kolmogorov-Smirnov test and its optional command identifying the distribution we are testing against. A special feature of this test is its ability to test against a known distribution. In our example, the distribution is defined as being a normal distribution. The following commands would accomplish this:

```
NPAR
USE DASET1
KS PRETEST*GROUP / NORMAL
```

On the other hand, if we wanted to test the question of whether the fall semester's grade equivalencies of third-graders on a standardized test are significantly different from the spring semester's grade equivalencies for the same students, this would be a dependent-sample test. The test most appropriate for this procedure would be the sign test (**A4-64**), which is equivalent to a dependent t-test but is used when the data is ordinal rather than interval or ratio. Assuming we have such a data set, the *SYSTAT* commands would be:

```
NPAR
SIGN FALL, SPRING
```

The results of this example would be in a z-statistic and would resemble the following:

SIGN TEST RESULTS

COUNTS OF DIFFERENCES (ROW VARIABLE GREATER THAN COLUMN)

	FALL	SPRING
FALL	0	71
SPRING	37	0

TWO-SIDED PROBABILITIES FOR EACH PAIR OF VARIABLES

	FALL	SPRING
FALL	1.000	
SPRING	0.001	1.000

The results of the above analysis indicate that there is a significant difference between our third-graders' grade equivalencies for the fall testing and the spring testing.*

The Wilcoxon Matched Pairs Signed Ranks Test (**A4-65**) utilizes a ranking strategy to find differences between matched pairs of dependent samples. *SYSTAT*

---

*The reader is cautioned against the tendency to draw an unwarranted inference from the results of this test. The test only indicates a significant difference exists between the two testings, not that one testing proved better or greater than the other.

here has an advantage over other statistical programs in that it will rank the data being tested rather than expecting it to be ranked by the user before analysis. The commands to perform a Wilcoxon Signed Ranks Test would be:

```
NPAR
USE DASET1
WILCOXON PRETEST,POSTTEST
```

A related test used quite often is the Chi-square test of independence, or test of homogeneity (**A4-61**). This test is printed with the TABLES module when the PRINT=LONG precedes the TABULATE command. For example, to test whether the assignment to a GROUP is independent of the TRSP (teacher rating of student potential) in the DASET1 data set, we would use the following set of commands:

```
TABLES
USE DASET1
PRINT=LONG
TABULATE GROUP*TRSP /SORT
```

The result of this set of commands is a two-by-two contingency table, with the appropriate statistics printed below the table.

Comparisons: Two or More Samples (Nonparametric) (A4-66 to A4-69)

SYSTAT offers the Kruskal-Wallis rank test for K-independent samples (**A4-68**) and the Friedman two-way analysis of variance by ranks for dependent samples (**A4-69**). The Kruskal-Wallis test may also be used with two-sample tests but, as noted earlier, it will print the results as the Mann-Whitney U-test (**A4-60**). Remember that the Kruskal-Wallis test is similar to an analysis of variance for ranked data, and that *SYSTAT* substitutes a Chi-square statistic instead of the H-statistic noted in **A4-68**.

As an example of the Kruskal-Wallis test, suppose we were interested in discovering if the variable of cost per pupil in our COOK1 data set was significantly different by county region. The following commands would obtain this information:

```
NPAR
USE COOK1
KRUSKAL COST*AREA$
```

Since *SYSTAT* automatically ranks data when performing a Friedman two-way analysis of variance, it may be implemented in much the same way as the previous

Kruskal-Wallis test. For example, the following commands would perform a Friedman test on the COOK1 data for GR3, GR6, and GR8:

```
NPAR
USE COOK1
FRIEDMAN GR3,GR6,GR8
```

The results would be the following*:

FRIEDMAN TWO-WAY ANALYSIS OF VARIANCE RESULTS FOR 111 CASES

VARIABLE	RANK SUM
GR3	252.000
GR6	214.500
GR8	199.500

FRIEDMAN TEST STATISTIC =   13.176
KENDALL COEFFICIENT OF CONCORDANCE =   0.059
PROBABILITY IS 0.001 ASSUMING CHI-SQUARE DISTRIBUTION WITH 2 DF

Comparisons: Two or More Samples—Analysis of Variance (ANOVA) (A4-70 to A4-90)

The utilization of analysis of variance (ANOVA) models in research paradigms is one of the most frequent techniques employed in quantitative research. The advent of computer-based statistical packages has had more to do with the reporting of results incorporating ANOVA models than has a true understanding of the models. When the researcher employs the ANOVA model, it is imperative that the model be employed with the correct elements and correct attributes. The variety of strategies available include factorial (**A4-71** to **A4-78**), repeated measures/split-plot (**A4-79** to **A4-83**) and nested or hierarchical (**A4-84** to **A4-86**). As in the chapter on *SAS*, only the most commonly occurring models will be illustrated below, with repetitious examples omitted.

FACTORIAL ANALYSIS OF VARIANCE (A4-71 TO A4-78)

The factorial ANOVA procedures will be illustrated with the same data set used in the *SAS* chapter, Data Set 3, p. 208 (stored as DASET3.SYS), which involves 18 subjects in a two-factor design. The factors are (A) treatment, with three levels (E1, E2, and C), and (B) gender, with two levels (F and M). The dependent variable is a

---

*The data in this example were originally interval; however they were converted to ordinal (ranked) data to illustrate the test.

test score. Before *SYSTAT* can properly execute our ANOVA commands, we must recode the treatment variables from E1, E2, and C to 1, 2, and 3, respectively, and gender must be recoded from F and M to 1 and 2, respectively. The transformation is required because *SYSTAT* employs a dummy coding strategy in its least-squares approximations that is not possible with character variables.

All ANOVA procedures are executed through the MGLH module within *SYSTAT*. The general form is the same, regardless of the finely tuned testing to be undertaken. For example, to run a one-way ANOVA (**A4-71**) with DASET3 the following commands would be issued:

```
MGLH
USE DASET3
ANOVA SCORE
CATEGORY GROUP=3
ESTIMATE
```

These command lines tell *SYSTAT*'s MGLH module that the program to be run is an ANOVA model with SCORE as its dependent variable. The categorical, or class, variable is GROUP, and it has three levels. While the ANOVA procedure in *SYSTAT* employs the *special* "ANOVA [dependent variable]" statement, this is merely a convenience to the user. The following command could just as easily have been used with the same results:

```
MODEL SCORE = CONSTANT + GROUP
```

Many users prefer to use the traditional model statement and should remember that it must precede the CATEGORY=line.

The output from *SYSTAT* is relatively straightforward:

DEP VAR: SCORE   N: 18   MULTIPLE R: .463   SQUARED MULTIPLE R: .215

ANALYSIS OF VARIANCE

SOURCE	SUM-OF-SQUARES	DF	MEAN-SQUARE	F-RATIO	P
GROUP	652.444	2	326.222	2.048	0.164
ERROR	2389.167	15	159.278		

This output identifies the independent variable, the number of subjects, the multiple R, the multiple-R squared, and the traditional analysis-of-variance table. When the ANOVA design is balanced, the sums-of-squares are Type I; when it is unbalanced, they are Type III. Additional information is printed when the user types a PRINT=LONG command before the ESTIMATE command.

Similar output is obtained for a two-way analysis-of-variance model (**A4-72** to **A4-74**), such as a treatment (or group) (3) by gender (2) in Data Set 3. The *SYSTAT* commands for this model would be:

```
MGLH
USE DASET3
ANOVA SCORE
CATEGORY GROUP=3, GENDER=2
ESTIMATE
```

The last three lines could be rewritten as:

```
MODEL SCORE = CONSTANT + GROUP + GENDER +
GROUP*GENDER
CATEGORY GROUP=3, GENDER=2
ESTIMATE
```

In either case the results, as listed below, would be the same:

DEP VAR: SCORE   N: 18   MULTIPLE R: .936   SQUARED MULTIPLE R: .876

ANALYSIS OF VARIANCE

SOURCE	SUM-OF-SQUARES	DF	MEAN-SQUARE	F-RATIO	P
GROUP	652.444	2	326.222	10.393	0.002
GENDER	1942.722	1	1942.722	61.892	0.000
GROUP*GENDER	69.778	2	34.889	1.112	0.361
ERROR	376.667	12	31.389		

The above is a summary table of the components of the ANOVA model and does not contain the overall model. Using a hand calculator, such a table may be readily constructed.

Should the user wish to obtain the cells means for this 3 × 2 table, the most effective way would be to use the STATISTICS module with a "by group, gender" command. Such a procedure would first require that the data set be sorted. The most effective way to produce the cell means would be the following:

```
SSORT
USE DASET3
SAVE DASET3A
SORT GROUP, GENDER
SWITCHTO STATS
USE DASET3A
STATISTICS SCORE / MEAN SD VARIANCE
```

Both of the preceding models represent fixed-effects models (**A4-72**); however, many studies employ random-effects (**A4-73**) or mixed-effects (**A4-74**) models. The analysis of these models is undertaken by using the HYPOTHESIS, EFFECT, ERROR, and TEST commands in conjunction with the general commands. Since *SYSTAT* is, in effect, an interactive model, the HYPOTHESIS, EFFECT, ERROR, and TEST commands are written after the model is first run with the ESTIMATE command. The HYPOTHESIS command has no visible effect on the model, but serves to alert MGLH that a set of commands related to the previous ESTIMATE command will follow. The EFFECT command identifies which of the effects (main or interaction) are to be tested, and the ERROR command identifies the terms to be used as the error term in the F-statistic. These commands are followed by the TEST command, which informs MGLH to begin the analysis.

Using the same procedure for mixed effects (**A4-74**) illustrated in the *SAS* chapter, the following ANOVA model uses Data Set 3 with the GROUP variable treated as a random effect and GENDER as a fixed effect. The EFFECT= command identifies GENDER as a tested effect, while the ERROR= command identifies the interaction term (GROUP*GENDER) to be used as the error. The complete command set would be:

```
MGLH
USE DASET3
ANOVA SCORE
CATEGORY GROUP=3, GENDER=2
ESTIMATE
HYPOTHESIS
EFFECT=GENDER
ERROR=GROUP*GENDER
TEST
```

The initial output would resemble the above; however, an additional table will be produced that contains the test of interest:

TEST FOR EFFECT CALLED: GENDER

TEST OF HYPOTHESIS

SOURCE	SS	DF	MS	F	P
HYPOTHESIS	1942.722	1	1942.722	55.683	0.017
ERROR	69.778	2	34.889		

The extension of the above procedures to three-way ANOVAs (**A4-75** to **A4-78**) is very logical. The user must identify the dependent variable in the ANOVA command and the categorical, or class, variables in the CATEGORY command. The major difference comes in the testing for special effects for mixed and random models. Unlike some other programs that permit the user to list multiple hypotheses to be tested, *SYSTAT* requires that each HYPOTHESIS set of commands be issued independently, each following after the other has been calculated.

To illustrate the *SYSTAT* procedure, we can use the same example of a three-way ANOVA that was developed in the *SAS* chapter, where the independent variables were A, B, and C, with A fixed and B and C random (**A4-76**). The statistical model calls for a mixed-effects analysis and a series of HYPOTHESIS sets to test the appropriate effects with the appropriate error terms. The MGLH command set would be as follows:

```
MGLH
USE DASET3
ANOVA Y
CATEGORY A=3, B=2, C=2
ESTIMATE
HYPOTHESIS
EFFECT=A
ERROR=A*B+A*C-A*B*C
TEST
HYPOTHESIS
EFFECT=B
ERROR=B*C
TEST
HYPOTHESIS
EFFECT=C
ERROR=B*C
TEST
HYPOTHESIS
EFFECT=A*B
ERROR=A*B*C
TEST
HYPOTHESIS
EFFECT=A*C
ERROR=A*B*C
TEST
```

While this command set may seem unwieldy, remember that the HYPOTHESIS sets are run consecutively and they are typed directly from the keyboard. Another strategy would be to save the above as a command file ([filename].CMD) and use the SUBMIT [file] command to run the analysis as a batch.

Repeated Measures/Split-Plot Analysis of Variance (A4-79 to A4-83)

As previously discussed, the one-way repeated measures (or within-subjects) design (**A4-79**) involves a single group of subjects measured across two or more different occasions. While a variety of inputting strategies are often employed by statistical packages, the process is simplified if the coding is direct. Such coding is illustrated below in our Data Set 4 (labeled DASET4.SYS). Again, the top two lines are presented only as column guides.

```
                          1         2         3
                 1234567890123456789012345678 90

                 1   21  25  30  35
                 2   17  22  28  33
                 3   19  24  27  31
                 4   25  27  33  36
                 5   18  21  23  25
                 6   19  21  23  24
                 7   20  23  25  29
                 8   11  16  19  24
                 9   17  20  25  27
                10   13  15  19  21
```

The DATA module commands for reading this ASCII file and creating a *SYSTAT* file would be:

```
DATA
GET DASET4
SAVE DASET4
INPUT SUBJECT, OCCAS(1-4)
RUN
```

One of the strengths of *SYSTAT* is its ability to handle special analyses, such as a repeated-measures model. The one-way repeated-measures model (**A4-79**) may be analyzed with the following commands:

```
MGLH
MODEL OCCAS(1-4) = CONSTANT / REPEAT
ESTIMATE
```

The special REPEAT option permits *SYSTAT* automatically to calculate the appropriate error term to test the model.

The next logical extension of the one-way repeated-measures model is the two-way repeated-measures model (**A4-80**). A model of this type has one categorical variable and a set of measures spread across several occasions. This type of design is often referred to as a type of split-plot design. The best strategy for inputting the data set identified in the *SAS* chapter as Data Set 4B (to be called DASET4B.SYS in this chapter) would be the following:

```
                         1         2         3
                1234567890123456789012345678790

                1  21  25  30  35
                1  17  22  28  33
                1  19  24  27  31
                1  25  28  33  36
                1  18  21  23  25
                2  19  21  23  24
                2  20  23  25  29
                2  11  15  19  24
                2  17  20  25  27
                2  13  15  19  21
```

The data, in ASCII file format (saved as DASET4B.DAT), would be inputted into *SYSTAT* with the following set of commands from within the DATA module:

```
DATA
GET DASET4B
SAVE DASET4B
INPUT GROUP, TEST(1-4)
RUN
```

The actual command format would resemble the previous commands for the one-way repeated-measures model. These commands would be:

```
MGLH
USE DASET5
CATEGORY GROUP=2
MODEL TEST(1-4) = CONSTANT + GROUP / REPEAT
ESTIMATE
```

Once again, since we are using the *SYSTAT* option for repeated measures, the testing of specific hypotheses for effects is unnecessary. The output (in edited form) is presented on the next page.

```
***********************************
* BETWEEN SUBJECTS EFFECTS *
***********************************
TEST FOR EFFECT CALLED:
                    GROUP      [This is the test for the between-subjects effect.]
TEST OF HYPOTHESIS
SOURCE        SS         DF      MS         F        P
HYPOTHESIS    308.025    1       308.025    7.297    0.027
  ERROR       337.700    8       42.213

***********************************
* WITHIN SUBJECTS EFFECTS  *
***********************************
TEST FOR EFFECT CALLED:               [This is the test for the tests effect, since
                    CONSTANT          CONSTANT*TESTS = TESTS.]
UNIVARIATE REPEATED MEASURES F-TEST
SOURCE        SS         DF      MS         F        P
HYPOTHESIS    623.475    3       207.825    98.379   0.000
  ERROR       50.700     24      2.113
TEST FOR EFFECT CALLED:       [This is the test for the TEST*GROUP effect.]
                    GROUP
UNIVARIATE REPEATED MEASURES F-TEST
SOURCE        SS         DF      MS         F        P
HYPOTHESIS    12.075     3       4.025      1.905    0.156
  ERROR       50.700     24      2.113
```

The above command statement could have been written with the following set of commands:

```
MGLH
ANOVA TEST(1-4) / REPEAT
CATEGORY GROUP=2
ESTIMATE
```

In either case, identical results are obtained. A key to utilizing the REPEAT option within the ANOVA model is the interpretation of the output.

The tables produced by *SYSTAT* resemble the typical tables produced for repeated-measures designs:

SOURCE	SS	df	MS	F
Between Subjects				
A (Group)	308.025	1	308.025	7.297*
Subjects w/i Groups	337.700	8	42.213	
Within Subjects				
B (Tests)	623.475	3	207.825	98.379*
AB	12.075	3	4.113	1.905
B × Subj w/i Groups	50.700	24	2.113	

*indicates P < .05

The above table is typical of the type of table that one would see in studies and reports. The output produced by *SYSTAT* is readily read when viewed from within the framework of this table. The first variable of interest is the between-subjects factor, GROUP, and it is the first one produced by *SYSTAT*. The next variable of interest is the within-subjects factor, TESTS. The final factor of interest is also a within-subjects factor, the GROUP × TESTS interaction (AB).

The next variation is a three-way split-plot ANOVA with one-between and two-within subject factors (**A4-81**). The model that will be illustrated is drawn from the corresponding example in the *SAS* chapter. The between-subjects factor is GENDER, the first within-subjects is OCCA, and the second within-subjects factor is the learning TRIAL, with four levels. All the factors are fixed effects. The dependent variable is the amount of information recalled.

The first step is to read the data (Data Set 5) in from the ASCII data set:

```
                    1         2         3
           1234567890123456789012345667890

        m  1 3 6 8   3 5 6 7   4 5 8 8
        m  2 4 4 6   2 4 6 6   3 6 6 8
        m  3 3 4 5   1 2 5 7   2 2 6 7
        f  4 5 6 7   4 4 7 8   4 5 6 7
        f  1 2 4 5   2 4 5 7   3 6 6 8
        f  1 3 3 4   2 4 4 5   2 4 5 7
```

The input commands within the DATA module are:

```
DATA
GET DASET5
SAVE DASET5
INPUT GENDER$, RECALL(1-12)
```

(continued on next page)

```
IF GENDER$ = 'M' THEN LET GENDER=1
IF GENDER$ = 'F' THEN LET GENDER=2
LET OCCA(1) = SUM(RECALL(1-4))/4
LET OCCA(2) = SUM(RECALL(5-8))/4
LET OCCA(3) = SUM(RECALL(9-12))/4
LET TRIAL(1) = SUM(RECALL(1),RECALL(5),RECALL(9))/3
LET TRIAL(2) = SUM(RECALL(2),RECALL(6),RECALL(10))/3
LET TRIAL(3) = SUM(RECALL(3),RECALL(7),RECALL(11))/3
LET TRIAL(4) = SUM(RECALL(4),RECALL(8),RECALL(12))/3
LET TOTAL = SUM(RECALL(1-12))/12
```

As with the earlier model, we let the program calculate our individual means. Another strategy would be to use the SSORT module to align the data and switch to the STATISTICS module to calculate the means. Because the output from *SYSTAT* for this analysis is quite extensive only the pertinent information from the ANOVA table is presented below:

SOURCE	SS	df	MS	F
Between Subjects				
A (Gender)	.014	1	.014	.001
Subjects w/i Groups	41.222	4	10.306	
Within Subjects				
B (Occasions)	28.694	2	18.347	18.283*
AB	.086	2	.431	.549
B × Subj w/i Groups	6.278	8	.785	
C (Trials)	189.375	3	63.125	82.139*
AC	1.819	3	.606	.789
C × Subj w/i Groups	9.222	12	.769	
BC	2.083	6	.347	.545
ABC	1.472	6	.245	.881
BC × Subj w/i Groups	15.278	24	.637	

*indicates P < .05

The ANOVA table clearly indicates that the main effects for the within-subjects factors, OCCASIONS and TRIALS, are significant. Application of the STATISTICS module would produce the means for each of these within-subject sets and then these could be subjected to appropriate *post hoc* tests (**A4-91** to **A4-97**) to establish the nature of these differences.

A final model to be illustrated for this split-plot design is a three-way model with two between-subjects factors and one within-subjects factor (**A4-82**). For purposes of consistency, the same data set used to illustrate this model in the *SAS* chapter will be used in this chapter. The data set will be called DASET6 and involves GENDER (two levels) and METHOD of instruction (three levels) as the between-subjects factors. The within-subjects factor is time of assessment, OCCASION (three levels: before, during, and after). The dependent variable is the number of mathematics problems solved correctly in a three-minute period.

For this model, we will use the direct keyboard-entry method of the DATA module (using Data Set 6).

```
DATA
SAVE DASET6
INPUT METHOD GENDER$ OCCAS (1-3) \
1 M 11 23 37    1 M 13 22 39    1 M  9 24 41
1 F  9 25 41    1 F  8 21 37    1 F 11 22 37
2 M  7 18 35    2 M 13 25 46    2 M  8 21 37
2 F 11 21 41    2 F  7 24 36    2 F  7 21 38
3 M  9 23 38    3 M  9 20 39    3 M 12 21 38
3 F 12 24 41    3 F  9 21 37    3 F  7 20 38
IF GENDER$ = 'M' THEN LET GENDER = 1
IF GENDER$ = 'F' THEN LET GENDER = 2
RUN
```

When data sets are as small as this illustrative one, the above strategy is perhaps the most efficient manner for data-set creation. It also provides another opportunity to illustrate the repeat-across-the-line function of the backslash (\) option.

The analysis format is very similar to the previous examples:

```
MGLH
USE DASET6
CATEGORY GENDER=2, METHOD=3
MODEL OCCAS(1-3) = CONSTANT + GENDER + METHOD +
GENDER*METHOD, /REPEAT
ESTIMATE
```

The output would be very similar to the previous model, with the exception that there would be two between-subjects factors (A and B) and only one within-subjects factor.

One final note of caution about the output from *SYSTAT* for repeated measures: There is a lot of it. It is advised that the user employ either the OUTPUT @ command to send the output to a printer or use the OUTPUT [filename] command to send the output to a data file. When finished with the output, remember to change the output back to the screen with the OUTPUT * command.

NESTED / HIERARCHICAL ANOVA (A4-84 TO A4-86)

The application of nested designs is often ignored or avoided as being too difficult or too complex to analyze. The fact of the matter is that a lot of research in education would be more powerful if these designs were employed as part of the analysis strategy. To restate, a nested design is used when some levels of one factor appear only under one level of another factor. Typical designs of this type may include studies of instructional strategies where teachers are nested within a one-instructional method, rather than being fully crossed with all instructional strategies. While this type of experimental design is closer to the real expectations (or limitations) of teachers, the statistical model employed to analyze it must be sensitive enough to detect significant differences. Such a design is a nested or hierarchical model.

To illustrate the model, assume that an experiment is being conducted to discover which of two reading series is most effective for developing reading comprehension. Because a reading series is a complete instructional system, it would be totally inappropriate to expect our ten teachers to teach with both series. Therefore, the only acceptable method of analysis is a nested design, with five teachers using one series and five teachers using the other. The following set of commands should be used within the DATA module to input Data Set 7 into the *SYSTAT* system:

```
DATA
SAVE DASET7
INPUT METHOD TEACHER SCORE \
1 1 25     1 1 30     1 1 29     1 2 27     1 2 28     1 2 24
1 3 26     1 3 25     1 3 24     1 4 27     1 4 30     1 4 24
1 5 18     1 5 24     1 5 23     2 6 10     2 6 8      2 6 12
2 7 9      2 7 9      2 7 10     2 8 7      2 8 6      2 8 9
2 9 7      2 9 10     2 9 9      2 10 9     2 10 10    2 10 11
IF METHOD = 1 THEN LET TEACH (2) = 0
IF METHOD = 2 THEN LET TEACH (1) = 0
IF TEACHER = 1 THEN LET TEACH (1) = 1
IF TEACHER = 2 THEN LET TEACH (1) = 2
IF TEACHER = 3 THEN LET TEACH (1) = 3
IF TEACHER = 4 THEN LET TEACH (1) = 4
IF TEACHER = 5 THEN LET TEACH (1) = 5
IF TEACHER = 6 THEN LET TEACH (2) = 1
IF TEACHER = 7 THEN LET TEACH (2) = 2
IF TEACHER = 8 THEN LET TEACH (2) = 3
IF TEACHER = 9 THEN LET TEACH (2) = 4
IF TEACHER = 10 THEN LET TEACH (1) = 5
RUN
```

The above set of DATA commands reflects a recoding strategy necessary to carry out our analysis within *SYSTAT*. Since the ten teachers are nested within the two methods of instruction, we must create two new variables, TEACH(1) and TEACH(2). These variables contain the nested teacher number (1 to 5) for their respective method and a value of zero for the other method.

The strategy of commands for our initial ANOVA analysis is exactly like the fully crossed model. It is the HYPOTHESIS commands that will test the EFFECTS.

```
DATA
CATEGORY TEACH(1)=5, TEACH(2)=5, METHOD=2
ANOVA SCORE
ESTIMATE
HYPOTHESIS
EFFECT = TEACH(1) & TEACH(2)
TEST
HYPOTHESIS
EFFECT METHOD
ERROR = TEACH(1) & TEACH(2)
TEST
```

The HYPOTHESIS commands are necessary if we are to test the correct EFFECTS with the correct ERROR terms. The first HYPOTHESIS set tells *SYSTAT* that the levels of TEACH(1) and TEACH(2) are to be treated as a single factor. The TEST statement without an ERROR command informs *SYSTAT* to employ the mean square error term as the ERROR. The second HYPOTHESIS set is designed to test the METHODS EFFECT, because we are assuming that this is a random effect (i.e., representative of all the possible reading series we could have selected). Since METHOD is treated as a random effect, the appropriate ERROR term is the mean square for the teacher-nested-within-method factor, TEACH(1) & TEACH(2). *SYSTAT* is able to perform this analysis because we used the ampersand (&) to indicate that the factors should be considered together.

The following is a reduced set of *SYSTAT*'s output for the above command set:

ANALYSIS OF VARIANCE

SOURCE	SUM-OF-SQUARES	DF	MEAN-SQUARE	F-RATIO	P
TEACH (1)	72.267	4	18.067	4.336	0.011
TEACH (2)	14.933	4	3.733	0.896	0.485
METHOD	2050.133	1	2050.133	492.032	0.000
ERROR	83.333	20	4.167		

TEST FOR EFFECT CALLED:
    TEACH(1)
  AND
    TEACH(2)

TEST OF HYPOTHESIS

SOURCE	SS	DF	MS	F	P
HYPOTHESIS	87.200	8	10.900	2.616	0.039
ERROR	83.333	20	4.167		

TEST FOR EFFECT CALLED:
    METHOD

TEST OF HYPOTHESIS:

SOURCE	SS	DF	MS	F	P
HYPOTHESIS	2050.133	1	2050.133	188.086	0.000
ERROR	87.200	8	10.900		

General ANOVA Designs (A4-87 to A4-88)

The set of designs known as randomized-blocks designs (**A4-87**) may be implemented within the *SYSTAT* system in a manner similar to the factorial ANOVA models described earlier in this chapter. The strategy requires the model to be stated without an interaction between the category or class variables. However, the full model statement must be used rather than the ANOVA command, since *SYSTAT* automatically assumes an interaction with two or more categories with this command. The following would implement a randomized-blocks model:

```
MGLH
USE BLOCK1
CATEGORY VAR1=3, BLOCK=2
MODEL DEPEND = CONSTANT + VAR1 + BLOCK
ESTIMATE
```

Decisions regarding appropriate HYPOTHESIS commands should be based upon the nature of the variable(s) to be tested; whether the factors are random or fixed.

While the complexity of a Latin square design (**A4-88**) limits its appropriateness in this chapter, *SYSTAT* is quite capable of performing such an analysis. The true value of a Latin square design is its ability to isolate cellular effects; however, its complexity limits its application. The *SYSTAT* manual describes a strategy of commands that will perform a Latin square model, but the real key to its use is in the input plan used to build the data set. The interested reader is advised to seek additional references such as Keppel (1982),* Kirk (1982),† and Winer (1971)‡ about building Latin square analysis models.

Related Designs (A4-89 to A4-90)

The purpose behind many statistical procedures is to employ a test powerful enough to detect significant differences or associations between or among the variables of interest. The purpose of the application of analysis of covariance (ANCOVA) (**A4-89**) is to reduce known sources of influence (bias) on the dependent variable in a factorial ANOVA. While many issues must be answered before the use of the ANCOVA model (see Keppel [1982],* Kirk [1982],† and Winer [1971]‡), once answered, its calculation is readily performed by *SYSTAT*.

ANCOVA requires that the general linear model employ techniques of both ANOVA models and regression models simultaneously. In other words, the model statement contains variables that are both categorical and continuous. Also, the critical assumption of homogeneity of regression (slopes) must be tested before implementing the analysis model. The first step in the process is to test for homogeneity of regression. Assuming that our data set (called ANCOVA1) has one

---

*G. Keppel. *Design and Analysis: A Researcher's Handbook*. 2d edition. Englewood Cliffs, NJ: Prentice-Hall, 1982.
†R. Kirk. *Experimental Designs: Procedures for the Behavioral Sciences*. 2d edition. Monterey, CA: Brooks/Cole Publishing Co., 1982.
‡B. J. Winer, *Statistical Principles in Experimental Design*. 2d edition. New York: McGraw-Hill, 1971.

categorical variable called VAR1, one continuous variable called CON1, and a dependent variable called DEPEN, the following would test for homogeneity of regression:

```
MGLH
GET ANCOVA1
CATEGORY VAR1=3
MODEL DEPEN = CONSTANT + VAR1 + CON1 + VAR1*CON1
ESTIMATE
```

The resultant ANOVA table should indicate a nonsignificant interaction between VAR1 and CON1 (VAR1*CON1). If the p-value for this interaction is nonsignificant, then the assumption of homogeneity of regression has been supported. The output from this model would resemble the following table:

	ANALYSIS OF VARIANCE				
SOURCE	SUM-OF-SQUARES	DF	MEAN-SQUARE	F-RATIO	P
VAR1	3.237	1	3.237	3.029	0.088
CON1	390.444	1	390.444	365.375	0.000
VAR1*CON1	1.552	1	1.552	1.452	0.234
ERROR	49.156	46	1.069		

Since the interaction between VAR1 and CON1 is not significant, the homogeneity-of-regression assumption has been confirmed and we may proceed with the full ANCOVA test.

```
CATEGORY VAR1 = 2
MODEL DEPEN = CONSTANT + VAR1 + CON1
ESTIMATE
```

Once the influence of CON1 is accounted for, the output from the ANCOVA model above confirms our suspicion that the two groups are different for the treatment VAR1.

	ANALYSIS OF VARIANCE				
SOURCE	SUM-OF-SQUARES	DF	MEAN-SQUARE	F-RATIO	P
VAR1	21.453	1	21.453	19.884	0.000
CON1	392.732	1	392.732	364.013	0.000
ERROR	50.708	47	1.079		

SYSTAT offers a special COVARIANT command that may be used in combination with the ANOVA command. For example:

```
CATEGORY VAR1=2
ANOVA DEPEN
COVARIANT CON1
ESTIMATE
```

The set of commands above would produce the ANOVA table, but would not test for the assumption of homogeneity of regression. This procedure may also be used for implementing models with more than one covariant variable. The reader is cautioned about indiscriminately employing this strategy without first testing the assumption of homogeneity of regression.

Multiple-Comparison Procedures (A4-91 to A4-97)

Multiple comparisons are performed in *SYSTAT* with either the STATISTICS module or the MGLH module. To perform pairwise comparison, *SYSTAT* offers the Duncan Multiple Range (**A4-97**), Newman-Keuls (**A4-96**), and Tukey HSD (**A4-95**) tests. The Duncan, Newman-Keuls, and Tukey tests require that the cell sizes be equal; however, the Tukey will employ a harmonic mean when testing unequal cells.

The implementation of any of these tests is performed through the STATISTICS module, and only one test at a time may be employed. (Note: *SYSTAT* purposely limits the testing to one test to avoid indiscriminate hunting for pairwise effects.) An example of this procedure, using the Tukey HSD (**A4-95**) and the data set previously identified as Data Set 3, is as follows:

```
STATS
USE DASET3
BY TREAT
STATISTICS SCORE / TUKEY
```

The output would include a full set of descriptive statistics for each of the BY variables, an ANOVA table, and a breakdown of the pairwise comparisons. Only the pairwise comparison will be illustrated.

MATRIX OF PAIRWISE ABSOLUTE MEAN DIFFERENCES			
	1	2	3
1	0.000		
2	8.667	0.000	
3	6.000	14.667	0.000

TUKEY HSD MULTIPLE COMPARISONS MATRIX OF PAIRWISE COMPARISON PROBABILITIES			
	1	2	3
1	1.000		
2	0.477	1.000	
3	0.695	0.143	1.000

The output for each of the other tests (**A4-96** and **A4-97**) is exactly the same. The probabilities are calculated using the appropriate test.

Contrasts are obtained with the MGLH module and the HYPOTHESIS command, much in the manner of effects testing described above (**A4-72** to **A4-77**). To test for contrasts with the same data set, the command set would be:

```
MGLH
USE DASET3
CATEGORY GROUP = 3
ANOVA SCORE
ESTIMATE
HYPOTHESIS
EFFECT = GROUP
CONTRAST
1 1 -2
TEST
HYPOTHESIS
EFFECT = GROUP
CONTRAST
-1 0 -1
TEST
HYPOTHESIS
EFFECT = GROUP
CONTRAST
1 -2 1
TEST
```

The table below represents an edited version of the output of the HYPOTHESIS commands above:

	GROUPS 1 & 2 VS 3				
TEST OF HYPOTHESIS					
SOURCE	SS	DF	MS	F	P
HYPOTHESIS	427.111	1	427.111	2.682	0.122
ERROR	2389.167	15	159.278		

(continued on next page)

	LINEAR				
TEST OF HYPOTHESIS					
SOURCE	SS	DF	MS	F	P
HYPOTHESIS	108.000	1	108.000	0.678	0.423
ERROR	2389.167	15	159.278		
	QUADRATIC				
TEST OF HYPOTHESIS					
SOURCE	SS	DF	MS	F	P
HYPOTHESIS	544.444	1	544.444	3.418	0.084
ERROR	2389.167	15	159.278		

To perform contrasts on factorial models the process is exactly the same; state HYPOTHESIS, identify the EFFECT variable, define the CONTRAST, and TEST the hypothesis. Remember, unlike the earlier example in the *SAS* chapter, *SYSTAT* will perform only one contrast at a time. A limitation of *SYSTAT* is its inability to perform interactive contrasts. *SYSTAT* will, however, perform simple contrasts; that is, contrasts for a factor while holding the one level of the other factor constant. This process requires an understanding of matrix construction and is beyond the scope of discussion here.

To perform contrasts for repeated-measures designs, the process is essentially the same. The user, however, must carefully decide which EFFECT is being tested before writing the contrast statement. For example, using Data Set 4, we can perform a contrast to test for significant differences between occasion 1 and occasion 4 for our repeated factor OCCAS(1-4). Rather than using a contrast statement, use the CMATRIX command, which permits testing the dependent variables across the four occasions (trials). The MGLH commands would be:

```
MGLH
USE DASET4
MODEL OCCAS(1-4) = CONSTANT /REPEAT
ESTIMATE
HYPOTHESIS
EFFECT = CONSTANT
CMATRIX
1 0 0 -1
TEST
```

A similar set of HYPOTHESIS commands could be used to test any combination of contrasts for the repeated factor. Contrasts testing for any effects in other repeated-measures models are carried out in essentially the same manner as regular contrasts, except it is recommended that the CMATRIX command be used instead of the CONTRAST command. Further information and examples are beyond the present scope of this book. The reader should consult the *SYSTAT* manual and have a working knowledge of matrix algebra.

CHAPTER 4

SPSS-X

Introduction

SPSS-X* is a well established, versatile, and comprehensive statistical package for researchers in the social and behavioral sciences. Applications of *SPSS-X,* Release 3.0, will be illustrated in this chapter. The microcomputer version of *SPSS* provides many of the same procedures with very similar coding. Complete documentation of the applications described in this chapter can be found in *SPSS-X User's Guide, 3rd Edition,* and *SPSS-X Advanced Statistics Guide.* These manuals provide helpful information about the statistical procedures they are designed to perform. Coding described in this chapter will run under *SPSS* release 4.0.

Text Format

Examples of *SPSS-X* commands and output will be shown throughout the remainder of this chapter. *SPSS-X* code will be printed in CAPITAL letters when referenced in a paragraph and will be printed in a shaded area between parallel lines when *SPSS-X* statements appear on separate lines that illustrate the code. In this chapter, *SPSS-X* code will be written as if the left-most character of the code is in the first column of the input screen. Examples of *SPSS-X* output will be displayed in boxes to set them apart from the text. *SPSS-X* examples will use data sets described in detail in Chapter 2.

Some SPSS-X Basics

A standard form for an *SPSS-X* job on a mainframe would have the features shown in the sample job below. System JCL comes first. The *SPSS* program is initiated and desired options (e.g., SET WIDTH=80) are given. The data are listed with variable names and column locations. VARIABLE LABELS assigns descriptive

---

*SPSS-X is a trademark of SPSS Inc., Chicago, IL.

labels to variables and VALUE LABELS assigns descriptive labels to discrete values of variables. BEGIN DATA signals the beginning of the data set; data are read according to the DATA LIST specifications; and END DATA marks the last of the data set. The TITLE provides a user-supplied legend on the printout.

The heart of the *SPSS-X* statistical activity is controlled by the *SPSS-X* procedures. Procedure DESCRIPTIVES, for example, computes univariate summary statistics. The *SPSS-X* procedure CORRELATIONS produces the correlation between TEST1 and TEST2. The subcommands or options coded as /MISSING and /STATISTICS control the treatment of missing data and the production of univariate summary statistics.

SAMPLE SPSS JOB

JCL System required Job Control Language, or JCL, usually provides an account number, password, and other system control information.

// EXEC SPSS {System specific syntax, initiates execution of SPSS-X, 3.0}
SET WIDTH=80 {Sets the line size to 80 characters, makes reading output on a CRT much easier}
DATA LIST FIXED / **TEST1 1-2 TEST2 4-5 GENDER 7** {Names three input variables and identifies column locations for data entry}
COMPUTE DIFF=TEST2-TEST1 {Defines a new variable, DIFF, as the difference between TEST2 and TEST1}
VARIABLE LABELS TEST1 'PRE TEST'
    TEST2 'POST TEST'
VALUE LABELS GENDER 1 'MALE' 2 'FEMALE'
BEGIN DATA {Signals the beginning of data}
17 20 1 {TEST1, TEST2 and GENDER for first observation}
15 26 2 {TEST1, TEST2 and GENDER for second observation}
13 29 2 {TEST1, TEST2 and GENDER for third observation}
END DATA {Signals the end of data}
TITLE 'EXAMPLE OF SPSS-X JOB' {Title printed on output}
DESCRIPTIVES VARIABLES=TEST1 TEST2 DIFF
  /STATISTICS=ALL
CORRELATIONS VARIABLES=TEST1 TEST2
  /MISSING=LISTWISE
  /STATISTICS=DESCRIPTIVES
FINISH
JCL System JCL to signal end of job.

SPSS-X procedures are generally referred to as "commands." There is some variation in how *SPSS-X* commands are written. In general, these commands take the following form:

```
COMMAND VARIABLES= [Variable list]
   /SUBCOMMAND=
   /SUBCOMMAND=
```

COMMAND names the procedure that is to be executed: for example, DESCRIPTIVES, CORRELATIONS, and REGRESSION; VARIABLES=Variable list, supplies the names of the variables to which the COMMAND is applied. The /SUBCOMMAND controls options available for different *SPSS-X* commands. At least one blank must precede the /SUBCOMMAND. Details about *SPSS-X* commands and subcommands will follow the description of data-entry and data-manipulation procedures.

Some Helpful SPSS-X Features

COMMENT LINES

Lines beginning with the word COMMENT or an asterisk (*) in *SPSS* are comment lines used to document the program. These lines are notes used to explain the flow and operation of the program. COMMENT lines are not executed and do not effect the operation of the program. A COMMENT can be extended over several lines by leaving blank the first column of each consecutive line used to continue the comment.

LIST

The *SPSS-X* LIST command is a useful tool for verifying the accuracy of data entry. LIST would generally be used with a VARIABLES= instruction to specify the variables to be listed. A subcommand can be used to control the number of cases listed as follows:

```
LIST VARIABLES=ALL
   /CASES FROM 1 TO 25
```

These commands would produce a listing of all variables for the first 25 cases in a data file.

PRINT / [Variable(s)]

The PRINT instruction provides a simple procedure for printing the values of some or all of the variables in a file. Unlike LIST, PRINT functions only as data are being read and is therefore not executed unless followed by the word EXECUTE or by any *SPSS-X* command that results in the data being read. PRINT must follow the END DATA card.

VARIABLES=

The application of *SPSS-X* commands can be restricted to a specific set of variables named on a VARIABLES= list. Individual variables or groups of variables can be named.

GROUPS OF VARIABLES

Groups or strings of variables can be referred to by specifying the first and last variables connected by the word TO. For example, if variables A, B, X, and Y have been entered into the DATA LIST, they can be referred as follows:

```
LIST VARIABLES= A TO Y.
```

The variables must have been read or created by *SPSS-X* in the order A, B, X, and Y.

TITLE

Titles are used to print descriptive information on output. TITLE precedes the command that it describes and is used throughout the analysis until another TITLE is encountered.

SPLIT FILE BY [Variable(s)]

The SPLIT FILE command subdivides the active file into adjacent categories of the variable(s) following the word BY. Data must first be sorted relative to the BY variable before SPLIT FILE can be applied. If data are not sorted in the active file, a SORT command must be used. For example, the following code sorts the data by GENDER, splits the file into the two gender groups, and applies the DESCRIPTIVES procedure to variables TEST1 and TEST2.

```
SORT CASES BY GENDER
SPLIT FILE BY GENDER
DESCRIPTIVES VARIABLES=TEST1 TEST2
SPLIT FILE OFF
```

The SPLIT FILE command remains in operation until it is turned off with the SPLIT FILE OFF command.

DO COMMANDS AND LOOP COMMAND

SPSS-X contains a variety of DO statements that initiate a repetitious sequence of executable steps. DO REPEAT starts a sequence that is completed with END

REPEAT. DO IF starts a sequence completed with END IF. DO IF commands are used for manipulations for a subset of cases.

The LOOP command starts a repetitious sequence applied to a single record or case and is completed with END LOOP.

MISSING DATA

Missing data are usually handled in one of two ways. One option involves use of the MISSING VALUES command. This command allows the user to specify the value for each or all variables that are to be treated as missing data. For example,

> MISSING VALUES ALL (0)

defines zero as a missing value for all numeric variables. The MISSING VALUES command is especially useful for free and list formats.

A second approach involves using the /MISSING subcommand as part of the statistical procedure being employed. Options for missing data include EXCLUDE, INCLUDE, LISTWISE, and PAIRWISE. Missing values defined by the user can be excluded or included in computations with the EXCLUDE or INCLUDE options, respectively. Cases with missing values on any variable named in a list are excluded with the LISTWISE option, and cases missing either one or both of a pair of values needed for a computation—for instance, correlation—are excluded with PAIRWISE. The /MISSING subcommand is used to control the way statistical procedures treat missing data.

The MISSING VALUES command and the /MISSING subcommand used with various procedures are somewhat different. The former defines values of variables that are to be considered missing; the latter specifies how cases with missing values for a variable or variables should be treated in various statistical procedures.

Data Entry (Input)

SPSS-X provides a full range of input formats to meet the needs of users with different types of data sets. Only the four most commonly used—fixed, free, list, and FORTRAN—input formats will be illustrated and described below. These input formats all begin with DATA LIST. Some *SPSS-X* input formats specify the fields or columns in which data are located. Data can be anywhere within the specified fields.

FORMAT 1: FIXED FORMAT

The following is an example of Format 1.

```
DATA LIST FIXED / STUDENT 1-2 GROUP 4 IQ 6-8 PRETEST10-11
     POSTTEST 13-14 TRSP 16

BEGIN DATA
 1  1  112  21  35  2
 2  1  109  17  33  2
 3  1  105  19  31  1
 4  1  118  25  36  2
 5  1   95  18  25  1
 6  2  111  19  24  1
 7  2  107  20  29  2
 8  2  102  11  24  1
 9  2   98  17  27  1
10  2  105  13  21  1
END DATA
```

Format 1, like all input procedures, begins with DATA LIST. The format is described as FIXED. The variables are then given names. In this hypothetical example, henceforth referred to as Data Set 1, the variables are defined as follows:

STUDENT = Student ID number (two-digit maximum)
GROUP = Two groups, coded "1" or "2"
IQ = Score on a standardized IQ test
PRETEST = Score on a class pretest
POSTTEST = Score on a class posttest
TRSP = Teacher rating of student potential (1 or 2)

The fields or columns containing the data for each variable are specified after the variable name. For example, data for the variable STUDENT are contained in columns 1 and 2. If data for a variable are written with letters, rather than numbers, place an A in parentheses after the column location. For example,

```
DATA LIST / GENDER 3 (A)
```

would be needed if data for the variable GENDER were coded "F" and "M." Variables with alphabetic values are called string variables.

FORMAT 2: LIST FORMAT

Format 2 is very similar to the first format except that the fields or columns are not specified. For example,

```
DATA LIST / LIST STUDENT GROUP IQ PRETEST POSTTEST
     TRSP
```

This format can be used as long as there is at least one space between the data for each variable and one record contains the data for one case. This format is quite flexible and convenient but must be used with care. *SPSS-X* expects to read data for six variables for each case or record, since six variables are listed in the input statement. Data are read and assigned to each variable, in order, for each record. Missing data must be dealt with carefully. Specifically, missing data should be represented in the data file with a particular character and that character should be defined as the character for missing data.

FORMAT 3: FREE FORMAT

Format 3 is very similar to Format 2 except that data are read continuously across each record such that data for more than one case is read from each record. This format is written as follows:

```
DATA LIST FREE / STUDENT GROUP IQ PRETEST POSTTEST
TRSP
```

Again, missing data must be handled carefully. For example, if the value of GROUP for STUDENT=1 were omitted, GROUP would be assigned a value of 112, IQ would have a value of 21, and so on. To avoid such difficulties, missing data should be represented in the data file with a specific character and that character should be defined as the character for missing data.

When both numeric and string variables are entered, the following approach can be used in both LIST and FREE formats.

```
DATA LIST LIST / STUDENT GROUP PRETEST * GENDER (A1)
DATA LIST FREE / STUDENT GROUP PRETEST * GENDER (A1)
```

The asterisk (*) indicates that the variables preceding it are read using the LIST or the FREE format; the string variable, GENDER, is read in a FORTRAN-like format in which the "A" indicates that a string variable is being read and the "1" following the "A" indicates the width of the string field. The string variable must be enclosed in single quotation marks if it contains blanks or punctuation marks. For example:

```
DATA LIST / LIST STUDENT * NAME (A8)
BEGIN DATA
1 'CHANG, L'
2 'HESS, R'
3 'JONES, J'
4 'SMITH, M'
END DATA
```

Missing Values. The definition of missing values needed in formats 2 and 3 can be specified immediately following the DATA LIST statement or before the command lines where the variables are included in the procedure. The values assigned for missing data must be a number for numeric variables and a letter for string variables. The following are examples of missing value assignments:

> MISSING VALUES IQ(0)
> MISSING VALUES ALL(99)
> MISSING VALUES GENDER ('U')
> MISSING VALUES IQ TO POSTTESTS(0)

FORMAT 4: FORTRAN FORMAT

A useful and efficient format can be used when variables names are repetitious. Such situations occur when several measures are taken over different occasions or when other types of multiple measures are used. Consider a situation in which there are four scores for each subject. Each subject is specified by a three-digit ID number in columns 1-2-3. Input formats 1, 2, and 3 for this example are as follows:

Format 1:

> DATA LIST FIXED /ID 1-3 SCORE1 5-6 SCORE2 8-9 SCORE3 11-12 SCORE4 14-15

Format 2:

> DATA LIST LIST / ID SCORE1 SCORE2 SCORE3 SCORE4

Format 3:

> DATA LIST FREE / ID SCORE1 SCORE2 SCORE3 SCORE4

Format 4 can be used as a modification of the other formats to eliminate the need to write out the names of all the variables. Input Format 4 contains two new features not seen in the other input conventions. First, the names of all variables are not explicitly listed. This can be done whenever the variable names have a repetitious literal portion with a unique and sequential numeric ending. Second, FORTRAN format specifications may be used to read all or part of the data. An explanation of FORTRAN code is beyond the scope of this book; however, a brief description of some features will be provided for the sake of illustration.

The modification can be used within the FIXED, LIST, or FREE formats as follows:

> DATA LIST FIXED / ID 1-3 SCORE1 TO SCORE4 4-15
> DATA LIST FIXED / ID 1-3 SCORE1 TO SCORE4 (T5,4 (F2.0,1X))
> DATA LIST FIXED / ID SCORE1 TO SCORE4 (F3.0,1X,4 (F2.0,1X))
> DATA LIST LIST   / ID SCORE1 TO SCORE4 (T5,4 (F2.0,1X))
> DATA LIST LIST   / ID SCORE1 TO SCORE4 (F3.0,1X,4 (F2.0,1X))
> DATA LIST FREE   / ID SCORE1 TO SCORE4 (F3.0,1X,4 (F2.0,1X)

In these examples, the FORTRAN code operates as follows. The "T5" functions like a tab key and moves control to column 5. The "4" indicates that the expression in the parentheses that follows will be used four times to read data. The "F2.0" contains three elements: it is a floating point variable (F), with a two-character field (2), and zero digits after the decimal point (.0). The "F3.0" indicates a floating point variable (F), with a three-character (3) field, and no digits (.0) after the decimal point. The "1X" indicates skipping one field.

Format 4 would be used with the LIST format as follows:

```
DATA LIST LIST / ID SCORE1 TO SCORE4 (T5,4 (F2.0,1X))
BEGIN DATA
001 21 25 30 35
002 17 22 28 33
003 19 24 27 31
004 25 28 33 36
005 18 21 23 25
END DATA
```

Format 4 would also be used with the FREE format when data are entered continually across the line of input. The input format for this approach is shown below.

```
DATA LIST FREE / ID SCORE1 TO SCORE4 (F3.0,1X,4 (F2.0,1X))
BEGIN DATA
001 21 25 30 35 002 17 22 28 33 003 19 24 27 31
004 25 28 33 36 005 18 21 23 25
END DATA
```

Data Manipulation (Transformation)

Data that have been inputted can be transformed or manipulated in many different ways with a variety of easy-to-use expressions or commands. A large array of arithmetic, algebraic, trigonometric, statistical, and logical manipulations are available and easily managed.

RECODE COMMAND

The RECODE command is central to data manipulation in *SPSS-X*. The RECODE command changes the values for a variable as the data are being read. For example, the code

```
RECODE GROUP (0=1) (1=2) (2=3)
```

changes the values for a variable named GROUP from 0, 1, and 2, to 1, 2, and 3, respectively. RECODE can be used to define discrete categories from a continuous variable. For example, the variable IQ in Data Set 1 can be recoded with the following code:

```
RECODE IQ (50 THRU 85=1) (86 THRU 115=2) (116 THRU 180=3)
(ELSE = 777)
```

This expression recodes the continuous variable, IQ, into a three-valued variable. The ranges for each category are defined using the keyword THRU. The keyword ELSE recodes all values of IQ that are not in the defined ranges, from 50 to 180, as equal to 777. The value 777 can be defined as missing data with the statement

```
MISSING VALUES IQ (777).
```

RECODE can be used to redefine a numeric variable as a string variable and to label the categories with alphabetic string names. For example, the variable IQ can be recoded into three categories as follows.

```
STRING IQLEVEL(A4)
RECODE IQ(LO THRU 85='LOW') (86 THRU 115='MED')
    (116 THRU HI='HIGH') INTO IQLEVEL
```

The STRING variable IQLEVEL is first defined. Then the RECODE command uses LO meaning the lowest value, HI meaning the highest value, and THRU to facilitate the recoding. The keyword INTO specifies the string variable into which the values defined in the RECODE are to be stored.

In some applications, a variable entered into a data file as a string variable must be redefined as a numeric variable. For example, a variable GENDER, entered as a string variable with the values 'F' and 'M', can be redefined as a numeric variable SEX, with the following RECODE command:

```
RECODE GENDER ('F'=1) ('M'=2) INTO SEX
```

COMPUTE COMMAND

The COMPUTE command is an instruction used to perform a variety of data manipulations. Two new variables can be created in Data Set 1 with the following simple arithmetic manipulations:

```
COMPUTE GAIN = POSTTEST - PRETEST
COMPUTE TOTAL = PRETEST + POSTTEST
```

The variable GAIN is the difference between each student's performance on POSTTEST and PRETEST. The variable TOTAL is the total number of points each student earns on PRETEST and POSTTEST combined. GAIN and TOTAL will be assigned missing values if either PRETEST or POSTTEST is missing.

SPSS-X includes an expression, or operator, for summing a set of variables, and this takes the form

> COMPUTE VARIABLE = SUM(list of variables to be added)

Using this notation, the variable TOTAL, the sum of the PRETEST and POSTTEST, can be written as

> COMPUTE TOTAL= SUM(PRETEST, POSTTEST)

This coding for summing a set of variables is especially convenient when variables have been named using the serial notation of the form SCORE1–SCORE4. The sum of SCORE1, SCORE2, SCORE3, and SCORE4 can be written:

> COMPUTE TOTAL=SUM(SCORE1 TO SCORE4)

Test scores defined as the number of items answered correctly can be changed to percentages by simply defining a new variable as the test score divided by the total number of items on the test. For example, the percentage of items each student answers correctly on PRETEST and POSTTEST can be named as PREPCT and POSTPCT, respectively. These variables are defined as students' test scores divided by 40 (the number of items on the test) using the following code:

> COMPUTE PREPCT = PRETEST/40
> COMPUTE POSTPCT= POSTTEST/40

The *SPSS-X* symbols for the four arithmetic operations are the plus sign (+) for addition, the minus sign (−) for subtraction, the asterisk (*) for multiplication, and the slash (/) for division. Exponents are indicated with a double asterisk (**). Thus the PRETEST scores could be squared and called PRESQRED with either of the following notations:

> COMPUTE PRESQRED = PRETEST*PRETEST
> COMPUTE PRESQRED = PRETEST**2

LOGICAL OPERATORS

Four basic logical operators used in conjunction with IF statements can be used to manipulate data in *SPSS-X*.
The logical operators are:

$$LT = \text{less than}$$
$$LE = \text{less than or equal to}$$
$$GT = \text{greater than}$$
$$GE = \text{greater than or equal to}$$

These logical operations are used in expressions that take the form:

```
IF (IQ GE 107) IQGROUP=2
IF (IQ LT 107) IQGROUP=1
```

This expression could also be written as

```
DO IF (IQ GE 107)
COMPUTE IQGROUP=2
ELSE
COMPUTE IQGROUP=1
END IF
```

These lines of code would define a new variable named IQGROUP. Students would be assigned to IQGROUP=2 if their IQ score was greater than or equal (GE) to 107. Otherwise (ELSE)—that is, if their IQ scores are less than 107—they are assigned to IQGROUP=1.

The power and flexibility for data manipulation in *SPSS-X* has only been briefly described in this section. Readers who anticipate working with this program should become more familiar with these data-manipulation capabilities through a careful examination of the *SPSS-X* manuals.

Descriptive Statistics (A4-1 to A4-13)

A variety of *SPSS-X* commands can be used to obtain basic descriptive statistics. The most common procedures used for descriptive statistics are DESCRIPTIVES, MEANS, FREQUENCIES, CROSSTABS, and PLOT. The following is a brief summary of the descriptive statistics each of these commands produces.

DESCRIPTIVES

DESCRIPTIVES calculates the mean (**A4-2**), standard deviation (**A4-8**), minimum, and maximum for numeric variables. Using the subcommand STATISTICS directs DESCRIPTIVES to calculate the mean (**A4-2**), standard error of the mean, standard deviation (**A4-8**), variance (**A4-7**), skewness (**A4-12**), kurtosis (**A4-13**), range (**A4-5**), minimum. maximum, and sum. The following code would produce the full set of descriptive statistics available with the DESCRIPTIVES command:

```
TITLE 'EXAMPLE OF DESCRIPTIVES'
DESCRIPTIVES VARIABLES=PRETEST POSTEST
  /STATISTICS=ALL
```

DESCRIPTIVES will produce only the specified statistics if some statistics are specified with the STATISTICS subcommand. DESCRIPTIVES can also produce the z-score transformation of each variable in the VARIABLES= list statement with the subcommand /SAVE. The transformed variables have a mean of 0, variance of 1, and are named automatically by the addition of the letter z to the first seven characters of the original variable name.

MEANS

The MEANS command calculates the mean (**A4-2**) and standard deviation (**A4-8**) for a dependent variable(s) for subgroups defined by levels of an independent variable(s). MEANS applied to the variables in Data Set 1 could be used as follows:

```
MEANS TABLES=PRETEST BY GROUP
```

This coding would produce the PRETEST mean and standard deviation for each level of the variable GROUP. The analysis can include finer levels of subdivision by adding additional independent variables. For example, the following code adds to the previous analysis:

```
MEANS TABLES=PRETEST BY GROUP BY TRSP
```

The mean and standard deviation for PRETEST is broken down by levels of GROUP and by levels of TRSP within group.

Multiple dependent variables can be specified with the keyword TO, as in the example:

```
MEANS TABLES=SCORE1 TO SCORE4 BY GENDER
```

Additional subcommands for MEANS include CROSSBREAK for printing the results in crosstabular form, CELL to modify the information about each cell, and STATISTICS for additional statistics including a one-way analysis of variance (**A4-71**).

FREQUENCIES

FREQUENCIES is a versatile command that produces basic frequency-distribution information, such as bar graphs, histograms, and a variety of descriptive statistics. The following code illustrates a number of FREQUENCIES' features:

```
TITLE 'FREQUENCIES COMMAND'
FREQUENCIES VARIABLES=IQ
   /HISTOGRAM=INCREMENT (20)
   /PERCENTILES=25 50 75
   /STATISTICS=ALL
```

The VARIABLES list specifies the variables to which FREQUENCIES is applied. The default information includes the raw frequency, percent, and cumulative percent for each value of the variables specified. Integer and continuous variables are treated somewhat differently. The HISTOGRAM subcommand, like the BARCHART subcommand, produces a visual representation of the frequency distribution. HISTOGRAM will plot 21 intervals as a default value. The option INCREMENT=20 overrides the default and instructs the program to organize the data into intervals 20 units wide. The option NORMAL with HISTOGRAM superimposes the normal curve over the histogram. For both HISTOGRAM and BARCHART, the options PERCENT(N) AND FREQ(N) can be used to specify that the horizontal axis be scaled in percentages or raw frequencies, respectively. The (N) in both cases is optional but can be used to specify the maximum value on the horizontal axis.

An example of BARCHART is illustrated using the TRSP variable from Data Set 1:

```
FREQUENCIES VARIABLES=TRSP
  /BARCHART
  /PERCENTILES=25 50 75
```

The PERCENTILES subcommand causes the program to print the values for the percentiles indicated after the equal sign. NTILES is a similar subcommand, which takes the form:

```
/NTILES=N
```

where N indicates that the sample is to be divided into N groups of equal size. NTILES=4, for example, produces the values of the variables at the quartile positions: 25th, 50th, and 75th percentiles.

The subcommand STATISTICS in FREQUENCIES is a powerful option that can be used to produce the full array of descriptive univariate statistics. The default statistics are mean (**A4-2**), standard deviation (**A4-8**), minimum, and maximum. The statistics available with the STATISTICS subcommand are mean (**A4-2**), standard error of the mean, median (**A4-3**), mode (**A4-4**), range (**A4-5**), variance (**A4-7**), standard deviation (**A4-8**), skewness (**A4-12**), standard error of the skewness statistic, kurtosis (**A4-13**), standard error of the kurtosis statistic, minimum, maximum, and sum. Any of these statistics can be specified after the /STATISTICS= code. All of these are produced by /STATISTICS=ALL.

CROSSTABS

CROSSTABS produces frequency information for two or more variables that have a joint distribution. Its most common application involves the examination of contingency tables. By default, CROSSTABS produces cell frequencies, row, column, and total frequencies and percentages. The subcommand CELL can be used to obtain cell, row, and column percentages; the expected cell frequency; and the residual or difference between the observed and expected cell frequency. These features are illustrated in the following example:

```
TITLE 'CROSSTABS EXAMPLE'
CROSSTABS TABLES=GROUP BY GENDER
  /CELLS=COUNT ROW COLUMN EXPECTED RESID
```

These commands will produce the contingency table for the levels of GROUP (rows) by the levels of GENDER (columns). Cell frequencies, cell row and column percentages, and expected and residual cell frequencies will be printed in response to the CELL subcommand. Marginal row and column frequencies and percentages will be automatically printed.

Additional variables can be employed as follows:

```
CROSSTABS TABLES=GROUP BY GENDER BY CLASSRM.
```

This command will produce the GROUP by GENDER frequency table for each level of CLASSRM (classroom).

The STATISTICS subcommand computes a number of summary statistics. The default with the STATISTICS subcommand is the Chi-square statistic used to test the hypothesis of homogeneity or independence (**A4-61**). Key words used with the STATISTICS subcommand can be used to obtain the Phi-coefficient (**A4-21**), contingency coefficient (**A4-23**), Kendall's Tau (**A4-18**), Eta-squared (**A4-29**), and the Pearson product-moment correlation coefficient (**A4-16**). The following illustrates the code needed to obtain this information:

```
TITLE 'CROSSTABS EXAMPLE'
CROSSTABS TABLES=GROUP BY GENDER
  /CELLS=COUNT ROW COLUMN EXPECTED RESID
  /STATISTICS=CHISQ PHI CC BTAU CTAU ETA CORR
```

All available statistics can be requested with the keyword ALL.

PLOT

Plots showing the joint distribution of continuous variables are produced by PLOT. PLOT takes the following form:

```
PLOT TITLE=' EXAMPLE OF PLOT COMMAND'
  /VERTICAL='POSTTEST SCORES' MIN (15) MAX (150)
   REFERENCE(100)
  /HORIZONTAL='PRETEST SCORES' MIN (15) MAX (150)
   REFERENCE(100)
  /PLOT=POSTTEST WITH PRETEST
```

This coding plots POSTEST scores against PRETEST scores. The vertical and horizontal axes are given descriptive labels. The VERTICAL and HORIZONTAL subcommands specify minimum and maximum values plotted on each axis and draw reference lines across the plot at the value 100 on both axes.

Measures of Relationships (A4-14 to A4-30)

Different measures of association or relationship can be obtained using a number of *SPSS-X* commands and subcommands. The measure of association chosen depends, of course, on the nature of the research situation and the type of data involved.

CORRELATIONS

Correlations can be obtained by using the CORRELATIONS procedure. The default for this command produces a Pearson product-moment correlation with a one-tailed test of significance. The CORRELATIONS procedure is applied to all numeric variables in the VARIABLES= list. A particularly useful option with CORRELATIONS is the use of a VARIABLES= list and a WITH statement. This produces the correlation of the variable(s) in the VAR list with the variable(s) listed in the WITH statement. The use of this *SPSS-X* command will be illustrated using data from Ryan (1977)* with 23 children (observations) on five variables. This data set will be referred to as Data Set 2. The variables are:

AGEMONTH = children's age in months
CYSN = children's score on a measure of syntactic (grammar) ability
MSYN = mothers' prediction of children's score on the syntax test
CVOCAB = children's score on a vocabulary test
MVOCAB = mothers' prediction of children's vocabulary score

The data are entered as follows:

```
DATA LIST FREE / AGEMONTH CSYN MSYN CVOCAB MVOCAB
BEGIN DATA
61  9  8 68 56    48  7 10 41 47    72 10 10 66 68    49  5  4 46 48
39  8  6 19 33    90  9 10 64 79    43  9 10 53 54    84 10 10 64 75
72 10 10 79 65    29  7  3 27 39    48  7  8 49 63    64  7  6 59 61
41  5  7 42 47    76  9 10 65 69    42  7  7 38 49    46  5 10 44 51
63 10  8 60 63    72  8 10 59 89    70  9 10 60 76    37  5  9 37 57
70  8 10 57 53    40  9 10 52 58    45  6  6 15 20
END DATA
TITLE 'PEARSON CORRELATIONS AMONG VARIABLES, DATA
   SET 2'
CORRELATIONS VARIABLES=AGEMONTH TO MVOCAB
  /PRINT=TWOTAIL
  /STATISTICS=DESCRIPTIVES
```

---

* J.P. Ryan. *Expansions in the Speech of Mothers to Their Young Children.* Doctoral dissertation, University of Chicago, 1977.

This code will produce the Pearson product-moment correlations (**A4-16**) for all pairs of variables as well as the mean (**A4-2**) and standard deviation (**A4-8**). The mean and standard deviation are produced because of the subcommand STATISTICS=DESCRIPTIVES. The printout of the correlations produced by CORRELATIONS includes the number of subjects in parentheses (23), and the two-tailed significance level in response to the subcommand TWOTAIL. A sample of the output produced by CORRELATIONS is shown below:

	AGEMONTH	CSYN	MSYN	CVOCAB	MVOCAB
AGEMONTH	1.0000	.6305	.5411	.7818	.7521
	(23)	(23)	(23)	(23)	(23)
	P= .	P= .001	P= .008	P= .000	P= .000
CSYN	.6305	1.0000	.4865	.6571	.5303
	(23)	(23)	(23)	(23)	(23)
	P= .001	P= .	P= .019	P= .001	P= .009
MSYN	.5411	.4865	1.0000	.6003	.6220
	(23)	(23)	(23)	(23)	(23)
	P= .008	P= .019	P= .000	P= .002	P= .002
CVOCAB	.7818	.6571	.6003	1.0000	.8121
	(23)	(23)	(23)	(23)	(23)
	P= .008	P= .011	P= .002	P= .	P= .000
MVOCAB	.7521	.5303	.6220	.8121	1.0000
	(23)	(23)	(23)	(23)	(23)
	P= .000	P= .009	P= .002	P= .000	P= .

The correlation between AGEMONTH and CSYN is .6305. Data from 23 subjects are used to compute this correlation. The significance level for this correlation, testing the hypothesis that the correlation differs from zero (**A4-44**), is shown below the number of subjects as P= .001. This is a two-tailed test in response to the subcommand TWOTAIL.

The correlation of AGEMONTH with the other variables, but not the intercorrelations among the other variables, could be produced with the code:

```
CORRELATIONS VARIABLES=AGEMONTH WITH CSYN TO
MVOCAB
```

The point biserial correlation (**A4-19**) and biserial correlation (**A4-20**) are special cases of the Pearson product-moment correlation and, thus, are produced by CORRELATIONS when data for the variables have the appropriate form. The point biserial correlation is obtained when one variable is discrete dichotomous and the other is continuous. The biserial correlation is obtained when one variable is dichotomous but has an underlying continuous normal distribution and the other variable is continuous.

NONPAR CORR

The Spearman rank order correlation (Rho) (**A4-17**) and Kendall's Tau (**A4-18**) can be calculated with the NONPAR CORR command. These correlation coefficients are obtained with the following command and subcommand:

```
NONPAR CORR VARIABLES=AGEMONTH TO MVOCAB
   /PRINT=BOTH
```

The PRINT=BOTH subcommand produces both the Spearman and Kendall coefficients. The default is only Spearman. The keyword WITH can be used in the VARIABLES= statement exactly as it was used in the CORRELATIONS procedure to obtain the correlations of one or more variables listed before the keyword WITH and one or more variables listed after the keyword WITH. The correlations among the variables preceding the WITH and the correlations among the variables following WITH are not computed. NONPAR CORR produces a one-tailed test of significance unless the subcommand PRINT=TWOTAIL is used.

CROSSTABS

The Phi-coefficient (**A4-21**) and contingency coefficient (**A4-23**) are produced by the CROSSTABS procedure using the subcommand STATISTICS. The Phi-coefficient and contingency coefficient for the variables GROUP and GENDER are produced by the following code:

```
TITLE 'EXAMPLE PHI AND CONTINGENCY COEFFICIENT'
CROSSTABS TABLES=GROUP BY GENDER
   /STATISTICS=PHI CC
```

PARTIAL CORR

Partial correlations (**A4-27**) describe the relationship between two variables while controlling for a third variable. Partial correlations are produced with the PARTIAL CORR procedure. An example of this procedure, using variables from Data Set 2, is as follows:

```
TITLE 'EXAMPLE OF PARTIAL CORRELATION'
PARTIAL CORR VARIABLES=CSYN TO MVOCAB BY
AGEMONTH(1)
   /STATISTICS=CORR DESCRIPTIVES
```

This command will compute the correlations among CSYN, MSYN, CVOCAB, and MVOCAB, while removing the influence of AGEMONTH. The keyword BY specifies the control variable to be partialed out of the correlations and the (1) following AGEMONTH specifies the first-order partial correlations. The subcommand STATISTICS=CORR DESCRIPTIVES produces the zero-order correlation matrix (CORR) and the means, standard deviations, and number of nonmissing cases.

Partial correlation (**A4-27**) and part correlation (**A4-26**) are also produced with the ZPP option in the STATISTICS subcommand of the REGRESSION procedure.

FACTOR ANALYSIS

SPSS-X provides a comprehensive approach to factor analysis (**A4-30**), with a wide range of subcommands and keywords in the FACTOR procedure. A variety of factor analyses, component analyses, and rotations can be carried out with this procedure. A useful description and illustration of FACTOR would be too lengthy for this chapter. The interested reader is referred to *SPSS-X User's Guide, 3rd Edition,* Chapter 28.

Regression Analysis (A4-31 to A4-39)

SPSS-X performs a wide range of regression analyses with the REGRESSION procedure. REGRESSION performs simple and multiple regression, with several options for selecting variables to be used in the equation. A variety of nonlinear regression analyses can be performed with the *SPSS-X* procedures CNLR (constrained nonlinear regression) and NLR (nonlinear regression). The description of these nonlinear analyses is beyond the scope of this chapter, since nonlinear analyses is not extensively covered in Section A.

A sample of *SPSS-X* code for a simple regression analysis (**A4-32**) is based on Data Set 2, shown below:

```
TITLE 'SIMPLE LINEAR REGRESSION'
REGRESSION VARIABLES=AGEMONTH CSYN
  /DEPENDENT=CSYN
  /METHOD=ENTER AGEMONTH
  /RESIDUALS
  /CASEWISE=DEFAULTS ALL
```

This is an analysis of the prediction of CSYN from AGEMONTH. The sample output using this REGRESSION procedure looks like this:

```
LISTWISE DELETION OF MISSING DATA
EQUATION NUMBER 1 DEPENDENT VARIABLE.. CSYN
BEGINNING BLOCK NUMBER 1. METHOD: ENTER    AGEMONTH
VARIABLE(S) ENTERED ON STEP NUMBER 1..          AGEMONTH

MULTIPLE R              .63054
R SQUARE                .39759
ADJUSTED R SQUARE       .36890
STANDARD ERROR          1.37507

ANALYSIS OF VARIANCE
                              SUM OF       MEAN
                    DF        SQUARES      SQUARE

REGRESSION          1         26.206       26.206

RESIDUAL            21        39.706       1.890

F=         13.859   SIGNIF F=  .0013


------------ VARIABLES IN THE EQUATION ------------
VARIABLE       B          SE B        BETA        t         SIG t

AGEMONTH      .064734    .017388     .630545    3.723      .0013
(CONSTANT)   4.120896   1.024510                4.022      .0006
```

The multiple R correlation (**A4-28**) is .63054, the R-squared (**A4-37**) is .39759. The standard error of estimate (**A4-35**) is 1.37507. The analysis of variance of the regression indicates that the independent variable accounts for a significant proportion of the variation in the dependent measure (R-squared =.397, F=13.85, P=.0013).

Estimates of the regression coefficient (**A4-33**) and Y-intercept (**A4-34**) are listed in the output above under the heading VARIABLE. The regression coefficient for AGEMONTH is .064734 and the Y-intercept is 4.120896. Thus the regression equation is

$$CSYN = .064734(AGEMONTH) + 4.12.$$

VARIABLES SUBCOMMAND

The VARIABLES subcommand following the REGRESSION command specifies the variables to be used in the analysis. The option VARIABLES=ALL may be used to include all variables. The VARIABLES=ALL option should be used with care, since the program calculates intercorrelations of all variables when the ALL option is used. If the VARIABLES subcommand is omitted, the default is an option named COLLECT, which includes all the variables in the DEPENDENT and METHOD subcommand.

DEPENDENT SUBCOMMAND

Subcommand DEPENDENT specifies the dependent variable(s). More then one dependent variable can be specified in the DEPENDENT subcommand and each dependent variable is analyzed separately. However, the dependent variables on a single DEPENDENT subcommand are not used as independent variables in the analysis of the other dependent variables listed on the same subcommand. Multiple DEPENDENT subcommands can be used with a single REGRESSION command.

METHOD SUBCOMMAND

The METHOD subcommand specifies the method by which the independent variable(s) are to be entered into the regression model and which independent variable or block of independent variables are selected by the method. The following are the options for selection with the METHODS subcommand:

FORWARD — Variables in the block are added to the equation, one per step, based on the order of smallest probability of the F-statistic, and only if they meet a specified criteria.

BACKWARD — All blocks of variables are included in the equation. Variables are then eliminated, one per step, based on the largest probability of the F-statistic.

STEPWISE — This method uses a combination of FORWARD and BACKWARD procedures. The variable with the largest probability of the F-statistic is removed from the equation if it exceeds the criterion. The remaining equation is recomputed to see if additional variables can be removed. After all variables have been removed in this fashion, the variable not in the equation with the smallest probability of F-statistic is entered. The equation is then recomputed for variables to be removed. The process is continued until all variables are removed or entered based on the criteria.

ENTER — All variables in a block are forced into the equation, one at a time, in a forward order based on the tolerance criteria.

REMOVE — All variables in a block are removed from the equation.

TEST — Tests the significance of the R-squared change based on the removal of a specified subset of items from the full model.

CRITERIA SUBCOMMAND

The CRITERIA subcommand is used to set statistical criteria for entering or removing variables in a regression analysis. The following are the criteria employed, with their default values listed in parentheses:

PIN(.05)	Probability of F needed to enter.
POUT(.10)	Probability of F needed to remove.
FIN(3.84)	Value of F-statistic needed to enter.
FOUT(2.71)	Value of F-statistic needed to remove.
MAXSTEPS(N)	Maximum number of steps. The default varies by METHOD subcommand.
TOLERANCE(.0001)	Proportion of variance for a variable independent of the variance attributable to other variables.

The CRITERIA subcommand must appear *after* the VARIABLES subcommand but *before* the DEPENDENT subcommand.

STATISTICS SUBCOMMAND

The STATISTICS subcommand appears before the DEPENDENT subcommand and is used to control the calculation and printing of statistics for the regression equation and independent variables. If STATISTICS is not specified the default is used, which includes the following options:

R	Multiple R, R-squared, adjusted R-squared, standard error of estimate.
ANOVA	Analysis of variance table.
COEFF	Unstandardized and standardized regression coefficients.
OUTS	Coefficients and statistics for variables not entered into the equation.

The ZPP option in the STATISTICS subcommand results in Pearson product-moment correlation (**A4-16**), part correlation (**A4-26**), and partial correlation (**A4-27**).

ANALYSIS OF RESIDUALS

The analysis of residuals is managed by the subcommands RESIDUALS, CASEWISE, and SCATTERPLOT. RESIDUALS is used to produce and print (1) statistical information about outliers and (2) histograms and normal probability plots for residuals and expected values from the last regression analysis performed. CASEWISE provides information similar to RESIDUALS and allows control of variables on a casewise basis. The casewise plot of standardized residuals, produced as one of the default options, can be especially helpful. SCATTERPLOT identifies variables used in bivariate plots and controls formating of the plots. The variables may include various forms of residuals produced by the RESIDUALS subcommand.

MULTIPLE REGRESSION

Multiple regression (**A4-36**), which includes curvilinear regression (**A4-38**) as a special case, is performed with the REGRESSION command using the subcommands and options described in the preceding sections. Multiple regression and stepwise multiple regression (**A4-39**) predicting CSYN from AGEMONTH, CVOCAB, MSYN, and MVOCAB from Data Set 2 is produced by the following code:

```
TITLE 'MULTIPLE REGRESSION'
REGRESSION VARIABLES=AGEMONTH TO MVOCAB
  /DEPENDENT=CSYN/ENTER
  /DEPENDENT=CSYN/STEPWISE
```

The subcommand /DEPENDENT=CSYN/ENTER specifies a multiple regression in which all independent variables are entered simultaneously in the order listed in the input. The regression coefficients (**A4-33**) using this model for AGEMONTH, CVOCAB, MSYN, and MVOCAB are, respectively, .033, .049, .116, and −.020. The Y-intercept (**A4-34**) is 3.547. The standard error of estimate (**A4-35**) is 1.37. The R-squared (**A4-37**) is .484. The multiple correlation (**A4-28**) is the square root of the R-squared, or .696.

The subcommand /DEPENDENT=CSYN/STEPWISE requests stepwise multiple regression.

Within the framework of multiple regression, part (**A4-26**) and partial (**A4-27**) correlations are produced as an option in the STATISTICS subcommand as follows:

```
TITLE 'PART AND PARTIAL CORRELATION'
REGRESSION VARIABLES=AGEMONTH CSYN CVOCAB
  /STATISTICS=DEFAULTS ZPP
  /DEPENDENT=CSYN/ENTER AGEMONTH, CVOCAB
```

The correlation between CSYN and CVOCAB is .657. The part correlation between CSYN and CVOCAB .263. This is the part correlation between children's scores on the syntax measure (CSYN) and scores on the vocabulary test, with the influence of age removed from the vocabulary scores. The partial correlation between children's vocabulary and syntax scores is .339.

Curvilinear regression can be performed by using nonlinear transformations of the independent variable(s). For example, the following line of code

```
COMPUTE AGESQ=AGEMONTH*AGEMONTH
```

would produce the quadratic form of the independent variable AGEMONTH, which would be of interest if the relationship between CSYN and children's age was thought to be curvilinear. The curvilinear regression of CSYN and AGEMONTH would be specified as follows:

```
TITLE 'CURVILINEAR REGRESSION'
REGRESSION VARIABLES=CSYN AGEMONTH AGESQ
  /DEPENDENT=CSYN
  /METHOD=ENTER AGEMONTH, AGESQ
```

This model would predict CSYN from two independent variables, namely, AGEMONTH and AGESQ, in which AGESQ is the quadratic version of AGEMONTH (AGEMONTH*AGEMONTH).

Comparison: One Sample (A4-40 to A4-47)

Hypotheses related to one sample involve the comparison of a mean (**A4-41**), variance (**A4-42**), proportion (**A4-43**), or correlation (**A4-44** and **A4-45**) to some value, or the comparison of a set of frequencies to some expected frequency values (**A4-46** and **A4-47**). Two general approaches can be used in applying *SPSS-X* to test such hypotheses. First, *SPSS-X* can be used to calculate the appropriate sample statistic, then the actual test statistic (z, Chi-square, t, or F) for testing the hypothesis (**A4-41** to **A4-47**) can be calculated with a hand calculator. This is often a practical and sensible approach when a small number of statistical tests are involved. DESCRIPTIVES can be used to obtain sample means and variances. Sample proportions and frequencies can be produced with FREQUENCIES. CORRELATIONS will calculate sample correlation coefficients and test the hypothesis that the correlations differ from zero (**A4-44**). *SPSS-X* performs a one-tailed test as the default, but the two-tailed test can be requested.

A second approach involves using *SPSS-X* to calculate the test statistics for hypotheses related to one sample. This is advisable if there are a large number of statistical tests. Hypotheses comparing the mean to some value (A) or the variance to some value (A) can be tested using a COMPUTE statement and the information obtained from DESCRIPTIVES. For example, the mean, standard error of the mean, and variance for the variable AGEMONTH can be obtained by using DESCRIPTIVES. If these three statistics for AGEMONTH are named MEANAGE, SEAGE, and VARAGE, respectively, then the following *SPSS-X* COMPUTE statements will calculate (1) the appropriate t-statistic for testing the mean against some hypothesized value A and (2) the appropriate Chi-square statistic for testing the variance against some hypothesized value A:

```
COMPUTE T=(MEANAGE−A)/SEAGE
COMPUTE CHISQ=((N−1)*VARAGE)/A
```

The *SPSS-X* code COMPUTE T=(MEANAGE−A)/SEAGE calculates the t-statistic described in **A4-41** for comparing the mean (MEANAGE) to some value (A). The *SPSS-X* code COMPUTE CHISQ=((N−1)*VARAGE)/A, calculates the Chi-square statistic described in **A4-42** for comparing the variance (VARAGE) to some value.

Hypotheses in which a sample proportion is compared to some population value (**A4-43**) are tested with a z-statistic. This z-statistic can be calculated in the *SPSS-X* data step. If PS represents the sample proportion, PP represents the population proportion, and N is the sample size, then the z-statistic is obtained by using the following *SPSS-X* coding:

```
COMPUTE Z=(PS − PP)/(SQRT((PP*(1−PP))/N))
```

Sample correlations are frequently examined to determine if they are significantly different from zero (**A4-44**). This test is automatically performed by the *SPSS-X* CORRELATIONS procedure and the exact significance level is reported for a one-tailed test. The significance level for a two-tailed test can be requested as an option. Hypotheses comparing sample correlations to some other hypothesized value (**A4-45**) may also be of interest. Such comparisons require the Fisher Z-transformation of both the sample correlation and the hypothesized correlation. This transformation has the following form:

```
ZR = .5*(LN((1+r)/(1−r)))
```

If the sample correlation is noted as SCORR, the hypothesized population value is noted as PCORR, and N is the sample size, the following *SPSS-X* coding will produce the z-statistic needed to test the hypothesis that the two correlations are equal. This coding would appear in the data step:

```
COMPUTE ZRSCORR = .5 * (LN((1+SCORR)/(1-SCORR)))
COMPUTE ZRPCORR = .5 * (LN((1+PCORR)/(1-PCORR)))
COMPUTE          Z = (ZRSCORR-ZRPCORR)/SQRT(1./(N-3))
```

It is useful to note that the *SPSS-X* coding for the one-sample hypotheses about a proportion (**A4-43**) or a correlation compared to some value (**A4-45**) basically employ *SPSS-X* as a calculator. As mentioned before, these calculations may be more efficiently performed with a hand calculator.

SPSS-X performs the Chi-square goodness-of-fit test (**A4-46**) and the Kolmogorov-Smirnov one-sample test (**A4-47**) with the NPAR TESTS (nonparametric tests) procedure, subcommands, and options. For example, consider a situation in which the proportion or number of subjects in each of four groups is of interest. The coding below provides a goodness-of-fit test examining whether the categories of GROUP have equal frequencies:

```
NPAR TESTS CHISQUARE=GROUP
```

Related, but slightly different, is the goodness-of fit test for which there are expected values for the frequencies in each category. For example, the following code

```
NPAR TESTS CHISQUARE=GROUP
   /EXPECTED=25,30,30,15
```

specifies the expected frequency in each category of the variable GROUP. These expected frequencies are summed, and each expected frequency is divided by the total to estimate the proportion of subjects expected in each category.

The Kolmogorov-Smirnov one-sample test (**A4-47**) can be used to test whether the observed distribution of some variable follows a uniform, normal, or Poisson distribution. The following coding would test the normality of a standardized variable X:

```
NPAR TESTS K-S (NORMAL,0,1) = X
```

The values 0 and 1 in this example are the mean and standard deviation of the hypothesized distribution. *SPSS-X* uses the observed mean and standard deviation if values for these parameters are not specified.

Comparisons: Two Samples (Parametric) *(A4-48 to A4-56)*

Comparisons involving two samples include both independent- and dependent-sample tests. Hypotheses for independent and dependent samples, respectively, examine questions about means (**A4-49** and **A4-50**), variances (**A4-51** and **A4-52**), proportions (**A4-53** and **A4-54**), and correlations (**A4-55** and **A4-56**). As in the one-sample case, two general approaches can be used in applying *SPSS-X* to test these hypotheses. First, *SPSS-X* can be used to calculate the appropriate statistics for each of the two samples. The actual test statistic (z, Chi-square, t, or F) for examining differences between the sample statistics (**A4-49** to **A4-56**) can then be determined easily with a hand calculator. This is often a practical and sensible approach when a small number of statistical tests are involved, especially if the tests relate to comparisons of variances, proportions, and correlations. A second approach involves using *SPSS-X* to calculate the actual test statistics for hypotheses related to two samples. This approach is advisable if there are a large number of statistical tests and is especially appropriate for comparisons of means.

MEANS

SPSS-X is especially suited for testing hypotheses about the means for two independent or dependent samples. The comparison of independent sample means (**A4-49**) is performed with the T-TEST procedure. The use of this procedure can be illustrated with Data Set 1 using the following *SPSS-X* code:

```
TITLE 'TWO-SAMPLE INDEPENDENT T-TEST'
T-TEST GROUPS=GROUP(1,2)/ VARIABLES = POSTTEST
```

The GROUPS subcommand specifies the independent variable with two levels (1,2). In this example, the specification of the categories as 1 and 2 is unnecessary because the values for the variable GROUP in Data Set 1 include only 1 and 2. Any two values can be used in the parentheses to define the two groups. The results of this independent-sample T-TEST procedure are shown in the output below.

VARIABLE	NUMBER OF CASES	MEAN	STANDARD DEVIATION	STANDARD ERROR	t-VALUE	2-TAIL PROB
POSTTEST						
GROUP 1	5	32.00	4.359	1.949		
					2.93	.019
GROUP 2	5	25.00	3.082	1.378		

For each group this includes the number of subjects, the mean, standard deviation, standard error, and the t-statistic testing the hypothesis that the means are statistically equivalent. The validity of this statistical test assumes that the variances for the two groups are equivalent (homogeneity of variances). T-TEST actually carries out the statistical test examining the assumption that the variances for the two groups are equal (**A4-51**). The F-statistic and significance level for the comparison of the independent-sample variances are produced by T-TEST. In this example, F=2.00, P=.51, indicating that the assumption of homogeneity of variance cannot be rejected.

A single value may be specified as a means for defining the two categories of the independent variable. In such a case, all subjects with a value for the independent variable equal to or greater than the specified value are placed on one category, all other subjects are in the second category. For example:

```
T-TEST GROUPS=RACE(2)/ VARIABLES =POSTTEST
```

In this example, all subjects with a value of 2 or greater for the variable RACE form one group, all subjects with a value for RACE less than 2 form the second group.

The comparison of means for two dependent samples (**A4-50**) is also performed by T-TEST. The following coding is used for the dependent-sample t-test:

```
TITLE 'DEPENDENT SAMPLE T-TEST'
T-TEST PAIRS=PRETEST POSTTEST
```

The PAIRS subcommand defines the two variables that are to be compared for each subject. The following sample of output for the dependent-sample T-TEST procedure was generated by this coding:

VARIABLE	NUMBER OF CASES	MEAN	STANDARD DEVIATION	STANDARD ERROR	t VALUE	2-TAIL PROB.
PRETEST		18.0	3.944	1.247		
	10				−9.84	.000
POSTTEST		28.5	5.126	1.621		

This procedure produces the mean, standard deviation, standard error for each variable, the t-statistic related to the hypothesis that the mean of the difference is zero, and the significance level for the t-statistic and appropriate degrees of freedom.

VARIANCES

Comparison of the variances for two independent samples (**A4-51**) is performed with the T-TEST procedure when the test of the assumption of homogeneity of variance is performed. The dependent-sample t-test comparing two variances (**A4-52**) is calculated using the variances of the two variables (PRETEST, POSTTEST), the correlation between the two variables, and the number of subjects involved. In most cases, it is relatively simple to calculate this t-statistic with a hand calculator using the variances, correlation, and sample size obtained from the *SPSS-X* procedures DESCRIPTIVES and CORRELATIONS.

PROPORTIONS

Procedures for comparing proportions for two independent or dependent samples are described in **A4-53** and **A4-54,** respectively. In most cases, these comparisons can be performed efficiently with a hand calculator using information from FREQUENCIES or CROSSTABS. With Data Set 1, for example, the coding

```
CROSSTABS TABLES = TRSP BY GENDER
    /CELLS
```

would produce a 2 × 2 contingency table. The information in this table could be used to compare the proportion of students in level 1 of TRSP for students in Group 1 compared to Group 2. This would be a comparison of proportions for independent samples (**A4-53**). In addition, the comparison of the proportion of students in Group 1 who are in levels 1 and 2 of TRSP also could be calculated directly from information contained in the 2 × 2 table. This would be a comparison of proportions for dependent samples (**A4-54**). (This example with Data Set 1 is used only for illustrative purposes. The small sample size of Data Set 1 would make the procedures in the illustration statistically inappropriate.)

CORRELATIONS

The comparison of independent-sample correlations (**A4-55**) requires that the correlation between two variables be calculated separately for each of two groups. This easily accomplished with the SORT CASES and SPLIT FILE command as follows:

```
SORT CASES BY GROUP
SPLIT FILE BY GROUP
TITLE 'SAMPLE CORRELATION BY GROUP'
CORRELATIONS VARIABLES=PRETEST POSTTEST
SPLIT FILE OFF
```

Data must first be sorted into the levels of the variable that defines the comparison categories or groups. This is accomplished with SORT CASES BY GROUP and SPLIT FILE BY GROUP. The CORRELATIONS procedure is then executed for each level of the variable in the BY statement. This produces the standard output for CORRELATIONS separately for the subjects in each level of GROUP. The correlation between PRETEST and POSTTEST for Group 1 and Group 2 can then be compared as described in **A4-55**. The SPLIT FILE command will remain in effect and will be used in other procedures unless negated with SPLIT FILE OFF.

Data needed for the comparison of two-sample correlation coefficients for dependent samples (**A4-56**) is generally produced with the CORRELATIONS procedure. A typical application of this test can be illustrated with Data Set 1. A researcher might be interested in knowing whether the correlation between PRETEST and IQ is significantly different from the correlation between POSTTEST and IQ. The correlations needed to calculate the t-statistic (**A4-56**) used to test whether the differences between the correlations is significant are produced by the following *SPSS-X* coding:

```
CORRELATIONS VARIABLES= PRETEST POSTTEST IQ
```

Comparisons: Two Samples (Nonparametric) (A4-57 to A4-65)

The *SPSS-X* command NPAR TESTS is a useful and comprehensive procedure for performing a wide range of nonparametric statistical tests. NPAR TESTS has subcommands for the following two-sample nonparametric statistics:

Independent-Sample Tests

(**A4-58**) Median Test
(**A4-59**) Wilcoxon Rank Sum Test
(**A4-60**) Mann-Whitney U-Test
(**A4-62**) Kolmogorov-Smirnov Two-Sample Test

Dependent-Sample Tests

(**A4-63**) McNemar's Test
(**A4-64**) Sign Test
(**A4-65**) Wilcoxon Matched-Pairs Signed-Rank Test

The *SPSS-X* coding for the median, Mann-Whitney, and Kolmogorov-Smirnov tests is as follows:

```
TITLE 'MEDIAN, MANN-WHITNEY AND K-S TWO SAMPLE
TESTS'
NPAR TESTS MEDIAN = POSTTEST BY GROUP (1,2)
NPAR TESTS M-W = POSTTEST BY GROUP (1,2)
NPAR TESTS K-S = POSTTEST BY GROUP (1,2)
```

The subcommands MEDIAN, M-W, and K-S specify the desired nonparametric test. The dependent variable or variables are listed after the equal sign. The variable following the keyword BY defines the independent variable. The two values of the independent variables used to define the two groups are specified in parentheses. The NPAR TESTS procedure produces a variety of descriptive statistics, the test statistic for the appropriate subcommand, and the associated significance level.

The MEDIAN subcommand calculates the median unless a median value appears in parentheses between the word MEDIAN and the equals sign. The M-W subcommand, in addition to producing the Mann-Whitney U-Test (**A4-60**), also produces the W-statistic for the Wilcoxon Rank Sum Test (**A4-59**).

The Chi-square test of independence for two samples (**A4-61**) is produced with the CROSSTABS procedure. This statistical test is illustrated using the following code based on Data Set 1 variables:

```
TITLE 'CHI-SQUARE TEST OF INDEPENDENCE'
CROSSTABS VARIABLES = GROUP (1,2) TRSP (1,2)
  /TABLES=GROUP BY TRSP
  /CELLS=COUNT ROW COLUMN TOTAL EXPECTED
  /STATISTICS=CHISQ
```

The use of CROSSTABS is described in some detail in the earlier section on descriptive statistics.

The nonparametric two-sample tests for dependent or correlated samples are produced by NPAR TESTS using the subcommands MCNEMAR, SIGN, or WILCOXON. McNemar's test (**A4-63**) is used when two dichotomous variables are involved. For example, if students' performance is recorded as pass = 1 or fail = 0 on both a pretest and a posttest, the following code would use McNemar's test to determine if there is a significant difference in the proportions of students with 1 or 0 on PRETEST compared to the POSTTEST:

```
TITLE 'EXAMPLE OF MCNEMARS TEST'
NPAR TESTS MCNEMAR = PRETEST POSTTEST
```

Variables used in the McNemar procedure must be dichotomous as entered or made dichotomous through a recoding or transformation.

The sign test for dependent samples (**A4-64**) is performed using the following code:

```
TITLE 'EXAMPLE OF SIGN TEST'
NPAR TESTS SIGN = PRETEST POSTTEST
```

For the SIGN test, the variables PRETEST and POSTTEST do not need to be dichotomous because the analysis counts the number of positive and negative differences between the variables over all the subjects.

The *SPSS-X* format for the Wilcoxon Matched-Pairs–Signed-Rank test (**A4-65**) that follows is virtually identical to that of the sign test (**A4-64**):

```
TITLE 'WILCOXON MATCHED-PAIRS SIGNED-RANK TEST'
NPAR TESTS WILCOXON = PRETEST POSTTEST
```

The Wilcoxon ranks the absolute value of the difference between the variables for each subject and then sums the positive and negative ranks. The test statistic z is produced along with its significance level.

Comparisons: Two or More Samples (Nonparametric) (A4-66 to A4-69)

The sign test for K-independent samples (or K-sample median test) (**A4-67**) and the Kruskal-Wallis rank test for independent samples (**A4-68**) are performed with NPAR TESTS. Examples of the *SPSS-X* coding used to obtain these procedures is as follows:

```
TITLE 'SIGN TEST (MEDIAN) AND KRUSKAL-WALLIS,
K-SAMPLES'
NPAR TESTS MEDIAN = POSTTEST BY GROUP (1,4)
NPAR TESTS K-W = POSTTEST BY GROUP (1,4)
```

The dependent variable for the K-independent sample median test, or sign test (**A4-67**), is POSTTEST. The keyword BY specifies the independent variables as having four levels. These instructions produce a two- (above and below the median) by-four (groups 1, 2, 3, and 4) table. The Kruskal-Wallis test, examining the ranks based on the total sample within each group, uses the same code as the median test.

The Friedman two-way analysis of variance by ranks (**A4-69**) is also a subcommand in NPAR TESTS. This test is used when each subject is measured on K-variables that are ordinal level or are treated as ordinal level. For example, consider a situation in which subjects are measured before (PRE), during (MID), and after (POST) an experiment. The Friedman analysis would be performed with the following code:

```
TITLE 'FRIEDMAN ANOVA FOR K-RELATED SAMPLES'
NPAR TESTS FRIEDMAN= PRE MID POST
```

NPAR produces the Chi-square test statistic for this analysis along with its significance level.

Comparisons: Two or More Samples—Analysis of Variance (ANOVA) (A4-70 to A4-90)

Analysis of variance, used to compare the means of two or more samples, is a powerful and general technique used to test hypotheses generated from a number of basic research designs. Factorial (**A4-71** to **A4-78**), repeated measures/split-plot (**A4-79** to **A4-83**), and nested or hierarchical (**A4-84** to **A4-86**) analyses are described in Section A, Chapter 4. The use of *SPSS-X* to perform these analyses will be demonstrated for the most commonly occurring factorial models, **A4-71** to **A4-78**. *SPSS-X* performs repeated-measures, split-plot, and nested analysis of variance in the context of multivariate analysis of variance. Since the multivariate approach was not included in Section A, these analyses will not be examined in the following discussion.

Three *SPSS-X* procedures can be used for factorial analysis of variance. These are ONEWAY, MEANS, and ANOVA. ONEWAY and MEANS are applicable only to one-way analysis of variance. ANOVA performs multiple-factor analysis of variance (and covariance). Each of these procedures, with relevant subcommands and options, will be illustrated below.

FACTORIAL ANALYSIS OF VARIANCE (**A4-71** TO **A4-78**)

Factorial ANOVA procedures will be illustrated with Data Set 3, which involves 18 subjects in a two-factor design. The factors are (A) treatment, with three levels (E1, E2, and C) and (B) gender, with two levels (F and M). The dependent measure is a test score. These data are read in *SPSS-X* as follows:

```
DATA LIST FREE / GROUP (A2) Gender (A) SCORE
BEGIN DATA
  E1  F  110
  E1  F   95
  E1  F   98
  E1  M   79
  E1  M   73
  E1  M   72
  E2  F  112
  E2  F  106
  E2  F   99
  E2  M   84
  E2  M   92
  E2  M   86
  C   F   95
  C   F   86
```

```
C  F  91
C  M  68
C  M  79
C  M  72
END DATA
RECODE GROUP ('E1'=1) ('E2'=2) ('C'=3) INTO TREAT/
     GENDER ('F'=1) ('M'=2) INTO SEX
```

The recoding of the string variables GROUP and GENDER into the numeric variables TREAT and SEX is necessary because the ONEWAY command requires numeric values for the levels of the independent variables. The recoded variables will be used for the examples using Data Set 3.

A one-way ANOVA (**A4-71**), with TREAT as the independent variable, is performed with ONEWAY using the following *SPSS-X* coding:

```
TITLE 'ONE-WAY ANOVA WITH THE ONEWAY PROCEDURE'
ONEWAY SCORE BY TREAT(1,3)
  /STATISTICS=DESCRIPTIVES HOMOGENEITY
```

The ONEWAY command specifies the dependent variable SCORE and the independent variable TREAT. The minimum and maximum values of TREAT are specified in the parentheses as 1 and 3, respectively. The STATISTICS subcommand uses the DESCRIPTIVES option to produce basic descriptive statistics for each cell in the analysis and for the total sample. The HOMOGENEITY option in the STATISTICS subcommand produces the test for homogeneity of variance within groups with the appropriate significance level. A sample of some of the output produced by *SPSS-X* using the ONEWAY ANOVA procedure follows:

VARIABLE SCORE
BY VARIABLE TREAT

ANALYSIS OF VARIANCE

SOURCE	DF	SUM OF SQUARES	MEAN SQUARE	F-RATIO	F-PROB
BETWEEN GROUPS	2	652.44	326.22	2.05	0.163
WITHIN GROUPS	15	2389.16	159.27		
TOTAL	17	3041.61			

GROUP	COUNT	MEAN	STANDARD DEVIATION	STANDARD ERROR
Grp 1	6	87.83	15.45	6.31
Grp 2	6	96.50	11.16	4.55
Grp 3	6	81.83	10.68	4.36
TOTAL	18	88.72	13.37	3.15

The ONEWAY subcommands POLYNOMIAL, CONTRAST, and RANGES can be used to examine a variety of multiple comparisons. These will be explained in a later section.

The MEANS command produces a one-way analysis of variance using the following coding:

```
TITLE 'EXAMPLE OF ANOVA WITH MEANS COMMAND'
MEANS TABLES=SCORE BY TREAT
  /STATISTICS=ANOVA
```

In this example, the dependent variable is specified as SCORE and the independent variable is TREAT. The MEANS procedure does not require that the independent variable be numeric.

The ANOVA procedure uses coding similar to that of ONEWAY to produce a one-way analysis of variance. The following coding applies the ANOVA procedure to the previous example.

```
TITLE 'EXAMPLE OF ONE-WAY ANOVA WITH ANOVA COMMAND'
ANOVA VARIABLES= SCORE BY TREAT(1,3)
  /STATISTICS=MEAN
```

The VARIABLES list defines the dependent variable SCORE, the independent variable TREAT, and specifies the levels of the independent variables as a minimum of 1 and maximum of 3. The levels of the independent variable must be integers. The MEANS option in the STATISTICS subcommand requests the cell sizes and means.

A two-way ANOVA (**A4-72**), sex (2) X treatment (3), is the appropriate analysis for the information in Data Set 3. The two-way analysis of variance, using the ANOVA procedure, is coded as follows:

```
TITLE 'TWO-WAY ANOVA, SEX BY TREATMENT'
ANOVA VARIABLES=SCORE BY SEX(1,2) TREAT(1,3)
  /STATISTIC=MEAN
```

The VARIABLES list defines the dependent variable as SCORE, and specifies two independent variables. The independent variables are SEX, with minimum and maximum values of 1 and 2, and TREAT, with minimum and maximum values of 1 and 3. The STATISTICS subcommand with the MEAN option produces the means and frequencies for each cell and for each level of the independent variables. A portion of the output from this two-way ANOVA analysis is shown on the next page:

	SCORE				
BY	SEX				
	TREAT				
SOURCE OF VARIATION	SUM OF SQUARES	DF	MEAN SQUARE	F	SIG of F
MAIN EFFECTS	2595.16	3	865.05	27.56	.000
SEX	1942.72	1	1942.72	61.89	.000
TREAT	652.44	2	326.22	10.39	.002
2-WAY INTERACTIONS	69.78	2	34.89	1.11	.361
SEX TREAT	69.78	2	34.89	1.11	.361
EXPLAINED	2664.94	5	532.99	16.98	.000
RESIDUAL	376.67	12	31.38		
TOTAL	3041.61	17	178.92		

The basic coding used for two-way ANOVAs applies to three-way ANOVA models and higher-order factorial designs.

It is critical to note that *SPSS-X* procedures for factorial analysis of variance assume that the data have come from a fixed-effects analysis of variance design. Researchers with random or mixed-effects designs need to use the appropriate models and select the appropriate error terms for the F-test examining different hypotheses. The *SPSS-X* procedures estimate all the required mean squares, leaving the researcher to select the correct effect and to carry out the F-test.

RELATED DESIGNS (A4-89)

Analysis of covariance (ANCOVA) (**A4-89**) improves on the statistical precision of analysis of variance by including a covariate(s) to reduce the proportion of variation attributed to error or residual. Data Set 1 will be used as an illustration. A simple one-way ANOVA is performed with POSTTEST as the dependent variable and GROUP as the independent variable. The data could also be analyzed as a one-way ANCOVA by including PRETEST as a covariate. A common procedure for analyzing such data would begin by testing whether the groups were significantly different on the covariate (PRETEST) and on the dependent variable (POSTTEST). This is followed by the one-way analysis of covariance with POSTTEST as the dependent variable, GROUP as the independent variable, and PRETEST as the covariate. The coding for these procedures is as follows:

```
TITLE 'ANOVA OF PRETEST AND POSTTEST'
ANOVA VARIABLES=PRETEST POSTTEST BY GROUP(1,2)
    /STATISTICS=MEAN
TITLE 'ANALYSIS OF COVARIANCE'
ANOVA VARIABLES=POSTTEST BY GROUP(1,2) WITH PRETEST
```

The first ANOVA procedure performs a one-way analysis of variance testing whether the groups are significantly different on the PRETEST and POSTTEST. The second ANOVA procedure actually performs the one-way ANCOVA with POSTTEST being the dependent variable, PRETEST being the covariate, and GROUP being the independent variable. The output from the analysis of covariance is very similar to the analysis of variance output previously displayed but adds the covariate to the model. The sum of squares, degrees of freedom, mean square, F-ratio, and significance level for the covariate are reported. The sum of squares, degrees of freedom, and mean square for the residual are changed in relationship to the proportion of variation accounted for by the covariate.

Multiple-Comparison Procedures (A4-91 to A4-97)

Multiple-comparison procedures are performed in *SPSS-X* within the ONEWAY procedure. Multiple-comparison procedures are specified with the subcommands POLYNOMIAL, CONTRAST, and RANGES. The one-way ANOVA for Data Set 3 is used below to illustrate the coding needed to perform all specific *a priori* contrasts, such as Groups 1 and 2 compared to Group 3, orthogonal contrasts (**A4-92**), orthogonal polynomial contrasts (**A4-93**), *post hoc* pairwise contrasts using Scheffé (**A4-94**), Tukey (**A4-95**), Newman-Keuls (coded as SNK) (**A4-96**), and the Duncan (**A4-97**) multiple-comparison procedure.

```
TITLE 'ONE-WAY ANOVA WITH MULTIPLE COMPARISONS'
ONEWAY SCORE BY GROUP(1,3)
  /CONTRAST = 1 1 -2
  /POLYNOMIAL = 2
  /RANGES=SCHEFFE
  /RANGES=TUKEY
  /RANGES=SNK
  /RANGES=DUNCAN
```

The subcommand CONTRAST is used to compare a group or combinations of groups to another group or combination of other groups. The contrast coefficients following the CONTRAST subcommand in the example above are 1 1 −2. This compares Groups 1 and 2 combined to Group 3. The POLYNOMIAL subcommand requests orthogonal polynomial contrast (**A4-93**). The POLYNOMIAL subcommand must specify the highest-order polynomial to be tested. This must be at least one less than the number of groups. In the example above with three groups, this is 2. The SCHEFFE, TUKEY, SNK (Student Neuman Keuls), and DUNCAN procedures are specified with RANGES subcommands.

CHAPTER

5

Minitab

Introduction

*Minitab**, as a mainframe statistical package, has been around for a long time. It is a package utilized as the basis for a number of introductory statistical texts and is available in forms for mainframe computers, minicomputers, and microcomputers. The most recent version is Release 7.1, which was delivered in mid 1989. The *Minitab* package is one of the most accessible packages on university campuses.

The following discussion is intended to be universal; that is, not specific to any single system. This means that the illustrated programs and models will be applicable regardless of the system environment available to the user. Documentation is available in a variety of forms, but the primary sources include the *Minitab Reference Manual*, the *Minitab Handbook* (2d edition, revised), and the *Minitab Primer*. Additional documentation is available depending on the system environment in which the user is working. All three of the above manuals are strongly recommended, regardless of the user's specific environment, particularly the *Minitab Reference Manual*.

Text Format

Examples of *Minitab* commands and output will be shown throughout this chapter. The *Minitab* code will be printed in CAPITAL letters when referenced in a paragraph, and it will be printed in a shaded area between parallel lines when *Minitab* statements appear on separate lines that illustrate the code. Examples in this chapter will be written as if the left-most *Minitab* code is in the first column of the input screen. Examples of *Minitab* output will be displayed in boxes to set them apart from the text.

---

*Minitab is a registered trademark of Minitab, Inc., State College, PA.

Some Minitab Basics

Minitab is generally accessed by logging onto (getting on) the computer facility where *Minitab* is stored (again, system specifics will generally dictate this process). Once in the proper setting, entry to the *Minitab* program is obtained by typing the word MINITAB. Depending on the environment, the user may or may not see several system-specific screens after which the user will see the *Minitab* prompt: MTB>. This prompt indicates that you are in the *Minitab* environment and ready to begin your data-analysis activity.

Minitab permits a variety of input strategies (including free format and fixed format), but did not introduce a full-screen editor (which is available only on the PC version) until Release 7.0. As a consequence, since not all users will have access to the screen editor, the data-input and editing strategies to be discussed will be limited to the basic editor and input strategies common to all systems.

Minitab employs the notion of storing data in the form of rows and columns, much in the manner of an electronic spreadsheet. Indeed, many of the arithmetic operations within *Minitab* operate in a fashion similar to an electronic spreadsheet. However, unlike spreadsheet programs, the only information contained in *Minitab* is in the form of data, either in alphanumeric (character) or numeric format. Labels are not directly a part of the row-by-column format and must be identified separately through the NAME command. The following is an illustration of the data-storage format of *Minitab*:

ROW	C1	C2	C3	C4	C5
1	61	9	8	68	56
2	48	7	10	41	47
3	72	10	10	66	68
4	49	5	4	46	48
5	39	8	6	19	33
6	90	9	10	64	79
7	43	9	10	53	54
8	84	10	10	64	75
9	72	10	10	79	65

The above data are drawn from the data identified throughout Section C as Data Set 2 and are stored in *Minitab* as DASET2.MTW. The extension identifies the file as a *Minitab* worksheet file. (The exact nature of file extensions is unique to each computer system so please check with your system.) The first thing to notice is that the columns do not have labels above them, and will not until the user tells *Minitab* to NAME C1 'AGEMONTH'. The previous command identifies the column of interest (C1) and the label, within single quotes, to be given the column. If the user completes the labeling process and directs *Minitab* to PRINT C1-C5, the output will resemble the following:

ROW	AGEMONTH	CSYN	MSYN	CVOCAB	MVOCAB
1	61	9	8	68	56
2	48	7	10	41	47
3	72	10	10	66	68
4	49	5	4	46	48
5	39	8	6	19	33

The ROW column is automatically generated by *Minitab* and is similar to the __N__ or CASE numbering system used by other statistical packages. Within *Minitab* each row is treated independently and can be directly referenced by an operation, argument, or statistical activity.

Data Entry (Input)

The entry of data into a *Minitab* program format may be accomplished in four primary ways: through the commands READ, SET, INSET, or RETRIEVE. Each command operates either upon an external file to move it into a *Minitab* worksheet or to combine data with an existing *Minitab* worksheet. The initial strategy often employed to create a *Minitab* worksheet is the READ command. The command is written as indicated below:

> READ 'FILENAME' into C1, C2, . . . Cx$_i$

If the user does not indicate a file name, then *Minitab* assumes that the data will be directly inputted through the keyboard.

As with many computer-based statistical programs, *Minitab* presupposes that any file read into it will be in ASCII format, and it will have a file-extension name of DAT (once again, the nature of the extension depends upon your local computer system). One of the idiosyncracies of *Minitab* is the need to identify the columns into which the data will be read. Unlike many of the other programs analyzed in this book, *Minitab* does not permit multiple-record sets on a single line during the input stage. This limitation often requires that the user have a clearly designed model of the data format (as it will later be needed for analysis) before beginning the read process. This does not mean you cannot read the data in through a multiple-set-of-records format and then later employ a CONCATENATE command to merge the data into proper data form for analysis, but it is often much easier to perform the data construction needs at the beginning than during the analysis stage.

If the user ends a command with a semicolon (;), this indicates to *Minitab* that a subcommand follows. The subcommand that typically follows a READ command is a FORMAT command. This command tells *Minitab* the arrangement of the data and designates which numbers or characters go into each column. When the user ends the READ command with a semicolon, *Minitab* begins the next line with a prompt (SUBC>) indicating a subcommand structure is in effect.

The FORMAT subcommand resembles a FORTRAN input style. Its application will typically reduce the time necessary to READ a file into the system (especially if the file is a large one). The format components are:

Fn	read an n-length numeric column
Fn.d	read an n-length numeric column with d decimal places
An	read an n-length character column*
X	skip a column
Tn	tab (skip) to column n
n	repeat factor
(	open parenthesis
)	close parenthesis
'	used to separate format lines
/	forward slash—move to next line

The following Data Set 1 should be read with the commands that follow it (the first two lines are column markers only):

```
             1         2         3
    12345678901234567890123456789 0
     1 1 112 21 35 2
     2 1 109 17 33 2
     3 1 105 19 31 1
     4 1 118 25 36 2
     5 1  95 18 25 1
     6 2 111 19 24 1
     7 2 107 20 29 2
     8 2 102 11 24 1
     9 2  98 17 27 1
    10 2 105 13 21 1
```

    MTB > READ 'DASET1' C1-C6;
    SUBC> FORMAT (F2,1X,F1,1X,F3,2(1X,F2),1X,F1).

The FORMAT command reads:

1. beginning in column 1 read a two variable-numeric figure and place in C1,
2. skip one space and read a one-variable numeric figure and place in C2,
3. skip one space and read a three-variable numeric figure and place in C3,

---

*Prior to Release 7.0 the maximum length of an alphanumeric (character) variable was 4 for READ and 80 for SET and INSERT. In Release 7.0 and above, all have a maximum length of 80.

4. for the next two figures, skip 1 space and read a set of two-variable numberic figures and place them in C4 and C5, and finally,
5. skip one space and read a one-variable figure and place it in C6.

The period (.) at the end of the FORMAT subcommand indicates the end of a SUBCOMMAND line. Failure to include this period would indicate to *Minitab* that another subcommand line follows. Should you fail to include the period, simply type it in by itself on the next command line. If you need more than one line to write your FORMAT subcommands, use the ampersand symbol (&) at the end of the line to indicate that the same set of information continues on the line that follows.

To further illustrate the flexible nature of the FORMAT command strategy, we can use a table similar to one on page 230 in Chapter 3 (*SYSTAT*). The table contains nine different variables but the input format is blocked into two sets: specifically, the first identifies the student's class (A or B), gender (M or F), and student number, while the second records scores on four quizzes and two unit tests. (Note that the first two lines are not part of the data; they are used only to identify column spaces.)

```
                    1          2          3
         12345678901234567890123456789 0
         AF01           293021294655
         AM02           303224296973
         BM03           293424299093
         BM04           353833349594
         AF05           283223285364
         BM25           222618258885
```

The correct READ and FORMAT commands for this data are:

```
MTB > READ 'SAMPLE' C1-C9;
SUBC> FORMAT (A1,A1,F2,T15,6(F2)).
```

The above FORMAT command reads the following information into the active memory of *Minitab:*

1. read a character variable of one place and place it into C1;
2. read a character variable of one place and place it into C2;
3. read a numeric variable of one place and place it into C3;
4. move to column 15, read six numeric variables of two places and place them into C4 through C9.

Once the last command has been entered and the period typed, the ENTER or RETURN key is pressed and *Minitab* begins its processing. The output to the screen following this activity is as follows:

```
     6 ROWS READ
ROW    C1     C2    C3    C4    C5    C6    C7    C8    C9
  1    A      F     1     29    30    21    29    46    55
  2    A      M     2     30    32    24    29    69    73
  3    B      M     3     29    34    24    29    90    93
  4    B      M     4     35    38    33    34    95    94
  .    .      .
```

As a means of validating the input process, *Minitab* will always identify the number of rows read into the file as well as printing the first four rows to verify a correct input. The data set, as it stands, does not possess any names for its columns. For later reference, the researcher will probably want to name the columns, one command line at a time. The following command lines would accomplish this:

```
MTB > name c1 'group'
MTB > name c2 'sex'
MTB > name c3 'stnum'
MTB > name c4 'quiz1'
MTB > name c5 'quiz2'
MTB > name c6 'quiz3'
MTB > name c7 'quiz4'
MTB > name c8 'test1'
MTB > name c9 'test2'
MTB > save 'sample'
```

The last command saves the data into a *Minitab* worksheet called SAMPLE.MTW.

A second method for inputting data into a *Minitab* file requires the use of the SET command. SET's purpose is to add a single column of data rather than a series or set of columns. The data in a SET command may be read from a data set or may be inputted directly from the keyboard. The general style of inputting is

```
SET 'filename' INTO Cxᵢ
```

where Cx_i is a column designation, such as C1 or C12. To add a column of data to our existing worksheet (SAMPLE), the following commands would be needed:

```
MTB > set into c10
DATA> 2 7 8 9 10 7
DATA> end
```

After instructing the program to initiate the SET command, notice that the pointer changes to DATA> from MTW>. Also, notice that after all the data of interest have been entered, the last line contains an END statement to inform the *Minitab* program that no further data will be entered. If we PRINT columns 6 through 10 the result would be:

```
MTB > print c6-c10
ROW        quiz3       quiz4       test1       test2       C10
  1          21          29          46          55          2
  2          24          29          69          73          7
  3          24          29          90          93          8
  4          33          34          95          94          9
  5          23          28          53          64         10
  6          18          25          88          85          7
```

Within the SET command, the input is either free-format or formatted. Since the information being inputted directly from the keyboard is only one column, no FORMAT command is required. However, if the input were being brought in from a file, the FORMAT subcommand might be necessary to tell *Minitab* how to organize, place decimal points, or interpret the new data.

One final comment concerning the SET command: sometimes the user requires only a certain number of data points from the data set. In order to limit the number of data points entering the worksheet the subcommand NOBS=x (where x = the number of data points) may be used.

The INSERT command works for rows of data in much the same fashion as SET does for a single column. One difference is that as many rows as necessary may be read into the existing worksheet. The command is:

> INSERT 'filename' BETWEEN ROWS X AND X_i OF Cn,
>   ..., Cn_i

The rows may be the first set (0 and 1) and anywhere within the worksheet. If the ROWS designation is left blank, *Minitab* appends the new information to the END of the current worksheet. Just as the user must designate the columns (e.g., C1 to C8) in the SET command, it is preferable that the rows be designated with the INSERT command.

The last strategy, RETRIEVE, is not precisely an input strategy. Correctly speaking, RETRIEVE actually works only with *Minitab* worksheets that have been previously created and saved (with the SAVE command) by *Minitab*. The command form is:

> RETRIEVE 'filename'

Minitab responds by providing the date when the file was saved. The nature of the file may be inspected by typing INFORMATION, after which *Minitab* prints a listing of the columns, the number of cases in each column, and the name of the column (if present). For example, if our active worksheet were the SAMPLE worksheet created earlier, the output would resemble the following:

```
MTB > retreive 'sample'
   WORKSHEET SAVED 12/29/1989

Worksheet retrieved from file:   sample.MTW
MTB > information
            COLUMN          NAME            COUNT
A           C1              group           6
A           C2              sex             6
            C3              stnum           6
            C4              quiz1           6
            C5              quiz2           6
            C6              quiz3           6
            C7              quiz4           6
            C8              test1           6
            C9              test2           6
CONSTANTS USED: NONE
```

OTHER INPUT STRATEGIES

The previous strategies were direct methods for data input and file creation within *Minitab*. For the microcomputer user, two additional techniques are available to move data directly into a *Minitab* worksheet. The first technique is to use the *Minitab*-supplied interface for moving data from *LOTUS 1-2-3* or *Symphony* spreadsheet files (or any program that writes a LOTUS-like or compatible spreadsheet file). The exact manner for setting up the interface is explained in the *Minitab User Guide: DOS Microcomputer Version* (Release 7).

The second method for moving data into a *Minitab* worksheet in the microcomputer version is to employ a translation program, such as *DBMS\COPY*, from Conceptual Software, Inc.* *DBMS\COPY* translates data files among more than 30 different programs, including *Minitab*. All the files in this book were created initially in an ASCII format and then translated via *DBMS\COPY* into the necessary microcomputer formats for *SYSTAT*, *Minitab*, *SAS*, and *SPSS-X*.

SAVING AND OUTPUTTING DATA

Minitab offers several formats for outputting and saving worksheets. The first strategy is the SAVE command, which has already been discussed. The technique is to create the worksheet and to issue the following command:

SAVE 'filename'

---
*Conceptual Software, Inc., P.O. Box 56627, Houston, TX 77256, (713) 667-4222.

Minitab will respond that the file has been saved. If the file previously existed, *Minitab* will ask if you wish to replace the file. Typing YES will tell *Minitab* to replace the existing file with the new file. When *Minitab* has completed the SAVE process, it will print a line of information stating the file name and file extension of the saved file.

If the user wishes to save the file in a format that can be read by another computer system (e.g., from mainframe to miniframe computer), the user must issue the PORTABLE subcommand following the SAVE command. The two lines necessary for this format are:

```
MTB > SAVE 'filename';
SUBC> PORTABLE.
```

The value of a PORTABLE file is its ability to be read by other computer systems through the following commands:

```
RETRIEVE 'filename';
PORTABLE
```

The portability of *Minitab* worksheets is a real value considering the ease of data transmission between computer systems available to the user. Further details concerning this process should be obtained from the user's local computer facility.

Another command of interest to the user is the PRINT command. PRINT is used within *Minitab* to display output on the screen. (Note: On some systems this may also direct output to other devices.) PRINT will produce output for columns, constants, or matrixes. The FORMAT subcommand may be employed to provide properly spaced output. If the user chooses not to use the FORMAT subcommand, the output is left justified and the rows are printed on the far left.

Similar to the PRINT command is the WRITE command. This command prints output to a disk file (or any file-storage system employed by the user's computer system). The general arrangement for this command is:

```
WRITE 'FILENAME' C1, C2, ..., Cxᵢ
```

The general purpose of the WRITE command is to create an output of the data set that may be read into or employed by another program, such as a word processor. The WRITE command may be followed by the FORMAT subcommand to organize the data in an arrangement preferred by the user. The resulting data are not labeled; that is, the data are placed in unlabeled columns. The output may be utilized by any program or storage device that stores data in ASCII form.

The SAVE command is crucial to the completion of any *Minitab* data-set manipulation. SAVE acts to close the activity and should be used as often as necessary to ensure that no data will be lost if the computer system fails (for whatever reason). The format for the SAVE command is:

```
SAVE 'filename'
```

The file created by the SAVE command is appended with a .MTW extension and is readable only by *Minitab*.

When a file is saved, *Minitab* stamps it with the current date. This permits the user to track generations of data sets if multiple sets are being used in an analysis. *Minitab* also checks to see if a previous version of the data set exists and asks the user if a replacement is desired. To replace the previous data set with the new one, the user should type Y and press the RETURN key. The following is an example of a SAVE procedure for our SAMPLE data set:

```
Worksheet retrieved from file:   sample.MTW
MTB > information
```

	COLUMN	NAME	COUNT
A	C1	group	6
A	C2	sex	6
	C3	stnum	6
	C4	quiz1	6
	C5	quiz2	6
	C6	quiz3	6
	C7	quiz4	6
	C8	test1	6
	C9	test2	6
	C10	grp	6
	C11	gender	6
	C12	zscore1	6

CONSTANTS USED: K1   K2

```
MTB > save 'sample'
Replace Existing File?  y

Worksheet saved into file:  sample.MTW
```

Working With *Minitab* Data Sets

The previous sections have dealt with the methods of data input and data-set creation within the *Minitab* system. One of the advantages of a full-function statistical-analysis system is its ability to manipulate the data to meet the needs of the user. While *Minitab*'s data manipulation is quite powerful, it does lack some of the more exotic features of other programs.

General Operations   *Minitab* begins most of its straight conversion operations with the LET command. This command informs the program that some type of manipulation is about to commence. The LET command operates very similar to the LET command of the BASIC programming languages for computers. For example, the following command would create a new variable from the sum of columns 1 and 2 times 50 divided by 100:

> LET Cx = ((C1 + C2) * 50)/100

Manipulating Data Sets *Minitab* permits a number of techniques for manipulating data once it has been entered into a data set. For example, data are often available in forms other than the traditional column format presented up to this point. Sometimes the data have been inputted in a linear fashion. As an illustration, the following data represent scores on three variables (VAR1, VAR2, and VAR3) for six subjects:

1	2	3	4	5	3	4	1	2	3
2	2	1	1	5	3	1	3	1	1

While some of the other programs described in this book can employ an input command strategy to read this as being three variables on six subjects, *Minitab* is unable to do this. However, the user may employ the STACK command to create a new set of columns completely structured in the manner needed by *Minitab*.

The set of commands, including the STACK command, needed to convert this data set would be as follows:

```
MTB > RETRIEVE 'SUBJECTS'
MTB > STACK (C1-C3) (C4-C6) (C12-C14)
MTB > STACK (C12-C14) (C7-C9) (C15-C17)
MTB > NAME C15 'VAR1'
MTB > NAME C16 'VAR2'
MTB > NAME C17 'VAR3'
MTB > SAVE 'SUBJECTS'
MTB > PRINT C15-C17
```

The output from these commands would be the following:

ROW	VAR1	VAR2	VAR3
1	2	3	4
2	2	1	1
3	5	3	4
4	5	3	1
5	1	2	3
6	3	1	1

The PRINT command following the SAVE command results in the above screen output. As can be seen, the data are now in a format more appropriate for use by *Minitab*.

Minitab also permits a number of special commands for the manipulation of data sets. The following are the essential commands that most users will need for their daily operations.

| COPY | COPY C1-C2 C12-C13 | Copies columns or sets to other columns. |

Several subcommands may be used with COPY to create appropriate subsets of the data:

SUBC>	USE 3:4	Copies only rows 3 and 4 to the new column(s).
SUBC>	OMIT 3:4	Copies all rows except 3 and 4.
SUBC>	USE 'VAR1' = 1	Copies only information when VAR1 equals 1.
SUBC>	OMIT 'VAR1' = '*'	Omits all missing data.

Other commands include:

ERASE	ERASE C1-C4	Erases columns or sets of columns.
CODE	CODE (0) 2 C23	Recodes all values of 0 to 2 in column 23.
DELETE	DELETE 3:8	Removes rows 3 through 8 from the data set (Note: the colon [:] is used to indicate continuous rows.)
INSERT	INSERT 'NEWFILE' 3:4 C12-C15	Inserts another file between rows 3 and 4 of the previous data set in columns C12 through C15. This command may be followed with a FORMAT and/or a NOBS = subcommand to specify the nature of the new file's data.
CONVERT	CONVERT C1 C2 C8 C9	Convert is used to translate alphanumeric data into numeric data through the use of a conversion table. In this example, C1 and C2 represent the conversion table, C8 is the data to be converted through the table, and C9 is the location where the conversion will be stored.

Variable Conversion Techniques *Minitab* permits a variety of arithmetic operations employing algebraic logic for its order of operations. In most cases *Minitab* permits comparison (e.g., greater than [>], less than [<], not equal [~=]) operations within the manipulation action. For example, the following statement asks *Minitab* to create a value of 1 (one) for column C5, when C1 is less than 5, otherwise the data are coded as 0 (zero):

> LET C5 = (C1 < 5)

Another example would be to create a column of z-scores for later plotting or analysis. Such a strategy would be to create the necessary constants (K1 = standard deviation and K2 = the mean) and employ these to create a new column of z-score values. For example:

> MTB > LET K1 = STDEV(C1)
> MTB > LET K2 = MEAN(C1)
> MTB > LET C5 = (C1-K2)/K2
> MTB > NAME C5 'ZSCORE'

The following example is drawn from the data set previously identified as SAMPLE. The example illustrates the full process of retrieving the data, creating the new constants for the variable TEST1, creating the z-score column (C12), and naming the column:

> MTB > retrieve 'sample'
>
> Worksheet retrieved from file: sample.MTW
>
> MTB > LET k1 = stdev(c8)
> MTB > LET k2 = mean(c8)
> MTB > LET c12 = (c8−k2)/k1
> MTB > name c12 'zscore1'
> MTB > print c12
>
> zscore1
>   −1.32880   −0.21744   0.79728   1.03888   −0.99056
>   0.70064

Substituting for If . . . Then . . . Else The LET command will permit the user to perform a variety of algebraic operations as well as comparison operations returning values of 0 (if false) or 1 (if true). However, *Minitab* will not permit the use of direct IF . . . THEN . . . ELSE statements as will some of the other programs. The user may create a set of columns or matrices of applicable arguments and by using the CONVERT command produce results quite similar to those of IF . . . THEN . . . ELSE. For example, suppose we wish to convert the data in column C8 of our SAMPLE file into alphanumeric values of HIGH or LOW. The following set of commands would accomplish this action and would create output for a column (C16) containing these new values:

```
MTB > retrieve 'sample'

Worksheet retrieved from file: sample.MTW

MTB > LET c13 = (c8 > 70)
MTB > set c15
DATA> 1 0
DATA> end
MTB > set c16;
SUBC> format (a4).
DATA> high
DATA> low
DATA> end
MTB > convert using table c15 c16, convert c13 to c17*
MTB > name c17 'category'
MTB > print c8 c13 c17

ROW    test1    C13    category
 1      46       0      low
 2      69       0      low
 3      90       1      high
 4      95       1      high
 5      53       0      low
 6      88       1      high
```

*The same statement could be written as:

CONVERT C15 C16, C13 C17

The result of the above would be the same as the following set of commands in *SAS*:

```
IF TEST1 > 70 THEN CATEGORY = 'HIGH';
ELSE CATEGORY = 'LOW'
```

In *SYSTAT*, the commands would be:

```
DATA
USE SAMPLE
IF TEST1 > 70 THEN LET CATEGORY$ = 'HIGH',
    ELSE LET CATEGORY$ = 'LOW'
```

While *Minitab* requires a few more steps, the end result is the same: the inability of *Minitab* to employ an IF . . . THEN . . . ELSE process is not a major hindrance in data manipulation.

ADDITIONAL LOGICAL OPERATIONS AND FUNCTIONS

Arithmetic Operations *Minitab*'s arithmetic operations are very similar to other statistical programs. For example:

* *   multiplication
* \+   addition
* \−   subtraction
* /   division
* **   raised to a power

Similarly, the general-comparison operations are:

=, EQ	equals
~=, NE	not equals
<, LT	less than
>, GT	greater than
<=, LE	less than or equal to
>=, GE	greater than or equal to

Minitab also permits several Boolean-logic arguments for comparisons and arguments. These arguments return a value of 1 when true and a value of 0 when false.

&, AND	and
|, OR	or
~, NOT	not

Furthermore, *Minitab* permits a number of row- and column-wise operations when creating constant variables. Some of the row-wise operations include ABSOLUTE, SQRT (square root), LOGTEN, LOGE, ROUND, and NSCORES. Operations on columns include COUNT, NMISS, N, SUM, MEDIAN, MIN (minimum), MAX (maximum), SORT, RANK, and MEAN. Each of these operations should be performed within a LET statement, and parentheses must be used to indicate the column or row. For example, to create a constant called MEAN1, the following command would be used:

```
LET MEAN1 = MEAN(C1)
```

Descriptive Statistics (A4-1 to A4-13)

The descriptive statistics explained in Section A, Chapter 4 may be obtained, for the most part, with the application of the DESCRIBE command. In fact, the user may be astonished to discover how much information *Minitab* will and will not provide with this command. The user may obtain the following: the number of cases, the mean (**A4-2**), the median (**A4-3**), the standard deviation (**A4-8**), the standard error of the mean, the trimmed mean (minus the smallest and largest 5% of the data set), the minimum value, the maximum value, the value at the first quartile, and the value at the third quartile. While the output from *Minitab* is revealing, *Minitab* will not directly produce statistics such as range, skewness, kurtosis, and variance.

The general format for obtaining descriptive statistics is to RETRIEVE the data set of interest and to type DESCRIBE followed by the variables of interest. Note that *Minitab* expects the user to employ column designations (e.g., C1 or C1-C12); however, the user may use variable labels surrounded by single quotation marks if the columns have been previously named. For example, using the previously created DASET1, the descriptive statistics for PRESTEST and POSTTEST could be obtained in the following manner:

```
MTB > describe 'pretest'- 'posttest'
                N      MEAN    MEDIAN   TRMEAN   STDEV   SEMEAN
PRETEST        10     18.00     18.50    18.00    3.94     1.25
POSTTEST       10     28.50     28.00    28.50    5.13     1.62

               MIN     MAX       Q1       Q3
PRETEST       11.00   25.00    16.00    20.25
POSTTEST      21.00   36.00    24.00    33.50
```

To obtain the variance (**A4-7**) the researcher would employ a set of LET commands:

```
MTB > LET k1=stdev('pretest')
MTB > LET k2=stdev('posttest')
MTB > LET k3=k1**2
MTB > LET k4=k2**2
MTB > print k3-k4
K3         15.5556
K4         26.2778
```

In the above output, K3 equals the variance for PRETEST and K4 equals the variance for POSTTEST. While the above is easy to obtain via *Minitab,* it is more readily accomplished with a hand calculator. In fact, if the user is working with a personal computer, it might be more advisable to use one of the many terminate-and-stay resident-shareware calculators that are available.

To obtain the range (**A4-5**), the user would create another constant variable with the MAX and MIN operations:

```
LET K5 = MAX('PRETEST')-MIN('PRETEST').
```

The skewness (**A4-12**) and kurtosis (**A4-13**) of the data set could be calculated in much the same manner by creating a series of constant variables. For example, to calculate the skewness of the PRETEST variable, the user would employ the formula (**A4-12**).

```
MTB > LET C7 = (C4-(MEAN(C4)))**2
MTB > LET C8 = (C4-(MEAN(C4)))**3
MTB > LET K6 = (SUM(C7)/(N(C7)))
MTB > LET K7 = (SUM(C8)/(N(C8)))
MTB > LET K8 = (K7)/(K6*(SQRT(K6)))
MTB > PRINT K8
     -0.171811
```

The result is the correct figure for the skewness for the variable PRETEST. A similar set of commands, with appropriate changes, would produce the kurtosis (**A4-13**) statistic.

Like many statistical packages, *Minitab* does not directly calculate a value for the mode (**A4-4**). An approximation of the modal value may be obtained by using the plotting operations of the package. The simplest and most direct procedure would be to use the DOTPLOT command to produce an elementary graph of the user's data. The following is an example using the DASET1 data described earlier:

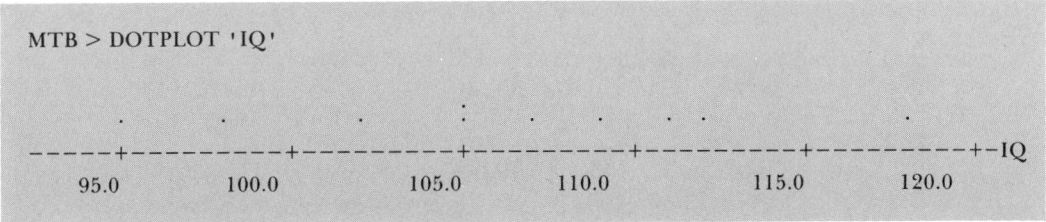

Based upon the above plot, the user may estimate that the modal value for IQ (IQ scores) is 105.

Minitab also permits the user to create frequency counts through the utilization of various tables. The tables may be created with either the TALLY, TABLE, or TABLES commands. The TALLY command produces a one-way table for each column listed in the command. Subcommands available include COUNTS, PERCENTS, CUMCOUNTS, CUMPERCENTS, and ALL. Using the SAMPLE data, the following commands would produce counts for the variables QUIZ1 and QUIZ2.

```
MTB > TALLY 'QUIZ1' 'QUIZ2';
SUBC> ALL.
```

The output from this command would be:

```
quiz1  COUNT  CUMCNT  PERCENT  CUMPCT   quiz2  COUNT  CUMCNT  PERCENT  CUMPC
  22     1       1     16.67    16.67     26     1       1     16.67    16.6
  28     1       2     16.67    33.33     30     1       2     16.67    33.3
  29     2       4     33.33    66.67     32     2       4     33.33    66.6
  30     1       5     16.67    83.33     34     1       5     16.67    83.3
  35     1       6     16.67   100.00     38     1       6     16.67   100.0
N=             6                         N=             6
```

The TABLE command will create a full contingency table for data within a data set. The command groups the data according to the numeric data being analyzed. The example below is from DASET1 and produces an output containing both the number of cases per cell and the percentage of the subjects for each cell.

```
MTB > TABLE 'GROUP' 'GENDER';
SUBC> COUNT;
SUBC> TOTPERCENTS.
```

```
ROWS: Group   COLUMNS: gender

                1           2      ALL
    1           2           1       3
              33.33       16.67   50.00

    2           1           2       3
              16.67       33.33   50.00

  ALL           3           3       6
              50.00       50.00  100.00

  CELL CONTENTS —
                    COUNT
                    % OF TBL
```

A variety of subcommands may be employed to obtain a full set of descriptive statistics for variables or sets of variables classified according to the variables indicated in the TABLE command. The statistics available are the same as those previously itemized in the DESCRIBE command, as well as the output of a special command that identifies the proportion of the data within the cells. Another subcommand that may be used with the TABLE command is CHISQUARE. This will produce a Chi-square test of independence or homogeneity (**A4-61**).

The full CHISQUARE command is similar to the subcommand for Chi-square but will produce a Chi-square test of association for a contingency table; however, the table must be formatted before the analysis. This limits the user somewhat, because the table will not automatically be based on the data in the columns. One advantage, however, is that many statistical programs will perform the Chi-square analysis only if the data *are not* in a contingency table. The suggested strategy for creating a contingency table is first to use the TABLE command to create the table and then to enter the table in appropriate columns using the READ and DATA commands described earlier.

A final strategy for describing data is the application of the PLOT command to produce a bivariate plotting of joint variables. The following command format will produce a plot of the variables for PRETEST and POSTTEST from the DASET1 data set:

```
MTB > PLOT 'PRETEST' 'POSTTEST';
SUBC> TITLE 'SAMPLE BIVARIATE PLOT';
SUBC> SYMBOL '*';
SUBC> XLABEL 'POSTTEST SCORES';
SUBC> YLABEL 'PRETEST SCORES'.
```

The three subcommands following the PLOT command are used to produce a more visually appealing plot and are not required for the plot. For additional subcommands and different types of plots available, refer to the *Minitab Reference Manual*.

In addition to the above commands, *Minitab* also provides some exploratory data-analysis commands to aid the user in describing the data. *Minitab*'s exploratory features include STEM-AND-LEAF plots and BOXPLOTS as well as condensed scatterplots (CPLOT), coded two-way tables (CTABLE), and others. The commands are very similar because the user must identify the appropriate column from which the data to be analyzed will be drawn. For example, the following commands would draw a boxplot for the COOK1 POP data, including notches for a confidence interval about the median of the data:

```
MTB > boxplot 'pop';
SUBC> notch.
```

The output from these commands would be:

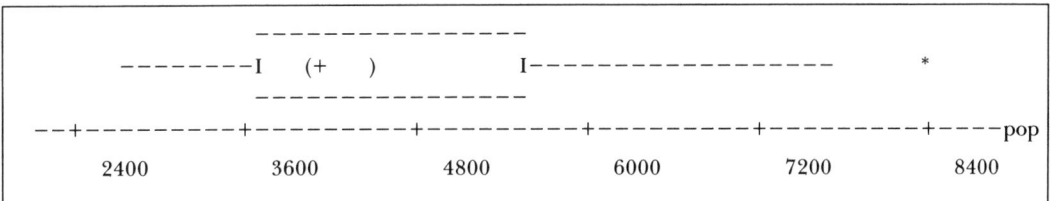

Measures of Relationships (A4-14 to A4-30)

The two most common measures of association, Pearson's product-moment coefficient (**A4-16**) and Spearman's Rho (**A4-17**) may be obtained with the CORRELATE command. For example, to calculate a correlation between the PRETEST and POSTTEST variables of DASAT1, the following commands would be used:

```
MTB > RETRIEVE 'DASET1'
MTB > CORRELATE 'PRETEST' 'POSTTEST'
```

If the data under analysis are ranked data, then the command RANK may be used first to create an ordered-by-rank data set and, the CORRELATE command will calculate the Spearman Rho value for the data set. Additional relationship measures that are directly available include point biserial (**A4-19**) and covariance (**A4-15**). To calculate the point biserial coefficient, the data for one of the variables must be in a 0,1 form while the other must be continuous. If the data are not in this form, the LET or the CONVERT command may be used to transform the data. The calculation of the covariance statistic requires only that the user employ the COVARIANCE FOR C_I C_{I+1} command (where C_I and C_{I+1} represent the column numbers), identifying the two columns from which the statistic will be calculated.

CORRELATIONS FOR CATEGORICAL DATA

Minitab does not directly support correlations for categorical data; however, most of the calculations may be done with applications of the TABLE command and the calculation of new columns and constants. For example, the Phi-coefficient (**A4-21**) may be calculated using the following strategy with the variables of GROUP and TRSP from DASET2:

```
MTB > RETRIEVE 'DASET2'
MTB > TABLE 'GROUP' 'TRSP';
SUBC> COUNTS.
MTB > READ C10-C11
MTB > LET K1 = 2
MTB > LET K2 = 3
MTB > LET K3 = 4
MTB > LET K4 = 1
MTB > LET K5 = ((K1*K2)-(K3*K4))/(SQRT((K1+K3)
*(K2+K4)*(K3+K4)*(K1+K2))
MTB > PRINT K5

-0.408
```

The contingency coefficient (**A4-23**) may be calculated in much the same manner as the above. First, using the TABLE command with COUNTS and CHISQUARE subcommands, obtain the appropriate Chi-square value. Second, use a LET command to define a constant. Finally, use another LET command to calculate the coefficient. The following would calculate a contingency coefficient for the same data as the above:

```
MTB > TABLE 'GROUP' 'TRSP';
SUBC> COUNTS;
SUBC> CHISQUARE.
MTB > LET K1 = 1.667
MTB > LET K2 = SQRT(K1/(10 + K1))
MTB > PRINT K2

0.378
```

The user is reminded that the contingency coefficient should be used only when both variables in the table are nominal and discrete and have two or more categories.

Regression Analysis (A4-31 to A4-39)

The majority of the regression analyses may be performed through the various operations of the REGRESS command. Some of the subcommands for this include:

NOCONSTANT	Removes the constant from the equation
WEIGHTS	Performs a weighted regression
MSE	Stores the mean square error
COEFFICIENTS	Stores the coefficients
HI	Stores the high leverage values
RESIDUALS	Stores the residual values
TRESIDUALS	Stores the studentized residuals
COOKD	Stores the Cook's-Distance values
PREDICT	Computes predicted Y-values for data range
VIF	Produces the variable-inflation-factor statistic
DW	Produces a Durbin-Watson statistic

To produce a simple linear regression (**A4-32**) with *Minitab* using the DASET2 data set, the appropriate commands would be as follows:

```
MTB > regress 'csyn' on 1 'agemonth';
SUBC> residuals c6;
SUBC> hi c7;
SUBC> tresiduals c8;
SUBC> cookd c9;
SUBC> predict 'agemonth'.
```

The above commands tell *Minitab* to create a regression equation with the variable CSYN as the predicted variable and AGEMONTH as the one predictor variable. The first subcommand tells *Minitab* to store the residuals in column C6. The second subcommand places the leverage values in C7, while the third subcommand places the studentized residuals in column C8. The fourth subcommand puts the associated Cook's-Distance value in column C9. The final subcommand, PREDICT 'CSYN', produces predicted values, standard deviation of fit, 95%-confidence intervals for the mean of the 'CSYN' responses, and the 95%-confidence intervals for each individual value.* This could be followed by NAME commands giving each a regular variable label. The result would be the following:

```
The regression equation is
CSYN  =  4.12  +  0.0647 AGEMONTH
```

Predictor	Coef	Stdev	t-ratio	P
Constant	4.121	1.025	4.02	0.001
AGEMONTH	0.06473	0.01739	3.72	0.001

S = 1.375   R-sq = 39.8%   R-sq(adj) = 36.9%

Analysis of Variance

SOURCE	DF	SS	MS	F	P
Regression	1	26.206	26.206	13.86	0.001
Error	21	39.707	1.891		
Total	22	65.913			

The estimate of the regression coefficient (**A4-33**) and the Y-intercept (**A4-34**) are listed in the identified regression equation (i.e., CSYN = 4.12 + 0.0647 AGEMONTH). The standard error of estimate (**A4-35**) is identified by "s" (1.375). The R-squared (**A4-37**) value is identified as "R-sq."

A PRINT command (PRINT C2 C6-C9) requesting a printing of the stored values in the *Minitab* data set would result in the following output (limited to the first five values, for illustration):

ROW	CSYN	residual	leverage	student	cookd
1	9	0.93031	0.046623	0.68407	0.011740
2	7	−0.22815	0.055209	−0.16670	0.000851
3	10	1.21823	0.081573	0.92111	0.037953
4	5	−2.29288	0.052630	−1.80257	0.081523
5	8	1.35446	0.092815	1.03598	0.054713

---

*If the command BRIEF = 1, 2, or 3 precedes the REGRESS command, the length of the output may be controlled. The higher the number associated with the BRIEF command, the longer the output. The level 3 produces a full output of predicted values as well as confidence intervals.

Each of the values has been relabeled with the appropriate name command.

In order to obtain the part correlation (**A4-26**) and partial correlation (**A4-27**) a little data manipulation is necessary (as it was with several of the other programs). Although the process appears to be rather complex, its nature is very similar to the strategy employed by other programs.

The first step, once again using the DASET2 data set, is to employ the appropriate REGRESS command with the variable CSYN predicted by the AGEMONTH variable to produce the appropriate residuals for the prediction. Step two is to save the data into another file with a new name. The third step is to run another regression model, using CVOCAB with AGEMONTH as the predictor, and to save the residuals. This step is followed by issuing a SAVE command to save the values for CVOCAB and residuals to a third file, once again with a different name. This is followed by retrieving the first file, using a READ command to merge the second file with the first and, finally, saving this combined file as a third new file.

Once these steps are completed, the user may employ the CORRELATE command to produce the necessary part correlation (**A4-26**) and partial correlation (**A4-27**). The major reason for all these steps is the inability of *Minitab* to work with multiple-prediction variables. The following is a summary of the commands necessary to produce part and partial correlations for CSYN, CVOCAB, and AGEMONTH:

```
MTB > RETRIEVE 'DASET2'
MTB > REGRESS 'CSYN' 1 'AGEMONTH'
SUBC> RESIDUAL C6.
MTB > NAME C6 'RESID1'
MTB > SAVE 'NEW1'
MTB > RETRIEVE 'DASET2'
MTB > REGRESS 'CVOCAB' 1 'AGEMONTH';
SUBC> RESIDUAL C6.
MTB > SAVE 'NEW2'
MTB > RETRIEVE 'NEW1'
MTB > READ 'NEW2' C7-C12
MTB > NAME C12 'RESID2'
MTB > SAVE 'BIGNEW'
MTB > CORRELATE 'RESID1' 'RESID2' 'CSYN' 'CVOCAB'
 'AGEMONTH'
```

The result of all this programming is the output below. As previously cautioned, the degrees of freedom are equal to N-3 and tests for statistical significance must be looked up in a table of correlation values.

	resid1	resid2	CSYN	CVOCAB
resid2	0.339			
CSYN	0.776	0.263		
CVOCAB	0.212	0.624	0.657	
AGEMONTH	−0.000	0.000	0.631	0.782

The correlation between CSYN and CVOCAB is .657, while the part correlation for CVOCAB and CSYN, which is the correlation between these variables with the influence of AGEMONTH removed from the CSYN scores, is .212. The part correlation for CVOCAB and CSYN with AGEMONTH removed from the CVOCAB scores is .263. The partial correlation between CSYN and CVOCAB with the influence of AGEMONTH removed from both is .339 (which is the correlation between RESID1 and RESID2). Although the manipulation is a little tricky, the result is far easier then to calculate with a hand calculator, especially for large data sets.

Multiple regression (**A4-36**) and curvilinear regression (**A4-38**) are logical extensions of the general REGRESS command described above. The *Minitab* commands for a multiple-regression model predicting CSYN with AGEMONTH, CVOCAB, MSYN, and MVOCAB would be the following:

```
MTB > brief 3
MTB > regress 'csyn' on 4 'agemonth' 'cvocab' 'msyn' 'mvocab'
```

The resulting output would be:

```
The regression equation is
CSYN  =  3.55  +  0.0337 AGEMONTH  +  0.0495 CVOCAB  +
0.177 MSYN  -  0.0201 MVOCAB
```

Predictor	Coef	Stdev	t-ratio	p
Constant	3.548	1.289	2.75	0.013
AGEMONTH	0.03367	0.02949	1.14	0.0269
CVOCAB	0.04953	0.03537	1.40	0.178
MSYN	0.1170	0.1787	0.65	0.521
MVOCAB	−0.02009	0.03543	−0.57	0.578

s = 1.375   R−sq = 48.4%   R−sq(adj) = 36.9%

Analysis of Variance

SOURCE	DF	SS	MS	F	p
Regression	4	31.905	7.976	4.22	0.014
Error	18	34.008	1.889		
Total	22	65.913			

SOURCE	DF	SEQ SS
AGEMONTH	1	26.206
CVOCAB	1	4.568
MSYN	1	0.523
MVOCAB	1	0.607

The only value missing from the data output is the value for the multiple correlation (**A4-28**); that is, the square root of the "R-sq" value ($.484^2 = .6957$). The regression equation is clearly identified at the top of the page, and the regression coefficients (**A4-33**) for each of the predictor variables are listed under the heading "Predictor."

The identification of a curvilinear relationship (**A4-38**) may be obtained by first identifying the possible source of the relationship (perhaps through use of the PLOT command to graph the relationship) and the recalculation of possible new variables (e.g., quadratic variables, log linear, and so forth). Using the previous example of a quadradic relationship of AGEMONTH, the process would be first to use the following command to create a new variable and to follow it with the adjusted REGRESS command:

```
MTB > LET C7 = 'AGEMONTH'**2
MTB > NAME C7 'AGEMOSQ'
MTB > REGRESS 'CSYN' ON 2 'AGEMONTH' 'AGEMOSQ'
```

The use of stepwise multiple-regression (**A4-39**) techniques in regression-model building has become very prevalent in the research literature. The reader is once again cautioned against indiscriminantly utilizing this technique without a sound rationale for its application. The STEPWISE command is a full-function model that employs a forward selection procedure by default and permits the user to use a variety of subcommands to fine-tune the model being built. The general procedure is similar to the REGRESS command, but the subcommands are unique to the STEPWISE command. Once the model is built to the satisfaction of the user, a subsequent analysis and output may be produced through the REGRESS command. The following is an illustration of the STEPWISE command using the data set called COOK1, previously created in the *SYSTAT* chapter:

```
MTB > RETRIEVE 'COOK1'
MTB > STEPWISE 'GRANDX' 'POP' 'COST' 'LOINCOM'
```

The results would indicate that the best model would include the variables of COST and LOINCOM, but would exclude POP from the model.

One final command available to the *Minitab* user is the BREG command, which produces a series of models with the best subsets of predictor variables. This command works in a fashion similar to the PROC RSQUARE described in the *SAS* chapter. The command format is:

```
MTB > BREG C1 C2-C6
```

Using Command Programs in Minitab

The maximum utilization of a statistical package is achieved when the package performs repetitive operations across a variety of data sets; that is, executes a program. While the command-driven interface of *Minitab* does not permit the more user-friendly operations of a menu-driven interface, it does permit the able user to develop the customized models and analyses as needed. The command operations of *Minitab* permit the development of stored programs called command programs (programs saved with an .MTB extension, such as SAMPLE.MTB). These programs may be interactive, requiring input from the user, looped to permit repetitive actions across columns or sets of columns, or even be nested and conditional. The limits of the command programs are circumscribed by the capabilities of the user.

The command files may be created in two manners. The first technique is somewhat like programming on the run. The user employs a STORE command, which initiates a command-program-writing sequence, writes the program steps, and terminates the writing session with an END command. Once the END command is issued, *Minitab* automatically saves the command file with the .MTB extension. The second technique is to use an editing program employed by your computer system to create, write, and save the program outside of the *Minitab* environment. Either strategy works equally well. Remember, the command file may have any name but must have a .MTB extension.

The following program illustrates the actions of a command file. First, it retrieves the COOK1 data file and performs an analysis of the descriptive statistics for the variables READ3, MATH3, POP, and LOINCOM. Next, it performs a series of plots to provide a visual picture of the data. In the third set of the analysis, it completes a multiple-regression analysis of the model READ3 = constant + MATH3 + POP + LOINCOM. The studentized residuals and predicted values are then stored in columns C10 and C11 and the residual and Cook's Distance values are stored in C12 and C13. The third set labels these new variables and saves the whole new set of data into a new file called COOK1A.

```
RETRIEVE 'COOK1'
DESCRIBE 'READ3' 'MATH3' 'POP' 'LOINCOM'
PLOT 'READ3' 'MATH3'
PLOT 'READ3' 'POP'
PLOT 'READ3' 'LOINCOM'
BRIEF = 3
REGRESS 'READ3' 3 'MATH3' 'POP' 'LOINCOM' C10 C11;
     RESIDUALS C12;
     COOKD C13.
NAME C10= 'STRESID' C11= 'PRED' C12= 'RESID'
C13= 'COOK'
SAVE 'COOK1A'
```

While the output of this program is directed totally to the user's computer screen, it easily could have been directed elsewhere. For example, to send the output to a printer, use the PAPER command. To send it to a file, use the OUTFILE 'FILENAME' command. The user, after sending the output, should remember to turn off the output by using either the NOPAPER or NOOUTFILE commands; otherwise, all output will continue to be directed as previously designated.

Comparison: One Sample (A4-40 to A4-47)

Hypotheses involving the comparison of a mean (**A4-41**), variance (**A4-42**), proportion (**A4-43**), or correlation (**A4-44, A4-45**) to some value, or the comparison of a set of frequencies to some expected frequency values (**A4-46, A4-47**), may be accomplished with the use of *Minitab*'s built-in functions. *Minitab* offers functions for 17 different distributions; specifically, uniform, normal, student's t, F, Chi-square, exponential, Gamma, Beta, binomial, Poisson, integer, discrete, Cauchy, Laplace, logistic, log normal, and Weibull. Each of these distributions is offered as a cumulative function, an inverse function, a discrete function, and a random-variate function.

While many of these functions are employed with all of the data sets as part of the *Minitab* system, the user may choose to implement them individually to calculate specific tests for one-sample cases. The user is cautioned once again, however, that to employ *Minitab* as a hand calculator is sometimes more complicated than simply using a hand calculator. Since *Minitab* can readily be used to calculate the sample statistic, it is recommended that the actual tests of significance of this type be performed using a hand calculator on all data sets except extremely large ones.

If the decision is to use *Minitab* to calculate the test statistics, the strategy is very similar to the one described in earlier chapters of Section C. The best strategy is to write a command file containing all the numerous commands and operations or to employ the constant operations of *Minitab* to create the necessary variables. Such a series of commands would resemble the following:

```
MTB > LET k1=n(c1)
MTB > LET k2=mean(c1)
MTB > LET k3=stdev(c1)
MTB > descr c1
```

	N	MEAN	MEDIAN	TRMEAN	STDEV	SEMEAN
AGEMONTH	23	56.57	49.00	56.29	16.86	3.52

	MIN	MAX	Q1	Q3
AGEMONTH	29.00	90.00	42.00	72.00

```
MTB > LET k4=3.52
MTB > PRINT k1-k4
K1      23.0000
K2      56.5652
K3      16.8599
K4       3.52000
MTB > LET k5 = (k2-32)/k4
MTB > PRINT k5
K5       6.97875
MTB > LET k6 = ((k1-1)*(k3**2))/32
MTB > PRINT k6
K6     195.427
```

The user needs to employ the DESCRIBE command because *Minitab* has no function for calculating the standard error of the mean, which is necessary for calculating the t-distribution. The value of "32" is arbitrary and was chosen simply to illustrate the process. The result indicated by K5 is the t-statistic, and the K6 result is the Chi-square statistic. The whole process could be incorporated within a command file using a special feature for making part of the file interactive. This is done as:

```
NOTE ENTER THE STANDARD ERROR OF MEASUREMENT
SET 'TERMINAL' C50;
    NOBS=1.
COPY C50 K50
NOTE ENTER THE NUMBER TO BE TESTED
SET 'TERMINAL' C51;
    NOBS=1.
COPY C51 K51
EXEC 'TTEST'
```

In the above example, the main program is stored in a command file called TTEST, and the above lines permit the user to interactively enter the value. The full command file would be:

```
RETRIEVE 'DASET2'
NOECHO
LET K1=N(C1)
LET K2=MEAN(C1)
LET K3=STDEV(C1)
ECHO
DESCRIBE 'AGEMONTH'
NOTE ENTER THE STANDARD ERROR OF MEASUREMENT
NOECHO
SET 'TERMINAL' C20;
    NOBS=1.
COPY C20 K20
NOTE ENTER THE NUMBER TO BE TESTED
SET 'TERMINAL' C21;
    NOBS=1.
COPY C21 K21
LET K5 = (K2-K21)/K20
PRINT K5
LET K6 = ((K1-1)*(K3**2))/K21
PRINT K6
ECHO
END
```

The use of the ECHO and NOECHO commands limits the actual output to the screen so that the user only sees what is necessary. The total output from this command file would be:

```
MTB > exec 'ttest'
MTB > RETRIEVE 'DASET2'
  WORKSHEET SAVED 12/20/1989

Worksheet retrieved from file: DASET2.MTW
MTB > DESCRIBE 'AGEMONTH'

              N    MEAN   MEDIAN   TRMEAN   STDEV   SEMEAN
AGEMONTH     23    56.57   49.00    56.29   16.86    3.52

             MIN    MAX      Q1       Q3
AGEMONTH   29.00  90.00   42.00    72.00

MTB > NOTE ENTER THE STANDARD ERROR OF MEASUREMENT
DATA> 3.52
MTB > NOTE ENTER THE NUMBER TO BE TESTED
DATA> 32
K5          6.97875
K6          195.427
MTB >END
```

A simpler strategy might be to use the DESCRIBE command and to employ a good hand calculator to perform the additional calculations.

A strategy similar to the above could also be employed if the user wanted to calculate z-statistics (**A4-41**) or to compare two correlations (**A4-44, A4-45**) with the use of a Fisher's Z-transformation. All of the transformations could be done for the z-test after getting the output from DESCRIBE and following these LET statements:

```
LET K1 = PS
LET K2 = PP
LET K3 = N(C1)
LET K4 = (K1−K2)/(SQRT((K1*(1−K1))/K3
PRINT K4
```

In the above command lines, PS would equal sample proportion, PP would equal the population proportion, K3 would equal the sample size, and K4 would be the z-statistic. Once again, this could be completely automated with the appropriate command file, including interactive responses to fill the necessary elements.

A more direct way to compute a z-statistic would be to use the special ZTEST command available in *Minitab*. The format of the command makes optional the specification of a population mean (assumed to be zero, if not specified), a required specification of the population standard deviation, and the required identification of the variable (or column) to be tested. Assuming our interest is in testing the READ3 variable of the COOK1 data set to see if it is significantly different from the expected mean of 250 and a population standard deviation of 50, the command format would be:

```
ZTEST 250 50 'READ3'
```

The output from this test would be:

```
TEST OF MU   =   250.000 VS MU N.E. 250.000
THE ASSUMED SIGMA   =   50.0

          N       MEAN      STDEV    SE MEAN     Z      P VALUE
read3    116    272.543    37.621     4.642    4.86    0.0000
```

The test indicates that the sample group is significantly different from the expected.

Sample correlations are often compared to see if they are significantly different from zero. Although this is not directly done by *Minitab*, the easiest procedure is to use a table of statistics to check the significance of the results of the CORRELATE command. On the other hand, many correlation comparisons are made between a sample correlation and a hypothesized correlation. In order to perform this comparison, the user must employ a Fisher's Z-transformation. After executing a CORRELATE command to get the necessary correlations—SC equals the sample correlation, HC equals the hypothesized correlation, and N equals the sample size in the correlation—the following command lines will perform a test of significance:

```
LET K1 = .5*(LOG((1+SC)/(1−SC)))
LET K2 = .5*(LOG((1+HC)/(1−HC)))
LET K3 = N
LET K4 = (K1−K2)/(SQRT(1/(K3−3))
```

Once again, the entire process would be more readily accomplished with the application of a command file having interactive facets to automate the task. (In fact, this is exactly what this writer has done for several of these one-sample tests.) The true advantage of a command file is its ability to perform repetitive tasks or make some tasks smoother. The calculation of one-sample tests is more easily accomplished with a set of command files that can be called up as necessary, rather than recreating them each time with an effort that would be more complex than a hand-calculated procedure.

Comparison: Two Samples (Parametric) (A4-48 to A4-56)

Minitab may be used to calculate a variety of two-sample parametric tests. The independent t-test (**A4-49**) is performed by envoking the TWOSAMPLE or the TWOT commands. When using the TWOSAMPLE command, *Minitab* assumes that the two samples are in different columns, and the user must identify the columns to be tested. The format for the command is:

> MTB > TWOSAMPLE C1 C2

The subcommands available are:

SUBC>   ALTERNATE = 1,−1 [where the 1 or −1 is used for directional hypotheses.]
SUBC>   POOLED [for applications requiring a pooled procedure.]

Related to the TWOSAMPLE command is the TWOT command for testing independent samples when the data to be tested is in one column and the grouping characteristic is in another. The format and subcommands are exactly the same.

The following is an example of output from the TWOT command for the IQ by GROUP data in the DASET1 data set.

```
TWOSAMPLE T FOR IQ
GROUP    N    MEAN    STDEV    SE MEAN
1        5    107.80   8.58     3.8
2        5    104.60   4.93     2.2

95 PCT CI FOR MU 1  −  MU 2: (−7.6, 14.0)

TTEST MU 1 = MU 2 (VS NE) : T=0.72  P=0.50  DF= 6
```

A similar command is the TTEST command for testing for two-sample dependent t-tests (**A4-50**). The TTEST also uses the ALTERNATE subcommand to specify a directional one-tailed test. Before calculating a dependent t-test, the user must first calculate a change variable. For example, suppose the researcher is interested in testing for differences between the PRETEST and POSTTEST scores in the DASET1 data set. The first step would be to create a new column containing the differences between these two variables. The following commands would accomplish this:

> NAME C9 'CHANGE'
> LET 'CHANGE' = 'POSTTEST' − 'PRETEST'

Once the new variable has been created, the TTEST command may be used to perform a dependent t-test under the assumption that the hypothesized mean difference is equal to 0. The result of this would be the following output:

```
TEST OF MU = 0.000 VS MU N.E.  0.000
         N    MEAN    STDEV    SE MEAN     T      P VALUE
change   10   10.500  3.375    1.067       9.84   0.0000
```

344 SOFTWARE APPLICATIONS IN STATISTICS

As previously noted, a critical assumption of the t-test is the assumption of homogeneity of variance. If a significant difference exists between the variances then the appropriate correction formula must be employed to adjust the resulting t-test. Use of a test such as **A4-51** for significant differences among variances must be undertaken if any doubt exists in the mind of the user. *Minitab,* unlike some programs, does not provide separate tests for any assumptions concerning significantly different variances. Should the user find that an independent t-test possesses significantly different variances, a formula such as Satterthwaite's correction formula* and an adjusted calculation for the standard error of difference** would be required to calculate the t-test.

The calculation of a two-sample variance for dependent samples (**A4-52**) is best calculated by using the DESCRIBE command and the CORRELATE command to obtain the necessary values. Once the values have been obtained the user should employ the formula in **A4-52** with a hand calculator.

The technique for comparing the proportions for two independent (**A4-53**) or two dependent (**A4-54**) samples is more readily accomplished by using the TABLE command with the PERCENTS subcommand. Once the data are obtained from the cells the application of the formulas is straightforward. For example, using DASET1 the commands would be as follows:

```
MTB > TABLE 'GROUP' 'TRSP';
SUBC> TOTPERCENTS.
```

The result of this two-by-two contingency table could be used to compare the proportion of students in Group 1 at level 1 of TRSP with the proportion of students in Group 2 at the same level. This would be a comparison of the proportion for two independent samples (**A4-53**). A second comparison of proportions of students in Group 1 who are in levels 1 and 2 of TRSP would be a dependent-sample proportion test (**A4-54**). Each of these calculations may be readily completed using the formulas and a hand calculator.

The tests of two sample correlations (**A4-55, A4-56**) requires that the correlations for each group must be available before completing the formulas. The first step in calculating an independent comparison of correlations (**A4-55**) requires that the data be grouped. In many programs this would simply require the use of a BY statement. Unfortunately, *Minitab* does not permit the use of such a statement and, consequently, the data must be manipulated before beginning the analysis. The following set of commands will take the data for the PRETEST in Data Set 1 and create two new columns for values, one for members of Group 1 and the other for members of Group 2:

---

* $df = \dfrac{(s_1^2/n_1 + s_2^2/n_2)}{\sqrt{[(s_1^2/n_1)^2/(n_1 - 1)] + [(s_2^2/n_2)^2/(n_2 - 1)]}}$

** $s_{x1-x2} = \sqrt{(s_1^2/n_1) + (s_2^2/n_2)}$

```
MTB > RETRIEVE 'DASET1'
MTB > COPY 'PRETEST' C9;
SUBC> USE 'GROUP' = 1.
MTB > COPY 'PRETEST' C10;
SUBC> USE 'GROUP' = 2.
MTB > NAME C9='GROUP1' C10='GROUP2'
```

Once the conversions have been completed, the CORRELATE 'GROUP1' 'GROUP2' and DESCRIBE commands may be issued, which would produce the necessary output for use with the previously developed Fisher's Z-transformation formula (or command file, if created).

The calculation of the comparisons of two-sample dependent correlations (**A4-56**) is first accomplished with application of the CORRELATE command. The results are best computed by a direct application of the formula with a hand calculator. Note that this, too, is a formula that can be readily converted into an interactive command file.

Significance Levels for z-, Chi-square, t-, and F-statistics

Minitab permits the user to calculate a variety of probability functions, using either the built-in cumulative distribution functions (CDF) or the inverse cumulative distribution functions (INVCDF). The CDF functions calculate the value associated with the area of a function. The INVCDF calculates the value below the associated area. *Minitab* offers functions for seventeen different distributions: uniform, normal, t, F, Chi-square, exponential, Gamma, Beta, binomial, Poisson, integer, discrete, Cauchy, Laplace, logistic, log normal, and Weibull. The following illustrate the use of the CDF to calculate appropriate significance levels for the z-, t-, F-, and Chi-square statistics. The numbers used in the examples are only for illustration purposes and may be substituted as needed with real numbers of interest. Additionally, any values created by the functions may be put into storage columns in the *Minitab* worksheet or may be drawn from the worksheet columns.

```
MTB > CDF 2.00;
SUBC> NORMAL 0 1.
    The proportion of a z-distribution that lies below the normal curve the
    value of 2.00 when the distribution has a mean of 0 and a standard
    deviation of 1.
```

```
MTB > CDF 3.5;
SUBC> T 30.
    The proportion of a t-distribution that lies below a value of 3.5 when the
    sample has 30 degrees of freedom.
```

```
MTB > CDF 3.5;
SUBC> F 2 18.
  The proportion of an F-distribution that lies below a value of 3.5 when the
  sample has 2 and 18 degrees of freedom.
```

```
MTB > CDF 3.5;
SUBC> CHISQUARE 8.
  The proportion of a Chi-square distribution that lies below a value of 3.5
  when the sample has 8 degrees of freedom.
```

Each of these functions may have their results stored in a column or as a constant. A constant could then be used to calculate a probability value for the statistic of interest, and this could be used to estimate the area of the distribution of interest above the value point. For example, the following is a command file designed to produce this outcome for a t-distribution test:

```
CDF 3.5 K1;
   T 30.
LET K2 = 1 - K1
PRINT K1
END
```

As with previous examples of command files, this file and its related functions could be stored as interactive files permitting the user to input the values of interest.

The INVCDF function may be used to calculate the value of a data point at an associated level of probability, given the appropriate critical elements. The user is advised to consult the *Minitab Reference Manual* for additional information concerning these functions.

Comparisons: Two Samples (Nonparametric) *(A4-57 to A4-65)*

Minitab's nonparametric tests include the sign test (**A4-64**) for two dependent samples and the Mann-Whitney U-test for the independent two-sample test (**A4-60**). The process for applying both of these tests is very similar: identify the command, identify the median or confidence level, and then identify the columns or variables for the test.

The sign test (**A4-64**) will be used in the example below to test for differences between a fall testing and a spring testing of students in a reading program. The question of interest: did any difference in median performance scores occur over the time period? The critical factors for employing the sign test are that the data must be ordinal in nature and the user must have an anchor median by which the comparison will be made. In our illustration below, the scores represent performance scales on a district-wide criterion-referenced reading test given at the beginning of the school year and again in April. The command and the output are given below.

```
MTB > Stest 266.0 'Fall' 'Spring'
```

```
SIGN TEST OF MEDIAN = 266.0 VERSUS  N.E.  266.0
          N     BELOW    EQUAL    ABOVE    P-VALUE    MEDIAN
Fall     116     57        1       58      1.0000     266.5
Spring   116     46        1       69      0.0402     274.0
```

The sign test asks if the same group of students performed differently in the spring than they did in the fall, based on a median for the fall testing of 266. A reminder: this test only indicates that a difference occurred in the test, not that one testing proved better or worse than the other.

The Mann-Whitney U-test (**A4-60**) is implemented in much the same manner, except that the user must indicate a confidence level (if none is indicated *Minitab* assumes a confidence level of 95%). The nature of this command is:

```
MANN-WHITNEY .05 'RANK1' 'RANK2'
```

The output for this comparison would be:

```
Mann-Whitney Confidence Interval and Test
C9    N =  5  Median =    19.000
C10   N =  5  Median =    17.000
Point estimate for ETA1-ETA2 is 4.000
96.3 pct. c.i. for ETA1-ETA2 is (-2.004, 9.996)
W = 34.0
Test of ETA1 = ETA2 vs. ETA1 n.e. ETA2 is significant at 0.2101
The test is significant at 0.2073 (adjusted for ties)

Cannot reject at alpha = 0.05
```

The results of our sample test indicate that we cannot reject a null hypothesis of significant difference between the medians of the two groups.

Comparisons: Two or More Samples (Nonparametric) (A4-66 to A4-69)

Minitab provides two tests to measure the nonparametric samples of two or more groups. The Kruskal-Wallis (**A4-68**) is used with independent samples; the Friedman two-way analysis of variance by ranks (**A4-69**) is used with dependent samples.

The Kruskal-Wallis test is treated by *Minitab* as a generalization of the Mann-Whitney test for two samples and will produce similar results when there are only two groups. The format of the command is to identify the column containing the variable of interest and the column containing the grouping variable. For example, the command below would produce results similar to the earlier example of the Mann-Whitney, but would not require the data to be in separate columns:

```
KRUSKAL C1 C2
```

```
LEVEL       NOBS      MEDIAN      AVE. RANK     Z VALUE
  1           5        19.00         6.8          1.36
  2           5        17.00         4.2         -1.36
OVERALL      10                      5.5

H = 1.84   d.f. = 1   p = 0.175
H = 1.87   d.f. = 1   p = 0.172 (adj. for ties)
```

As with the previous example, we see that there is no significant difference between the groups.

The Friedman test in *Minitab* assumes several critical points: first, that all the necessary data are in one column; second, that the first levels assigned subjects are in another column; and, finally, that the second level assigned subjects is in a third column. The command line must include all this information. Additional columns may be used to store residuals and fit statistics if desired by the user. The example below will be based upon data that are actually interval in nature, but will be treated as ranked for purposes of illustrating the FRIEDMAN command.

```
Friedman test of SCORE by C6 blocked by C7

S =  9.75   d.f. = 2   p = 0.008
S = 10.17   d.f. = 2   p = 0.006 (adjusted for ties)

                    Est.         Sum of
    C6      N      Median        RANKS
     1      6      89.667         10.5
     2      6      97.333         18.0
     3      6      82.000          7.5

Grand median = 89.667
```

The results of the test indicate a significant difference among subjects according to the levels of the grouping variable (C6) and blocking variable (C7). The reader is cautioned that the Friedman is most often used to test correlated samples when the assumptions of a repeated-measures parametric model cannot be met. The reader should also review the description in **A4-69** before undertaking this analysis.

Comparisons: Two or More Samples—Analysis of Variance (ANOVA) (A4-70 to A4-90)

The analysis of variance technique, used to compare the means of two or more samples, is a powerful and generic procedure used to test hypotheses derived from a number of fundamental research designs. The variety of strategies available range from factorial (**A4-71** to **A4-78**) to repeated measures/split-plot (**A4-79** to **A4-83**) to nested or hierarchical (**A4-84** to **A4-86**). As with the previous chapters, only the most commonly occurring models will be illustrated. Repetitious examples will be omitted.

FACTORIAL ANALYSIS OF VARIANCE (A4-71 TO A4-78)

The factorial ANOVA procedures will be illustrated with the same data set used in SAS, Data Set 3, p. 208 (stored in *Minitab* as DASET3.) This data set involves 18 subjects in a two-factor design. The factors are (A) treatment, with three levels (E1, E2, and C) and (b) gender, with two levels (F and M).

While many statistical packages can perform ANOVA analyses on the categorical variables stored in alphanumeric form, *Minitab* cannot. This limitation requires that the data be recoded into numeric values before initiating the analysis. The first part of our analysis would be the recoding of the data using the CONVERT command. Assuming that the values for A are stored in column 2 and the values for B are stored in column 3, we first must build a conversion table in our worksheet to perform the action. The table below is created in columns 4, 5, 6, and 7:

C4	C5	C6	C7
E1	1	F	1
E2	2	M	2
C	3		

Using the table in C4 and C5 the next step is to complete the conversion with the application of the CONVERT command:

```
CONVERT C4 C5 C2 C10
```

The above command tells *Minitab* to convert, using the table in C4 and C5, the data in C2 and to put the new values in C10. A similar command will put the correct values for gender into C11. The NAME command is issued to give the user labels for the variables:

```
CONVERT C6 C7 C3 C11
NAME C10='TRT' C11='SEX'
```

After these manipulations the data are in a form that *Minitab* may use in one of its several commands available to perform ANOVA analyses. In a fashion similar to many statistical programs, *Minitab* has available several different commands according to the needs and expertise of the user. The following is a short description of these commands:

AOVONEWAY   performs a one-way analysis of variance with the requirement that each of the categorical data be in an independent column. Does not require a balanced model.

ONEWAY   performs a one-way analysis of variance. The dependent variable is expected to be in one column and the categorical data in another column. Requires a balanced model.

TWOWAY   performs a two-way analysis of variance with the expectation that the dependent measure is in one column and the categorical levels are each in different columns (i.e., one column for levels of A and another column for levels of B). Requires that the cells be completely balanced.

ANOVA   employs a model statement (i.e., Y = A + B + A*B) to perform a multiway analysis of variance. For multiway models the design must be balanced. May be used for unbalanced one-way ANOVAs.

GLM   employs a model statement to perform balanced and unbalanced ANOVAs as well as ANCOVA and regression models. GLM is available only in version 7.0 and above. The output is similar to that found in *SAS, SPSS-X,* and *SYSTAT*.

The GLM and ANOVA commands will be used to illustrate most of the ANOVA models described in this chapter. To point out some of the differences, however, the initial example will employ both the ONEWAY and GLM commands.

The one-way analysis of variance (**A4-71**) example has as its dependent variable SCORE. There are three levels of the treatment, called TRT 1, 2, and 3 (as designated in the conversion from above). The command for the ONEWAY is:

```
ONEWAY 'SCORE' 'TRT'
```

On the command line, the user may also designate a storage location for residuals and fit statistics. The output from this is presented below.

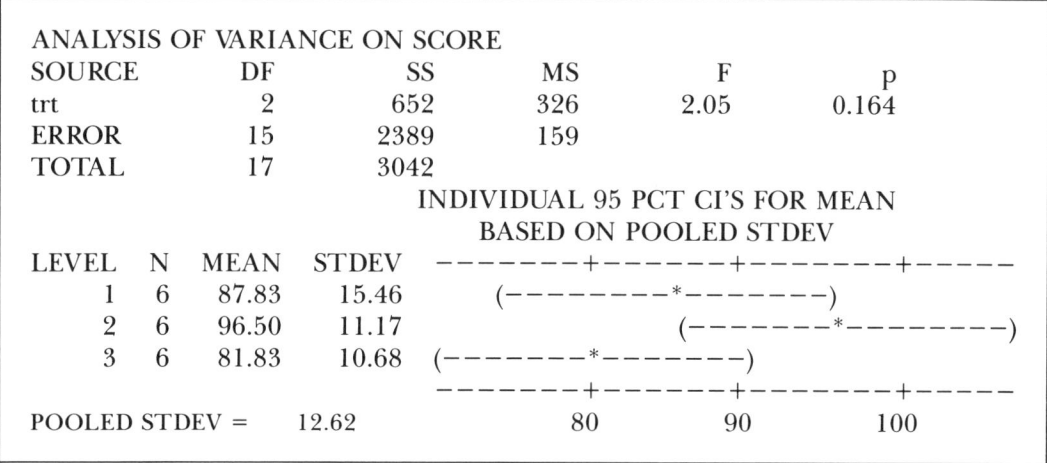

```
ANALYSIS OF VARIANCE ON SCORE
SOURCE          DF          SS           MS           F           p
trt              2         652          326         2.05       0.164
ERROR           15        2389          159
TOTAL           17        3042
                                    INDIVIDUAL 95 PCT CI'S FOR MEAN
                                        BASED ON POOLED STDEV
LEVEL    N    MEAN    STDEV    -------+-------+-------+-----
    1    6   87.83    15.46        (--------*-------)
    2    6   96.50    11.17                      (-------*--------)
    3    6   81.83    10.68    (-------*-------)
                                 -------+-------+-------+-----
POOLED STDEV =     12.62            80          90         100
```

The command for GLM would be:

```
GLM 'SCORE' = 'TRT'
```

The output would be:

```
Factor   Levels   Values
trt         3        1      2      3

Analysis of Variance for SCORE

Source          DF      Seq SS       Adj SS       Adj MS         F          P
trt              2       652.4        652.4        326.2       2.05      0.164
Error           15      2389.2       2389.2        159.3
Total           17      3041.6
```

These outputs differ in several critical ways. First, with ONEWAY the output is rounded to the one place, although the F-value and probability are carried at least to the hundredths place. Also, with ONEWAY the output contains a listing of the treatment means and a graphic representation of the confidence intervals based upon the pooled standard deviations.

At first glance the output from GLM appears to provide more precision in its presentation. The sums of squares and mean squares are carried out to the tenths place, and the output provides the sequential as well as the adjusted sums of squares. On the other hand, the initial output does *not* provide a test or listing of the cell means. GLM actually offers several subcommands that not only elaborate the output but provide additional information for the user. The following is an example of a simple elaboration that can be used:

```
MTB > BRIEF = 3
MTB > GLM 'SCORE' = 'TRT';
SUBC> MEANS 'TRT'.
```

The output from this set of commands is:

```
Factor   Levels  Values
trt         3      1     2     3
```

Analysis of Variance for SCORE

Source	DF	Seq SS	Adj SS	Adj MS	F	P
trt	2	652.4	652.4	326.2	2.05	0.164
Error	15	2389.2	2389.2	159.3		
Total	17	3041.6				

Term	Coeff	Stdev	t-value	P
Constant	88.722	2.975	29.83	0.000
trt				
1	−0.889	4.207	−0.21	0.836
2	7.778	4.207	1.85	0.084

Means for SCORE

trt	Mean	Stdev
1	87.83	5.152
2	96.50	5.152
3	81.83	5.152

The addition of the MEANS statement provides the additional means table. By preceding the GLM with BRIEF=3, the user obtains a full table of the coefficients—not necessary for ANOVA, but necessary for regression uses of GLM.

While the output of GLM is fairly complete, some of the additional information often provided by other statistical packages is not printed. For example, the printout does not provide an estimate of R-squared; nevertheless, this can be readily obtained by dividing the sums of squares for TRT (the between factor) by the sums of squares for the total (i.e., 652.4 / 3041.6 = .21449). The square root of R-squared can be calculated to get the multiple-R value.

The output for a two-way analysis of variance (**A4-72, A4-73, A4-74**) is obtained in a similar fashion. The model is designated in the GLM command, and the MEANS subcommand provides the means for the cells. The commands may be written in two ways:

```
MTB > GLM 'SCORE' = 'TRT' + 'SEX' + 'TRT'*'SEX';
SUBC> MEANS 'TRT' 'SEX' 'TRT'*'SEX'.
```

or

```
MTB > GLM 'SCORE' = 'TRT'| 'SEX';
SUBC> MEANS 'TRT' | 'SEX'.
```

The use of the straight line (|) in the second set of commands tells *Minitab* to treat each of the variables as independent and crossed. The results are exactly the same for either format. Remember, if the BRIEF=3 command is issued before the GLM command, *Minitab* will also print the regression coefficients.

```
Factor      Levels   Values
trt             3      1    2    3
sex             2      1    2

Analysis of Variance for SCORE

Source       DF     Seq SS     Adj SS     Adj MS        F        P
trt           2     652.44     652.44     326.22    10.39    0.002
sex           1    1942.72    1942.72    1942.72    61.89    0.000
trt*sex       2      69.78      69.78      34.89     1.11    0.361
Error        12     376.67     376.67      31.39
Total        17    3041.61

Means for SCORE

trt        Mean      Stdev
1         87.83      2.287
2         96.50      2.287
3         81.83      2.287
sex
1         99.11      1.868
2         78.33      1.868
trt*sex
1   1    101.00      3.235
1   2     74.67      3.235
2   1    105.67      3.235
2   2     87.33      3.235
3   1     90.67      3.235
3   2     73.00      3.235
```

The above ANOVA table does not contain the overall model with the between factor having five degrees of freedom. However, the quick use of a hand calculator should permit the user to easily construct it. Also, since the R-squared value is not printed, the user will have to calculate it using the formula described earlier. Another point of interest is the table of means produced by the MEANS command. The values represent the adjusted cell means for each listed term.

If the user chooses to use the ANOVA command, the output is very similar; however, the user may choose to store the cell means in worksheet columns through the FITS subcommand. The sequence for the ANOVA commands are listed below.

```
MTB > ANOVA 'SCORE' = 'TRT' | 'SEX';
SUBC> MEANS 'TRT' | 'SEX';
SUBC> FITS C14-C16.
```

The output from ANOVA would be:

Factor	Type	Levels	Values		
trt	fixed	3	1	2	3
sex	fixed	2	1	2	

Analysis of Variance for SCORE

Source	DF	SS	MS	F	P
trt	2	652.44	326.22	10.39	0.002
sex	1	1942.72	1942.72	61.89	0.000
trt*sex	2	69.78	34.89	1.11	0.361
Error	12	376.67	31.39		
Total	17	3041.61			

MEANS

trt	N	SCORE
1	6	87.833
2	6	96.500
3	6	81.833

sex	N	SCORE
1	9	99.111
2	9	78.333

trt	sex	N	SCORE
1	1	3	101.00
1	2	3	74.67
2	1	3	105.67
2	2	3	87.33
3	1	3	90.67
3	2	3	73.00

Note that the output is essentially the same for GLM and ANOVA. The motivation to choose one over the other is based more upon need and design type. GLM permits balanced as well as unbalanced designs and provides the same output as the other ANOVA commands, but it will not handle random or mixed models. While GLM does not have a RANDOM subcommand, both GLM and ANOVA permit the user to write customized TEST equations that allow the proper identification of an error term for an effect. ANOVA, on the other hand, while limited only to balanced models, will directly test for random models with the use of a RANDOM subcommand, as well as print expected mean squares with the EMS subcommand. Either command serves equally well; the choice between the two essentially depends upon the preferences and skills of the user. The recommendation is that a user should become familiar with both of these primary commands.

The random-effects and mixed-effects models (**A4-73, A4-74**) require the user correctly to test the treatment effects through the correct error term. Unless specified through the RANDOM subcommand in ANOVA or the TEST command in either GLM or ANOVA, *Minitab* assumes that the model being tested is a fixed-effects model and calculates its tests accordingly. The general format of the TEST command is:

> TEST name of effect(s) to test / error term

If we employ the same procedure used in previous chapters and use the TRT variable as a random effect and SEX as a fixed effect, the resulting model is a mixed-effects model (**A4-74**). The TEST command identifies the effect to be tested as SEX and the error term to be TRT*SEX. The complete command set would be:

> MTB > ANOVA 'SCORE' = 'TRT' | 'SEX';
> SUBC> TEST 'SEX' / 'TRT'*'SEX'.

Since the design is balanced, the critical output will be the same as the previous output for our fixed effect, regardless of whether the command is GLM or ANOVA. The only addition to either table will be the test of the SEX effect listed below:

```
F-test with denominator: trt*sex
Denominator MS = 34.889 with 2 degrees of freedom
Numerator        DF           MS            F            P
sex              1            1943          55.68        0.017
```

If the RANDOM (RANDOM 'TRT') subcommand had been with the ANOVA command instead of the TEST command, the ANOVA table would be printed only as a mixed-effects table. The result of this procedure is printed below:

Factor	Type	Levels	Values		
trt	random	3	1	2	3
sex	fixed	2	1	2	

Analysis of Variance for SCORE

Source	DF	SS	MS	F	P
trt	2	652.44	326.22	9.35	0.097
sex	1	1942.72	1942.72	55.68	0.017
trt*sex	2	69.78	34.89	1.11	0.361
Error	12	376.67	31.39		
Total	17	3041.61			

Once again, the results are equal. If all the effects in the ANOVA model are random, then the use of the RANDOM subcommand in the ANOVA command is more expedient than writing a series of test commands. If the model is not balanced, the only option would be to use the GLM command with the appropriate TEST subcommand.

The extension of the above procedures to three-way ANOVAs (**A4-75** to **A4-78**) is rather straightforward. However, once into the higher-order models the user is strictly limited to the ANOVA or the GLM commands. The primary consideration in making the extension to three-way models is specifying the appropriate error terms for random- and mixed-effects models. If the design is balanced, the user may readily depend upon the ANOVA command to perform the necessary tests; however, if the model is not balanced, the user will have to build the correct tests using the TEST subcommand and the guides provided by the appropriate models.

The following is an example of a three-way ANOVA first introduced in the *SAS* chapter. In this example, the independent variables are A, B, and C, with A fixed and B and C random (**A4-76**). The statistical model calls for a mixed-effects analysis with appropriate error terms to test the model.

```
MTB > GLM 'SCORE' = 'A'|'B'|'C';
SUBC> TEST A / A*B+A*C−A*B*C;
SUBC> TEST B C / B*C;
SUBC> TEST A*B A*C / A*B*C;
SUBC> MEANS 'A'|'B'|'C'.
```

While the use of the RANDOM term to specify the random effects would be possible in the ANOVA command, the above data set represents an unbalanced design and ANOVA, therefore, cannot be used. Since the design is unbalanced, the appropriate command to use is GLM with the utilization of a set of TEST commands to set up the proper testing for the effects. The same type of strategy may be employed with any of the three-way models using the strategies described in **A4-75** to **A4-78**.

Repeated-Measures/Split-Plot ANOVA (A-79 to A4-83)

One of the limitations of *Minitab* is its lack of a special command or subcommand for performing a repeated-measures analysis. By employing appropriate TEST statements and nesting strategies, however, most of the major types of repeated measures may be accomplished. Although most of the techniques to be described in this chapter have been developed by the authors, some were also drawn from Graf and Lapp (1987)* and Aubuchon (1989).**

---

*R. Graf and D.M. Lapp. *Repeated Measures Designs Using TWOWAY, REGRESS, and ANOVA*. State College, PA: Minitab Labs, 1987.
**J.C. Aubuchon. *Repeated measures Designs Using ANOVA*. State College, PA: Minitab Labs, 1989.

A one-way repeated measures (or within-subjects) design **(A4-79)** involves a single group of subjects measured across several different occasions. The illustration will be drawn from Data Set 4, (p. 212) stored as DASET4.MTW in *Minitab*. The initial data format had to be adjusted to meet the requirements of calculations with *Minitab*.

The format required the use of the STACK command to create a column containing only the scores of the ten subjects, creation of a column containing the repetitively stacked subjects, (1 through 10, times 4 occasions), and the use of the SET command to set a 40-element column containing 10 1s, 2s, 3s, and 4s—representing each subject on each of the four occasions. The following is a partial listing of the restructured Data Set 4:

ROW	sub	score	occas
1	1	21	1
2	2	17	1
3	3	19	1
4	4	25	1
5	5	18	1
6	6	19	1
7	7	20	1
8	8	11	1
9	9	17	1
10	10	13	1
11	1	25	2
12	2	22	2
13	3	24	2
14	4	27	2
15	5	21	2
16	6	21	2
17	7	23	2
18	8	16	2
19	9	20	2
20	10	15	2
21	1	30	3

This pattern was continued until all 40 scores had been presented.

The procedure for analyzing a one-way repeated measure is similar to the strategy employed in a two-way mixed-effects model. In this case the fixed effect, OCCASION, is tested against the interactive effect, OCCASION*SUBJECT. The output from *Minitab* clearly warns the user that the identified error term equals zero and, therefore, none of the effects are testable. However, since a TEST subcommand is employed, the effect of interest is established as OCCASION and the lack of testable effects on the main table is not of concern. The commands to perform this analysis are:

```
MTB > GLM 'SCORES' = 'SUBJECT'|'OCCASION';
SUBC> TEST 'OCCASION' / 'SUBJECT'*'OCCASION'.
```

The output is listed at the top of the next page.

```
Factor    Levels        Values
sub       10    1     2    3    4    5    6    7    8    9    10
occas      4    1     2    3    4
```

Analysis of Variance for score

Source	DF	Seq SS	Adj SS	Adj MS	F	P
subject	9	619.725	619.725	68.858	**	
occasion	3	623.475	623.475	207.825	**	
sub*occas	27	64.775	64.775	2.399	**	
Error	0	0.000	0.000	0.000		
Total	39	1307.975				

** denominator of F=test is zero

F=test with denominator: subject*occasion
Denominator MS = 2.3991 with 27 degrees of freedom

Numerator	DF	Seq MS	F	P
occasion	3	207.8	86.63	0.000

The results indicate that a significant difference exists across the OCCASIONS factor.

The two-way split-plot (repeated measures) design (**A4-80**) is a logical extension of the one-way model. In this design, one factor is a categorical variable (GROUP), one factor is the subjects, and the remaining factor is the occasions. As with the earlier example, the limitations of *Minitab* require that the data set, called DASET4, be reconstructed into a format that can be applied to test our model. The STACK and SET commands are here again employed to create a four-column worksheet: one column for the variable SCORE, one for GROUP, one for SUBJECT, and one for OCCASION. The model may be treated in two fashions. Since it is balanced, one strategy could employ the ANOVA command with a RANDOM statement for the subjects-nested-within-occasion factor. The commands for this would be:

```
MTB > ANOVA SCORE = OCCASION SUBJECT(OCCASION)
GROUP
OCCASION*GROUP;
SUBC> RANDOM SUBJECT(OCCASION).
```

The other technique would be to employ the less restrictive GLM command and build the proper error terms from the model in **A4-80**. These commands would be:

```
MTB > GLM SCORE = GROUP SUBJECT(GROUP) OCCASION
GROUP*OCCASION
OCCASION*SUBJECT(GROUP);
SUBC> TEST GROUP / SUBJECT(GROUP);
SUBC> TEST OCCASION GROUP*OCCASION /
OCCASION*SUBJECT(GROUP).
```

The results of the ANOVA table are flawed because of the use of an incorrect error term, but the TEST subcommands establish the proper terms and error terms. It is the results of the TEST subcommands that are of interest in this analysis. The following is an edited version of the output illustrating the hypotheses tested in the TEST subcommands:

```
F-test with denominator: S(GROUP)
Denominator MS = 42.213 with 8 degrees of freedom

Numerator      DF    Seq MS       F        P
GROUP           1     308.0     7.30    0.027

F-test with denominator: OCCASION*S (GROUP)
Denominator MS = 2.1125 with 24 degrees of freedom

Numerator          DF   Seq MS      F       P
OCCASION            3   207.825   98.38   0.000
GROUP*OCCASION      3     4.025    1.91   0.156
```

The extension to a three-way split-plot (repeated measures) model, whether that model is one-between and two-within (**A4-81**) or two-between and one-within (**A4-82**) is a logical extension of the models previously presented. The complexity of the analysis is more limited by the need to recode and restructure the data within the *Minitab* worksheet. The recoding strategies are rather complex, although the actual analyses follow the strategies previously detailed. The user is advised to create the data set in the worksheet initially in a way that meets the requirements of *Minitab* rather than attempting to employ STACK, SET, and INDICATOR commands later.

NESTED/HIERARCHICAL ANOVA (**A4-84** TO **A4-86**)

A nested design is used when some levels of one factor appear only under one level of another factor. Typical designs of this type include studies of instructional strategies with teachers nested within a given instructional method rather than being fully crossed; that is, each teacher uses only one instructional method, rather than all teachers teaching all methods. The statistical model used in these designs (**A4-84** to **A4-86**) to detect significant differences is the nested, or hierarchical, design.

The example that will be used is an experiment being conducted to discover which of two reading series is most effective for developing reading comprehension. Since a basal reading series is a complete instructional system, it would be inappropriate to expect all ten teachers in the study to teach with both series. The only acceptable method of analysis is a nested design, with five teachers using one series and five teachers using the other. This would not be a factorial design unless all teachers were crossed with the reading series.

The coding strategy for use with the data set, called DASET7 (p. 218), requires the creation of a single variable, TEACHER, which repeats for each of the methods. Thus the variable contains the nested teacher number 1 through 5 for the respective method. This strategy is necessary in order to nest properly the factor within the method and to create the proper TEST subcommand. The best way to verify the proper coding format is to use the following commands to create a table with the values for the teachers within the cells of the table:

```
MTB > TABLE 'METHOD' 'TEACHER';
SUBC> DATA 'SCORE'.
```

The result is the following table:

```
ROWS: METHOD         COLUMNS: teacher
                 1          2          3          4          5
       1     25.000     27.000     26.000     27.000     18.000
             30.000     28.000     25.000     30.000     24.000
             29.000     24.000     24.000     24.000     23.000

       2     10.000      9.000      7.000      7.000      9.000
              8.000      9.000      6.000     10.000     10.000
             12.000     10.000      9.000      9.000     11.000

       CELL CONTENTS —
                SCORE: DATA
```

Although it may appear that the variable TEACHER is a crossed effect, the GLM or ANOVA command will treat it as a nested effect. The proper error term for testing the METHOD variable is TEACHER(METHOD), which is read "teacher nested within method." The command lines for ANOVA are:

```
MTB > ANOVA SCORE = METHOD TEACHER(METHOD);
SUBC> TEST METHOD / TEACHER(METHOD).
```

The output, with the proper effect tested, is illustrated on the next page:

```
Factor              Levels         Values
METHOD              2       1   2
teacher (METHOD)    5       1   2   3   4   5
```

Analysis of Variance for SCORE

```
Source              DF    Seq SS    Adj SS    Adj MS        F        P
METHOD               1   2050.13   2050.13   2050.13   492.03    0.000
teacher (METHOD)     8     87.20     87.20     10.90     2.62    0.039
Error               20     83.33     83.33      4.17
Total               29   2220.67
```

F-test with denominator: teacher (METHOD)
Denominator MS = 10.900 with 8 degrees of freedom

```
Numerator    DF    Seq MS       F        P
METHOD        1      2050   188.09    0.000
```

GENERAL ANOVA DESIGNS (A4-87 AND A4-88)

The set of designs known as randomized-blocks designs (**A4-87**) may be implemented within *Minitab* in a manner similar to the factorial ANOVA models described earlier in this chapter. The strategy requires the model to be stated without an interaction between the category or class variables. The full model statement must be used, however, rather than using the straight line (|), since *Minitab* automatically assumes an interaction with two or more categories with this command. The following would implement a randomized-blocks model:

> MTB > ANOVA DEPENDENT = CATEGORY BLOCK;
> SUBC> MEANS CATEGORY;

Decisions regarding appropriate TEST commands (if any) should be based upon the nature of the variable(s) to be tested; that is, whether the factors are random or fixed.

While the complexity of a Latin square design (**A4-88**) limits it appropriateness in this chapter, *Minitab* is quite capable of performing such an analysis. The true value of a Latin square design is its ability to isolate cellular effects; however, its complexity limits its application. The *Minitab Reference Manual* does not describe a strategy of commands that will perform a Latin square model; however, the key to its use is in the input plan used to build the data set. The interested reader is advised to seek additional references such as Keppel (1982),* Kirk (1982),** and Winer (1971)*** about building Latin square analysis models.

---

*G. Keppel. *Design and Analysis: A Researcher's Handbook,* 2d edition. Englewood Cliffs, NJ: Prentice-Hall, 1982.
**R. Kirk. *Experimental Designs: Procedures for the Behavioral Sciences.* 2d edition. Monterey, CA: Brooks/Cole Publishing Co., 1982.
***B.J. Winer. *Statistical Principles in Experimental Design.* 2d edition. New York: McGraw-Hill, 1971.

Related Designs (A4-89)

The purpose of analysis of covariance, ANCOVA, is to reduce known sources of influence or bias on the dependent variable in a factorial ANOVA. The calculation of an ANCOVA model is accomplished through the use of the ANCOVA or GLM commands. Both commands essentially produce the same results, with the restriction placed on the ANCOVA command that the design must be balanced while the GLM does not have this constraint. The number of covariant variables ranges from 24 for DOS-based personal computers to 50 for miniframe and mainframe systems.

Using Data Set 1 (DASET1.MTW) for our sample, the first step is to check for homogeneity of variance. This is accomplished by first computing a less-than-full-rank model followed by the regular full-rank model. The difference in error sums of squares divided by the difference in degrees of freedom provides the mean square due to the homogeneity of slopes. A ratio of this value to the mean square for error for the full-rank model is the suggested F-test for the question of homogeneity of slopes.* The previously identified statistical texts (Keppel [1982]; Kirk [1982]; and Winer [1971]) also provide a variety of hand-calculable tests for homogeneity of slopes. The following commands provide an ANCOVA for POSTTEST, with GROUP as the categorical variable and PRETEST as the covariant. The MEANS subcommand produces the adjusted means for the analysis:

```
MTB > ANCOVA POSTTEST = GROUP;
SUBC> COVARIANT PRETEST;
SUBC> MEANS GROUP.
```

The output for the model is as follows:

```
Factor          Levels  Values
GROUP              2      1    2

Analysis of Covariance for POSTTEST

Source              DF      ADJ SS       MS         F         P
Covariates           1      44.890     44.890      4.55     0.070
GROUP                1      33.326     33.326      3.38     0.109
Error                7      69.110      9.873
Total                9     236.500

Covariate         Coeff       Stdev      t-value       P
PRETEST          0.6700       0.314       2.132      0.070

          ADJUSTED MEANS
GROUP        N        POSTTEST
  1          5         30.660
  2          5         26.340
```

Minitab Reference Manual, Version 6.1. State College, PA: Minitab Labs, 1988. Pp. 166–167.

The following output is the ANOVA model for the same data without the covariant, and the means are the means without the adjustment for the covariant:

```
Factor     Type   Levels  Values
GROUP      fixed    2      1   2

Analysis of Variance for POSTTEST

Source      DF         SS         MS         F        P
GROUP        1      122.50     122.50      8.60    0.019
Error        8      114.00      14.25
Total        9      236.50

            MEANS
GROUP        N      POSTTEST
  1          5       32.000
  2          5       25.000
```

As can be readily seen from the two outputs, the GROUP factor is statistically different for POSTTEST before the introduction of the covariant, but not after the covariant is introduced.

Multiple-Comparison Procedures (A4-91 to A4-97)

Minitab does not possess any documented built-in multiple-comparison methods, although Tukey's HSD (**A4-95**), Fisher's Least Significant Difference, Dunnett's test for control groups, and Hsu's multiple comparisons with the best test are available in version 7.0 and higher as undocumented features for the ONEWAY command.* While only these few tests are currently available, most of the tests (**A4-92** to **A4-97**) may be accomplished with the storing of adjusted cell means (by using a FIT subcommand) and through some calculations with a hand calculator. The following is an example using Data Set 3 (DASET3) with ONEWAY and its TUKEY option.

```
MTB > oneway c3, c4;
SUBC> tukey.
```

---

*J.C. Aubuchon. *Experimental Multiple Comparison Capability: ONEWAY Subcommands.* State College, PA: Minitab Labs, 1988.

```
ANALYSIS OF VARIANCE ON SCORE
SOURCE    DF     SS      MS       F       P
trt        2    652     326     2.05    0.164
ERROR     15   2389     159
TOTAL     17   3042

                        INDIVIDUAL 95 PCT CI'S FOR MEAN
                        BASED ON POOLED STDEV
LEVEL  N  MEAN  STDEV  -------+-------+-------+---
  1    6  87.83 15.46    (-------+-------)
  2    6  96.50 11.17              (-------+-------)
  3    6  81.83 10.68  (------+-------)
                        -------+-------+-------+---
POOLED STDEV =  12.62         80      90     100

CONTINUE?
TUKEY'S multiple comparison procedure

        Nominal level = 0.0500
    Family error rate = 0.0500
Individual error rate = 0.0202

Critical value = 3.68

Intervals for (mean of column group) - (mean of row group)

          1         2
  2    -27.61
         10.28

  3    -12.94     -4.28
         24.94    33.61
```

Orthogonal contrasts (**A4-92**) and orthogonal polynomial contrasts (**A4-93**) are not directly available through *Minitab*. The application of appropriate formulas to cell means and marginal means may be obtained through the various commands, such as ANOVA, GLM, and so forth. These means may be tested according to appropriate statistical tests. Another strategy, for the sophisticated user, would be to employ the matrix functions of *Minitab* to calculate the comparisons, but this strategy is beyond the scope of this present text.

Appendixes

APPENDIX A

Normal Distribution Curve: Ordinates and Areas

$\frac{x}{\sigma}$	Area	Ordinate	$\frac{x}{\sigma}$	Area	Ordinate	$\frac{x}{\sigma}$	Area	Ordinate	$\frac{x}{\sigma}$	Area	Ordinate
.00	.0000	.3989	.30	.1179	.3814	.60	.2257	.3332	.90	.3159	.2661
.01	.0040	.3989	.31	.1217	.3802	.61	.2291	.3312	.91	.3186	.2637
.02	.0080	.3989	.32	.1255	.3790	.62	.2324	.3292	.92	.3212	.2613
.03	.0120	.3988	.33	.1293	.3778	.63	.2357	.3271	.93	.3238	.2589
.04	.0160	.3986	.34	.1331	.3765	.64	.2389	.3251	.94	.3264	.2565
.05	.0199	.3984	.35	.1368	.3752	.65	.2422	.3230	.95	.3289	.2541
.06	.0239	.3982	.36	.1406	.3739	.66	.2454	.3209	.96	.3315	.2516
.07	.0279	.3980	.37	.1443	.3725	.67	.2486	.3187	.97	.3340	.2492
.08	.0319	.3977	.38	.1480	.3712	.68	.2517	.3166	.98	.3365	.2468
.09	.0359	.3973	.39	.1517	.3697	.69	.2549	.3144	.99	.3389	.2444
.10	.0398	.3970	.40	.1554	.3683	.70	.2580	.3123	1.00	.3413	.2420
.11	.0438	.3965	.41	.1591	.3668	.71	.2611	.3101	1.01	.3438	.2396
.12	.0478	.3961	.42	.1628	.3653	.72	.2642	.3079	1.02	.3461	.2371
.13	.0517	.3956	.43	.1664	.3637	.73	.2673	.3056	1.03	.3485	.2347
.14	.0557	.3951	.44	.1700	.3621	.74	.2703	.3034	1.04	.3508	.2323
.15	.0596	.3945	.45	.1736	.3605	.75	.2734	.3011	1.05	.3531	.2299
.16	.0636	.3939	.46	.1772	.3589	.76	.2764	.2989	1.06	.3554	.2275
.17	.0675	.3932	.47	.1808	.3572	.77	.2794	.2966	1.07	.3577	.2251
.18	.0714	.3925	.48	.1844	.3555	.78	.2823	.2943	1.08	.3599	.2227
.19	.0753	.3918	.49	.1879	.3538	.79	.2852	.2920	1.09	.3621	.2203
.20	.0793	.3910	.50	.1915	.3521	.80	.2881	.2897	1.10	.3643	.2179
.21	.0832	.3902	.51	.1950	.3503	.81	.2910	.2874	1.11	.3665	.2155
.22	.0871	.3894	.52	.1985	.3485	.82	.2939	.2850	1.12	.3686	.2131
.23	.0910	.3885	.53	.2019	.3467	.83	.2967	.2827	1.13	.3708	.2107
.24	.0948	.3876	.54	.2054	.3448	.84	.2995	.2803	1.14	.3729	.2083
.25	.0987	.3867	.55	.2088	.3429	.85	.3023	.2780	1.15	.3749	.2059
.26	.1026	.3857	.56	.2123	.3410	.86	.3051	.2756	1.16	.3770	.2036
.27	.1064	.3847	.57	.2157	.3391	.87	.3078	.2732	1.17	.3790	.2012
.28	.1103	.3836	.58	.2190	.3372	.88	.3106	.2709	1.18	.3810	.1989
.29	.1141	.3825	.59	.2224	.3352	.89	.3133	.2685	1.19	.3830	.1965

$\frac{x}{\sigma}$	Area	Ordinate	$\frac{x}{\sigma}$	Area	Ordinate	$\frac{x}{\sigma}$	Area	Ordinate	$\frac{x}{\sigma}$	Area	Ordinate
1.20	.3849	.1942	1.65	.4505	.1023	2.10	.4821	.0440	2.55	.4946	.0154
1.21	.3869	.1919	1.66	.4515	.1006	2.11	.4826	.0431	2.56	.4948	.0151
1.22	.3888	.1895	1.67	.4525	.0989	2.12	.4830	.0422	2.57	.4949	.0147
1.23	.3907	.1872	1.68	.4535	.0973	2.13	.4834	.0413	2.58	.4951	.0143
1.24	.3925	.1849	1.69	.4545	.0957	2.14	.4838	.0404	2.59	.4952	.0139
1.25	.3944	.1826	1.70	.4554	.0940	2.15	.4842	.0395	2.60	.4953	.0136
1.26	.3962	.1804	1.71	.4564	.0925	2.16	.4846	.0387	2.61	.4955	.0132
1.27	.3980	.1781	1.72	.4573	.0909	2.17	.4850	.0379	2.62	.4956	.0129
1.28	.3997	.1758	1.73	.4582	.0893	2.18	.4854	.0371	2.63	.4957	.0126
1.29	.4015	.1736	1.74	.4591	.0878	2.19	.4857	.0363	2.64	.4959	.0122
1.30	.4032	.1714	1.75	.4599	.0863	2.20	.4861	.0355	2.65	.4960	.0119
1.31	.4049	.1691	1.76	.4608	.0848	2.21	.4864	.0347	2.66	.4961	.0116
1.32	.4066	.1669	1.77	.4616	.0833	2.22	.4868	.0339	2.67	.4962	.0113
1.33	.4082	.1647	1.78	.4625	.0818	2.23	.4871	.0332	2.68	.4963	.0110
1.34	.4099	.1626	1.79	.4633	.0804	2.24	.4875	.0325	2.69	.4964	.0107
1.35	.4115	.1604	1.80	.4641	.0790	2.25	.4878	.0317	2.70	.4965	.0104
1.36	.4131	.1582	1.81	.4649	.0775	2.26	.4881	.0310	2.71	.4966	.0101
1.37	.4147	.1561	1.82	.4656	.0761	2.27	.4884	.0303	2.72	.4967	.0099
1.38	.4162	.1539	1.83	.4664	.0748	2.28	.4887	.0297	2.73	.4968	.0096
1.39	.4177	.1518	1.84	.4671	.0734	2.29	.4890	.0290	2.74	.4969	.0093
1.40	.4192	.1497	1.85	.4678	.0721	2.30	.4893	.0283	2.75	.4970	.0091
1.41	.4207	.1476	1.86	.4686	.0707	2.31	.4896	.0277	2.76	.4971	.0088
1.42	.4222	.1456	1.87	.4693	.0694	2.32	.4898	.0270	2.77	.4972	.0086
1.43	.4236	.1435	1.88	.4699	.0681	2.33	.4901	.2064	2.78	.4973	.0084
1.44	.4251	.1415	1.89	.4706	.0669	2.34	.4904	.0258	2.79	.4974	.0081
1.45	.4265	.1394	1.90	.4713	.0656	2.35	.4906	.0252	2.80	.4974	.0079
1.46	.4279	.1374	1.91	.4719	.0644	2.36	.4909	.0246	2.81	.4975	.0077
1.47	.4292	.1354	1.92	.4726	.0632	2.37	.4911	.0241	2.82	.4976	.0075
1.48	.4306	.1334	1.93	.4732	.0620	2.38	.4913	.0235	2.83	.4977	.0073
1.49	.4319	.1315	1.94	.4738	.0608	2.39	.4916	.0229	2.84	.4977	.0071
1.50	.4332	.1295	1.95	.4744	.0596	2.40	.4918	.0224	2.85	.4978	.0069
1.51	.4345	.1276	1.96	.4750	.0584	2.41	.4920	.0219	2.86	.4979	.0067
1.52	.4357	.1257	1.97	.4756	.0573	2.42	.4922	.0213	2.87	.4979	.0065
1.53	.4370	.1238	1.98	.4761	.0562	2.43	.4925	.0208	2.88	.4980	.0063
1.54	.4382	.1219	1.99	.4767	.0551	2.44	.4927	.0203	2.89	.4981	.0061
1.55	.4394	.1200	2.00	.4772	.0540	2.45	.4929	.0198	2.90	.4981	.0060
1.56	.4406	.1182	2.01	.4778	.0529	2.46	.4931	.0194	2.91	.4982	.0058
1.57	.4418	.1163	2.02	.4783	.0519	2.47	.4932	.0189	2.92	.4982	.0056
1.58	.4429	.1145	2.03	.4788	.0508	2.48	.4934	.0184	2.93	.4983	.0055
1.59	.4441	.1127	2.04	.4793	.0498	2.49	.4936	.0180	2.94	.4984	.0053
1.60	.4452	.1109	2.05	.4798	.0488	2.50	.4938	.0175	2.95	.4984	.0051
1.61	.4463	.1092	2.06	.4803	.0478	2.51	.4940	.0171	2.96	.4985	.0050
1.62	.4474	.1074	2.07	.4808	.0468	2.52	.4941	.0167	2.97	.4985	.0048
1.63	.4484	.1057	2.08	.4812	.4059	2.53	.4943	.0163	2.98	.4986	.0047
1.64	.4495	.1040	2.09	.4817	.0449	2.54	.4945	.0158	2.99	.4986	.0046
									3.00	.4987	.0044

Source: J.E. Wert. *Educational Statistics*. New York: McGraw-Hill, 1938.

APPENDIX B

t-Distribution: Critical Values

df	Level of significance for one-tailed test					
	.10	.05	.025	.01	.005	.0005
	Level of significance for two-tailed test					
	.20	.10	.05	.02	.01	.001
1	3.078	6.314	12.706	31.821	63.657	636.619
2	1.886	2.920	4.303	6.965	9.925	31.598
3	1.638	2.353	3.182	4.541	5.841	12.941
4	1.533	2.132	2.776	3.747	4.604	8.610
5	1.476	2.015	2.571	3.365	4.032	6.859
6	1.440	1.943	2.447	3.143	3.707	5.959
7	1.415	1.895	2.365	2.998	3.499	5.405
8	1.397	1.860	2.306	2.896	3.355	5.041
9	1.383	1.833	2.262	2.821	3.250	4.781
10	1.372	1.812	2.228	2.764	3.169	4.587
11	1.363	1.796	2.201	2.718	3.106	4.437
12	1.356	1.782	2.179	2.681	3.055	4.318
13	1.350	1.771	2.160	2.650	3.012	4.221
14	1.345	1.761	2.145	2.624	2.977	4.140
15	1.341	1.753	2.131	2.602	2.947	4.073
16	1.337	1.746	2.120	2.583	2.921	4.015
17	1.333	1.740	2.110	2.567	2.898	3.965
18	1.330	1.734	2.101	2.552	2.878	3.922
19	1.328	1.729	2.093	2.539	2.861	3.883
20	1.325	1.725	2.086	2.528	2.845	3.850
21	1.323	1.721	2.080	2.518	2.831	3.819
22	1.321	1.717	2.074	2.508	2.819	3.792
23	1.319	1.714	2.069	2.500	2.807	3.767
24	1.318	1.711	2.064	2.492	2.797	3.745
25	1.316	1.708	2.060	2.485	2.787	3.725

(continued)

	Level of significance for one-tailed test					
	.10	.05	.025	.01	.005	.0005
df	Level of significance for two-tailed test					
	.20	.10	.05	.02	.01	.001
26	1.315	1.706	2.056	2.479	2.779	3.707
27	1.314	1.703	2.052	2.473	2.771	3.690
28	1.313	1.701	2.048	2.467	2.763	3.674
29	1.311	1.699	2.045	2.462	2.756	3.659
30	1.310	1.697	2.042	2.457	2.750	3.646
40	1.303	1.684	2.021	2.423	2.704	3.551
60	1.296	1.671	2.000	2.390	2.660	3.460
120	1.289	1.658	1.980	2.358	2.617	3.373
∞	1.282	1.645	1.960	2.326	2.576	3.291

Source: Taken from Table III, p. 46 of Ronald A. Fisher and Frank Yates: *Statistical Tables for Biological, Agricultural and Medical Research*, published by Longman Group Ltd., London (previously published by Oliver and Boyd, Edinburgh), and by permission of the authors and publishers.

APPENDIX C
Percentage Points for F-Distribution

df for denominator	α	\multicolumn{9}{c}{df for numerator}								
		1	2	3	4	5	6	7	8	9
1	.25	5.83	7.50	8.20	8.58	8.82	8.98	9.10	9.19	9.26
	.10	39.9	49.5	53.6	55.8	57.2	58.2	58.9	59.4	59.9
	.05	161	200	216	225	230	234	237	239	241
2	.25	2.57	3.00	3.15	3.23	3.28	3.31	3.34	3.35	3.37
	.10	8.53	9.00	9.16	9.24	9.29	9.33	9.35	9.37	9.38
	.05	18.5	19.0	19.2	19.2	19.3	19.3	19.4	19.4	19.4
	.01	98.5	99.0	99.2	99.2	99.3	99.3	99.4	99.4	99.4
3	.25	2.02	2.28	2.36	2.39	2.41	2.42	2.43	2.44	2.44
	.10	5.54	5.46	5.39	5.34	5.31	5.28	5.27	5.25	5.24
	.05	10.1	9.55	9.28	9.12	9.01	8.94	8.89	8.85	8.81
	.01	34.1	30.8	29.5	28.7	28.2	27.9	27.7	27.5	27.3
4	.25	1.81	2.00	2.05	2.06	2.07	2.08	2.08	2.08	2.08
	.10	4.54	4.32	4.19	4.11	4.05	4.01	3.98	3.95	3.94
	.05	7.71	6.94	6.59	6.39	6.26	6.16	6.09	6.04	6.00
	.01	21.2	18.0	16.7	16.0	15.5	15.2	15.0	14.8	14.7
5	.25	1.69	1.85	1.88	1.89	1.89	1.89	1.89	1.89	1.89
	.10	4.06	3.78	3.62	3.52	3.45	3.40	3.37	3.34	3.32
	.05	6.61	5.79	5.41	5.19	5.05	4.95	4.88	4.82	4.77
	.01	16.3	13.3	12.1	11.4	11.0	10.7	10.5	10.3	10.2
6	.25	1.62	1.76	1.78	1.79	1.79	1.78	1.78	1.78	1.77
	.10	3.78	3.46	3.29	3.18	3.11	3.05	3.01	2.98	2.96
	.05	5.99	5.14	4.76	4.53	4.39	4.28	4.21	4.15	4.10
	.01	13.7	10.9	9.78	9.15	8.75	8.47	8.26	8.10	7.98
7	.25	1.57	1.70	1.72	1.72	1.71	1.71	1.70	1.70	1.69
	.10	3.59	3.26	3.07	2.96	2.88	2.83	2.78	2.75	2.72
	.05	5.59	4.74	4.35	4.12	3.97	3.87	3.79	3.73	3.68
	.01	12.2	9.55	8.45	7.85	7.46	7.19	6.99	6.84	6.72
8	.25	1.54	1.66	1.67	1.66	1.66	1.65	1.64	1.64	1.63
	.10	3.46	3.11	2.92	2.81	2.73	2.67	2.62	2.59	2.56
	.05	5.32	4.46	4.07	3.84	3.69	3.58	3.50	3.44	3.39
	.01	11.3	8.65	7.59	7.01	6.63	6.37	6.18	6.03	5.91
9	.25	1.51	1.62	1.63	1.63	1.62	1.61	1.60	1.60	1.59
	.10	3.36	3.01	2.81	2.69	2.61	2.55	2.51	2.47	2.44
	.05	5.12	4.26	3.86	3.63	3.48	3.37	3.29	3.23	3.18
	.01	10.6	8.02	6.99	6.42	6.06	5.80	5.61	5.47	5.35

df for denominator	α	df for numerator								
		10	11	12	15	20	24	30	40	50
1	.25	9.32	9.36	9.41	9.49	9.58	9.63	9.67	9.71	9.74
	.10	60.2	60.5	60.7	61.2	61.7	62.0	62.3	62.5	62.7
	.05	242	243	244	246	248	249	250	251	252
2	.25	3.38	3.39	3.39	3.41	3.43	3.43	3.44	3.45	3.45
	.10	9.39	9.40	9.41	9.42	9.44	9.45	9.46	9.47	9.47
	.05	19.4	19.4	19.4	19.4	19.4	19.5	19.5	19.5	19.5
	.01	99.4	99.4	99.4	99.4	99.4	99.5	99.5	99.5	99.5
3	.25	2.44	2.45	2.45	2.46	2.46	2.46	2.47	2.47	2.47
	.10	5.23	5.22	5.22	5.20	5.18	5.18	5.17	5.16	5.15
	.05	8.79	8.76	8.74	8.70	8.66	8.64	8.62	8.59	8.58
	.01	27.2	27.1	27.1	26.9	26.7	26.6	26.5	26.4	26.4
4	.25	2.08	2.08	2.08	2.08	2.08	2.08	2.08	2.08	2.08
	.10	3.92	3.91	3.90	3.87	3.84	3.83	3.82	3.80	3.80
	.05	5.96	5.94	5.91	5.86	5.80	5.77	5.75	5.72	5.70
	.01	14.5	14.4	14.4	14.2	14.0	13.9	13.8	13.7	13.7
5	.25	1.89	1.89	1.89	1.89	1.88	1.88	1.88	1.88	1.88
	.10	3.30	3.28	3.27	3.24	3.21	3.19	3.17	3.16	3.15
	.05	4.74	4.71	4.68	4.62	4.56	4.53	4.50	4.46	4.44
	.01	10.1	9.96	9.89	9.72	9.55	9.47	9.38	9.29	9.24
6	.25	1.77	1.77	1.77	1.76	1.76	1.75	1.75	1.75	1.75
	.10	2.94	2.92	2.90	2.87	2.84	2.82	2.80	2.78	2.77
	.05	4.06	4.03	4.00	3.94	3.87	3.84	3.81	3.77	3.75
	.01	7.87	7.79	7.72	7.56	7.40	7.31	7.23	7.14	7.09
7	.25	1.69	1.69	1.68	1.68	1.67	1.67	1.66	1.66	1.66
	.10	2.70	2.68	2.67	2.63	2.59	2.58	2.56	2.54	2.52
	.05	3.64	3.60	3.57	3.51	3.44	3.41	3.38	3.34	3.32
	.01	6.62	6.54	6.47	6.31	6.16	6.07	5.99	5.91	5.86
8	.25	1.63	1.63	1.62	1.62	1.61	1.60	1.60	1.59	1.59
	.10	2.54	2.52	2.50	2.46	2.42	2.40	2.38	2.36	2.35
	.05	3.35	3.31	3.28	3.22	3.15	3.12	3.08	3.04	3.02
	.01	5.81	5.73	5.67	5.52	5.36	5.28	5.20	5.12	5.07
9	.25	1.59	1.58	1.58	1.57	1.56	1.56	1.55	1.55	1.54
	.10	2.42	2.40	2.38	2.34	2.30	2.28	2.25	2.23	2.22
	.05	3.14	3.10	3.07	3.01	2.94	2.90	2.86	2.83	2.80
	.01	5.26	5.18	5.11	4.96	4.81	4.73	4.65	4.57	4.52

df for numerator							df for denominator
60	100	120	200	500	∞	α	
9.76	9.78	9.80	9.82	9.84	9.85	.25	
62.8	63.0	63.1	63.2	63.3	63.3	.10	1
252	253	253	254	254	254	.05	
3.46	3.47	3.47	3.48	3.48	3.48	.25	
9.47	9.48	9.48	9.49	9.49	9.49	.10	2
19.5	19.5	19.5	19.5	19.5	19.5	.05	
99.5	99.5	99.5	99.5	99.5	99.5	.01	
2.47	2.47	2.47	2.47	2.47	2.47	.25	
5.15	5.14	5.14	5.14	5.14	5.13	.10	3
8.57	8.55	8.55	8.54	8.53	8.53	.05	
26.3	26.2	26.2	26.2	26.1	26.1	.01	
2.08	2.08	2.08	2.08	2.08	2.08	.25	
3.79	3.78	3.78	3.77	3.76	3.76	.10	4
5.69	5.66	5.66	5.65	5.64	5.63	.05	
13.7	13.6	13.6	13.5	13.5	13.5	.01	
1.87	1.87	1.87	1.87	1.87	1.87	.25	
3.14	3.13	3.12	3.12	3.11	3.10	.10	5
4.43	4.41	4.40	4.39	4.37	4.36	.05	
9.20	9.13	9.11	9.08	9.04	9.02	.01	
1.74	1.74	1.74	1.74	1.74	1.74	.25	
2.76	2.75	2.74	2.73	2.73	2.72	.10	6
3.74	3.71	3.70	3.69	3.68	3.67	.05	
7.06	6.99	6.97	6.93	6.90	6.88	.01	
1.65	1.65	1.65	1.65	1.65	1.65	.25	
2.51	2.50	2.49	2.48	2.48	2.47	.10	7
3.30	3.27	3.27	3.25	3.24	3.23	.05	
5.82	5.75	5.74	5.70	5.67	5.65	.10	
1.59	1.58	1.58	1.58	1.58	1.58	.25	
2.34	2.32	2.32	2.31	2.30	2.29	.10	8
3.01	2.97	2.97	2.95	2.94	2.93	.05	
5.03	4.96	4.95	4.91	4.88	4.86	.01	
1.54	1.53	1.53	1.53	1.53	1.53	.25	
2.21	2.19	2.18	2.17	2.17	2.16	.10	9
2.79	2.76	2.75	2.73	2.72	2.71	.05	
4.48	4.42	4.40	4.36	4.33	4.31	.01	

df for denominator	α	df for numerator											
		1	2	3	4	5	6	7	8	9	10	11	12
10	.25	1.49	1.60	1.60	1.59	1.59	1.58	1.57	1.56	1.56	1.55	1.55	1.54
	.10	3.29	2.92	2.73	2.61	2.52	2.46	2.41	2.38	2.35	2.32	2.30	2.28
	.05	4.96	4.10	3.71	3.48	3.33	3.22	3.14	3.07	3.02	2.98	2.94	2.91
	.01	10.0	7.56	6.55	5.99	5.64	5.39	5.20	5.06	4.94	4.85	4.77	4.71
11	.25	1.47	1.58	1.58	1.57	1.56	1.55	1.54	1.53	1.53	1.52	1.52	1.51
	.10	3.23	2.86	2.66	2.54	2.45	2.39	2.34	2.30	2.27	2.25	2.23	2.21
	.05	4.84	3.98	3.59	3.36	3.20	3.09	3.01	2.95	2.90	2.85	2.82	2.79
	.01	9.65	7.21	6.22	5.67	5.32	5.07	4.89	4.74	4.63	4.54	4.46	4.40
12	.25	1.46	1.56	1.56	1.55	1.54	1.53	1.52	1.51	1.51	1.50	1.50	1.49
	.10	3.18	2.81	2.61	2.48	2.39	2.33	2.28	2.24	2.21	2.19	2.17	2.15
	.05	4.75	3.89	3.49	3.26	3.11	3.00	2.91	2.85	2.80	2.75	2.72	2.69
	.01	9.33	6.93	5.95	5.41	5.06	4.82	4.64	4.50	4.39	4.30	4.22	4.16
13	.25	1.45	1.55	1.55	1.53	1.52	1.51	1.50	1.49	1.49	1.48	1.47	1.47
	.10	3.14	2.76	2.56	2.43	2.35	2.28	2.23	2.20	2.16	2.14	2.12	2.10
	.05	4.67	3.81	3.41	3.18	3.03	2.92	2.83	2.77	2.71	2.67	2.63	2.60
	.01	9.07	6.70	5.74	5.21	4.86	4.62	4.44	4.30	4.19	4.10	4.02	3.96
14	.25	1.44	1.53	1.53	1.52	1.51	1.50	1.49	1.48	1.47	1.46	1.46	1.45
	.10	3.10	2.73	2.52	2.39	2.31	2.24	2.19	2.15	2.12	2.10	2.08	2.05
	.05	4.60	3.74	3.34	3.11	2.96	2.85	2.76	2.70	2.65	2.60	2.57	2.53
	.01	8.86	6.51	5.56	5.04	4.69	4.46	4.28	4.14	4.03	3.94	3.86	3.80
15	.25	1.43	1.52	1.52	1.51	1.49	1.48	1.47	1.46	1.46	1.45	1.44	1.44
	.10	3.07	2.70	2.49	2.36	2.27	2.21	2.16	2.12	2.09	2.06	2.04	2.02
	.05	4.54	3.68	3.29	3.06	2.90	2.79	2.71	2.64	2.59	2.54	2.51	2.48
	.01	8.68	6.36	5.42	4.89	4.56	4.32	4.14	4.00	3.89	3.80	3.73	3.67
16	.25	1.42	1.51	1.51	1.50	1.48	1.47	1.46	1.45	1.44	1.44	1.44	1.43
	.10	3.05	2.67	2.46	2.33	2.24	2.18	2.13	2.09	2.06	2.03	2.01	1.99
	.05	4.49	3.63	3.24	3.01	2.85	2.74	2.66	2.59	2.54	2.49	2.46	2.42
	.01	8.53	6.23	5.29	4.77	4.44	4.20	4.03	3.89	3.78	3.69	3.62	3.55
17	.25	1.42	1.51	1.50	1.49	1.47	1.46	1.45	1.44	1.43	1.43	1.42	1.41
	.10	3.03	2.64	2.44	2.31	2.22	2.15	2.10	2.06	2.03	2.00	1.98	1.96
	.05	4.45	3.59	3.20	2.96	2.81	2.70	2.61	2.55	2.49	2.45	2.41	2.38
	.01	8.40	6.11	5.18	4.67	4.34	4.10	3.93	3.79	3.68	3.59	3.52	3.46
18	.25	1.41	1.50	1.49	1.48	1.46	1.45	1.44	1.43	1.42	1.42	1.41	1.40
	.10	3.01	2.62	2.42	2.29	2.20	2.13	2.08	2.04	2.00	1.98	1.96	1.93
	.05	4.41	3.55	3.16	2.93	2.77	2.66	2.58	2.51	2.46	2.41	2.37	2.34
	.01	8.29	6.01	5.09	4.58	4.25	4.01	3.84	3.71	3.60	3.51	3.43	3.37
19	.25	1.41	1.49	1.49	1.47	1.46	1.44	1.43	1.42	1.41	1.41	1.40	1.40
	.10	2.99	2.61	2.40	2.27	2.18	2.11	2.06	2.02	1.98	1.96	1.94	1.91
	.05	4.38	3.52	3.13	2.90	2.74	2.63	2.54	2.48	2.42	2.38	2.34	2.31
	.01	8.18	5.93	5.01	4.50	4.17	3.94	3.77	3.63	3.52	3.43	3.36	3.30
20	.25	1.40	1.49	1.48	1.46	1.45	1.44	1.43	1.42	1.41	1.40	1.39	1.39
	.10	2.97	2.59	2.38	2.25	2.16	2.09	2.04	2.00	1.96	1.94	1.92	1.89
	.05	4.35	3.49	3.10	2.87	2.71	2.60	2.51	2.45	2.39	2.35	2.31	2.28
	.01	8.10	5.85	4.94	4.43	4.10	3.87	3.70	3.56	3.46	3.37	3.29	3.23

df for numerator													df for denominator
15	20	24	30	40	50	60	100	120	200	500	∞	α	
1.53	1.52	1.52	1.51	1.51	1.50	1.50	1.49	1.49	1.49	1.48	1.48	.25	
2.24	2.20	2.18	2.16	2.13	2.12	2.11	2.09	2.08	2.07	2.06	2.06	.10	10
2.85	2.77	2.74	2.70	2.66	2.64	2.62	2.59	2.58	2.56	2.55	2.54	.05	
4.56	4.41	4.33	4.25	4.17	4.12	4.08	4.01	4.00	3.96	3.93	3.91	.01	
1.50	1.49	1.49	1.48	1.47	1.47	1.47	1.46	1.46	1.46	1.45	1.45	.25	
2.17	2.12	2.10	2.08	2.05	2.04	2.03	2.00	2.00	1.99	1.98	1.97	.10	11
2.72	2.65	2.61	2.57	2.53	2.51	2.49	2.46	2.45	2.43	2.42	2.40	.05	
4.25	4.10	4.02	3.94	3.86	3.81	3.78	3.71	3.69	3.66	3.62	3.60	.01	
1.48	1.47	1.46	1.45	1.45	1.44	1.44	1.43	1.43	1.43	1.42	1.42	.25	
2.10	2.06	2.04	2.01	1.99	1.97	1.96	1.94	1.93	1.92	1.91	1.90	.10	12
2.62	2.54	2.51	2.47	2.43	2.40	2.38	2.35	2.34	2.32	2.31	2.30	.05	
4.01	3.86	3.78	3.70	3.62	3.57	3.54	3.47	3.45	3.41	3.38	3.36	.01	
1.46	1.45	1.44	1.43	1.42	1.42	1.42	1.41	1.41	1.40	1.40	1.40	.25	
2.05	2.01	1.98	1.96	1.93	1.92	1.90	1.88	1.88	1.86	1.85	1.85	.10	13
2.53	2.46	2.42	2.38	2.34	2.31	2.30	2.26	2.25	2.23	2.22	2.21	.05	
3.82	3.66	3.59	3.51	3.43	3.38	3.34	3.27	3.25	3.22	3.19	3.17	.01	
1.44	1.43	1.42	1.41	1.41	1.40	1.40	1.39	1.39	1.39	1.38	1.38	.25	
2.01	1.96	1.94	1.91	1.89	1.87	1.86	1.83	1.83	1.82	1.80	1.80	.10	14
2.46	2.39	2.35	2.31	2.27	2.24	2.22	2.19	2.18	2.16	2.14	2.13	.05	
3.66	3.51	3.43	3.35	3.27	3.22	3.18	3.11	3.09	3.06	3.03	3.00	.01	
1.43	1.41	1.41	1.40	1.39	1.39	1.38	1.38	1.37	1.37	1.36	1.36	.25	
1.97	1.92	1.90	1.87	1.85	1.83	1.82	1.79	1.79	1.77	1.76	1.76	.10	15
2.40	2.33	2.29	2.25	2.20	2.18	2.16	2.12	2.11	2.10	2.08	2.07	.05	
3.52	3.37	3.29	3.21	3.13	3.08	3.05	2.98	2.96	2.92	2.89	2.87	.01	
1.41	1.40	1.39	1.38	1.37	1.37	1.36	1.36	1.35	1.35	1.34	1.34	.25	
1.94	1.89	1.87	1.84	1.81	1.79	1.78	1.76	1.75	1.74	1.73	1.72	.10	16
2.35	2.28	2.24	2.19	2.15	2.12	2.11	2.07	2.06	2.04	2.02	2.01	.05	
3.41	3.26	3.18	3.10	3.02	2.97	2.93	2.86	2.84	2.81	2.78	2.75	.01	
1.40	1.39	1.38	1.37	1.36	1.35	1.35	1.34	1.34	1.34	1.33	1.33	.25	
1.91	1.86	1.84	1.81	1.78	1.76	1.75	1.73	1.72	1.71	1.69	1.69	.10	17
2.31	2.23	2.19	2.15	2.10	2.08	2.06	2.02	2.01	1.99	1.97	1.96	.05	
3.31	3.16	3.08	3.00	2.92	2.87	2.83	2.76	2.75	2.71	2.68	2.65	.01	
1.39	1.38	1.37	1.36	1.35	1.34	1.34	1.33	1.33	1.32	1.32	1.32	.25	
1.89	1.84	1.81	1.78	1.75	1.74	1.72	1.70	1.69	1.68	1.67	1.66	.10	18
2.27	2.19	2.15	2.11	2.06	2.04	2.02	1.98	1.97	1.95	1.93	1.92	.05	
3.23	3.08	3.00	2.92	2.84	2.78	2.75	2.68	2.66	2.62	2.59	2.57	.01	
1.38	1.37	1.36	1.35	1.34	1.33	1.33	1.32	1.32	1.31	1.31	1.30	.25	
1.86	1.81	1.79	1.76	1.73	1.71	1.70	1.67	1.67	1.65	1.64	1.63	.10	19
2.23	2.16	2.11	2.07	2.03	2.00	1.98	1.94	1.93	1.91	1.89	1.88	.05	
3.15	3.00	2.92	2.84	2.76	2.71	2.67	2.60	2.58	2.55	2.51	2.49	.01	
1.37	1.36	1.35	1.34	1.33	1.33	1.32	1.31	1.31	1.30	1.30	1.29	.25	
1.84	1.79	1.77	1.74	1.71	1.69	1.68	1.65	1.64	1.63	1.62	1.61	.10	20
2.20	2.12	2.08	2.04	1.99	1.97	1.95	1.91	1.90	1.88	1.86	1.84	.05	
3.09	2.94	2.86	2.78	2.69	2.64	2.61	2.54	2.52	2.48	2.44	2.42	.01	

df for denominator	α	df for numerator											
		1	2	3	4	5	6	7	8	9	10	11	12
22	.25	1.40	1.48	1.47	1.45	1.44	1.42	1.41	1.40	1.39	1.39	1.38	1.37
	.10	2.95	2.56	2.35	2.22	2.13	2.06	2.01	1.97	1.93	1.90	1.88	1.86
	.05	4.30	3.44	3.05	2.82	2.66	2.55	2.46	2.40	2.34	2.30	2.26	2.23
	.01	7.95	5.72	4.82	4.31	3.99	3.76	3.59	3.45	3.35	3.26	3.18	3.12
24	.25	1.39	1.47	1.46	1.44	1.43	1.41	1.40	1.39	1.38	1.38	1.37	1.36
	.10	2.93	2.54	2.33	2.19	2.10	2.04	1.98	1.94	1.91	1.88	1.85	1.83
	.05	4.26	3.40	3.01	2.78	2.62	2.51	2.42	2.36	2.30	2.25	2.21	2.18
	.01	7.82	5.61	4.72	4.22	3.90	3.67	3.50	3.36	3.26	3.17	3.09	3.03
26	.25	1.38	1.46	1.45	1.44	1.42	1.41	1.39	1.38	1.37	1.37	1.36	1.35
	.10	2.91	2.52	2.31	2.17	2.08	2.01	1.96	1.92	1.88	1.86	1.84	1.81
	.05	4.23	3.37	2.98	2.74	2.59	2.47	2.39	2.32	2.27	2.22	2.18	2.15
	.01	7.72	5.53	4.64	4.14	3.82	3.59	3.42	3.29	3.18	3.09	3.02	2.96
28	.25	1.38	1.46	1.45	1.43	1.41	1.40	1.39	1.38	1.37	1.36	1.35	1.34
	.10	2.89	2.50	2.29	2.16	2.06	2.00	1.94	1.90	1.87	1.84	1.81	1.79
	.05	4.20	3.34	2.95	2.71	2.56	2.45	2.36	2.29	2.24	2.19	2.15	2.12
	.01	7.64	5.45	4.57	4.07	3.75	3.53	3.36	3.23	3.12	3.03	2.96	2.90
30	.25	1.38	1.45	1.44	1.42	1.41	1.39	1.38	1.37	1.36	1.35	1.35	1.34
	.10	2.88	2.49	2.28	2.14	2.05	1.98	1.93	1.88	1.85	1.82	1.79	1.77
	.05	4.17	3.32	2.92	2.69	2.53	2.42	2.33	2.27	2.21	2.16	2.13	2.09
	.01	7.56	5.39	4.51	4.02	3.70	3.47	3.30	3.17	3.07	2.98	2.91	2.84
40	.25	1.36	1.44	1.42	1.40	1.39	1.37	1.36	1.35	1.34	1.33	1.32	1.31
	.10	2.84	2.44	2.23	2.09	2.00	1.93	1.87	1.83	1.79	1.76	1.73	1.71
	.05	4.08	3.23	2.84	2.61	2.45	2.34	2.25	2.18	2.12	2.08	2.04	2.00
	.01	7.31	5.18	4.31	3.83	3.51	3.29	3.12	2.99	2.89	2.80	2.73	2.66
60	.25	1.35	1.42	1.41	1.38	1.37	1.35	1.33	1.32	1.31	1.30	1.29	1.29
	.10	2.79	2.39	2.18	2.04	1.95	1.87	1.82	1.77	1.74	1.71	1.68	1.66
	.05	4.00	3.15	2.76	2.53	2.37	2.25	2.17	2.10	2.04	1.99	1.95	1.92
	.01	7.08	4.98	4.13	3.65	3.34	3.12	2.95	2.82	2.72	2.63	2.56	2.50
120	.25	1.34	1.40	1.39	1.37	1.35	1.33	1.31	1.30	1.29	1.28	1.27	1.26
	.10	2.75	2.35	2.13	1.99	1.90	1.82	1.77	1.72	1.68	1.65	1.62	1.60
	.05	3.92	3.07	2.68	2.45	2.29	2.17	2.09	2.02	1.96	1.91	1.87	1.83
	.01	6.85	4.79	3.95	3.48	3.17	2.96	2.79	2.66	2.56	2.47	2.40	2.34
200	.25	1.33	1.39	1.38	1.36	1.34	1.32	1.31	1.29	1.28	1.27	1.26	1.25
	.10	2.73	2.33	2.11	1.97	1.88	1.80	1.75	1.70	1.66	1.63	1.60	1.57
	.05	3.89	3.04	2.65	2.42	2.26	2.14	2.06	1.98	1.93	1.88	1.84	1.80
	.01	6.76	4.71	3.88	3.41	3.11	2.89	2.73	2.60	2.50	2.41	2.34	2.27
∞	.25	1.32	1.39	1.37	1.35	1.33	1.31	1.29	1.28	1.27	1.25	1.24	1.24
	.10	2.71	2.30	2.08	1.94	1.85	1.77	1.72	1.67	1.63	1.60	1.57	1.55
	.05	3.84	3.00	2.60	2.37	2.21	2.10	2.01	1.94	1.88	1.83	1.79	1.75
	.01	6.63	4.61	3.78	3.32	3.02	2.80	2.64	2.51	2.41	2.32	2.25	2.18

				df for numerator									df for denominator
15	20	24	30	40	50	60	100	120	200	500	∞	α	
1.36	1.34	1.33	1.32	1.31	1.31	1.30	1.30	1.30	1.29	1.29	1.28	.25	
1.81	1.76	1.73	1.70	1.67	1.65	1.64	1.61	1.60	1.59	1.58	1.57	.10	22
2.15	2.07	2.03	1.98	1.94	1.91	1.89	1.85	1.84	1.82	1.80	1.78	.05	
2.98	2.83	2.75	2.67	2.58	2.53	2.50	2.42	2.40	2.36	2.33	2.31	.01	
1.35	1.33	1.32	1.31	1.30	1.29	1.29	1.28	1.28	1.27	1.27	1.26	.25	
1.78	1.73	1.70	1.67	1.64	1.62	1.61	1.58	1.57	1.56	1.54	1.53	.10	24
2.11	2.03	1.98	1.94	1.89	1.86	1.84	1.80	1.79	1.77	1.75	1.73	.05	
2.89	2.74	2.66	2.58	2.49	2.44	2.40	2.33	2.31	2.27	2.24	2.21	.01	
1.34	1.32	1.31	1.30	1.29	1.28	1.28	1.26	1.26	1.26	1.25	1.25	.25	
1.76	1.71	1.68	1.65	1.61	1.59	1.58	1.55	1.54	1.53	1.51	1.50	.10	26
2.07	1.99	1.95	1.90	1.85	1.82	1.80	1.76	1.75	1.73	1.71	1.69	.05	
2.81	2.66	2.58	2.50	2.42	2.36	2.33	2.25	2.23	2.19	2.16	2.13	.01	
1.33	1.31	1.30	1.29	1.28	1.27	1.27	1.26	1.25	1.25	1.24	1.24	.25	
1.74	1.69	1.66	1.63	1.59	1.57	1.56	1.53	1.52	1.50	1.49	1.48	.10	28
2.04	1.96	1.91	1.87	1.82	1.79	1.77	1.73	1.71	1.69	1.67	1.65	.05	
2.75	2.60	2.52	2.44	2.35	2.30	2.26	2.19	2.17	2.13	2.09	2.06	.01	
1.32	1.30	1.29	1.28	1.27	1.26	1.26	1.25	1.24	1.24	1.23	1.23	.25	
1.72	1.67	1.64	1.61	1.57	1.55	1.54	1.51	1.50	1.48	1.47	1.46	.10	30
2.01	1.93	1.89	1.84	1.79	1.76	1.74	1.70	1.68	1.66	1.64	1.62	.05	
2.70	2.55	2.47	2.39	2.30	2.25	2.21	2.13	2.11	2.07	2.03	2.01	.01	
1.30	1.28	1.26	1.25	1.24	1.23	1.22	1.21	1.21	1.20	1.19	1.19	.25	
1.66	1.61	1.57	1.54	1.51	1.48	1.47	1.43	1.42	1.41	1.39	1.38	.10	40
1.92	1.84	1.79	1.74	1.69	1.66	1.64	1.59	1.58	1.55	1.53	1.51	.05	
2.52	2.37	2.29	2.20	2.11	2.06	2.02	1.94	1.92	1.87	1.83	1.80	.01	
1.27	1.25	1.24	1.22	1.21	1.20	1.19	1.17	1.17	1.16	1.15	1.15	.25	
1.60	1.54	1.51	1.48	1.44	1.41	1.40	1.36	1.35	1.33	1.31	1.29	.10	60
1.84	1.75	1.70	1.65	1.59	1.56	1.53	1.48	1.47	1.44	1.41	1.39	.05	
2.35	2.20	2.12	2.03	1.94	1.88	1.84	1.75	1.73	1.68	1.63	1.60	.01	
1.24	1.22	1.21	1.19	1.18	1.17	1.16	1.14	1.13	1.12	1.11	1.10	.25	
1.55	1.48	1.45	1.41	1.37	1.34	1.32	1.27	1.26	1.24	1.21	1.19	.10	120
1.75	1.66	1.61	1.55	1.50	1.46	1.43	1.37	1.35	1.32	1.28	1.25	.05	
2.19	2.03	1.95	1.86	1.76	1.70	1.66	1.56	1.53	1.48	1.42	1.38	.01	
1.23	1.21	1.20	1.18	1.16	1.14	1.12	1.11	1.10	1.09	1.08	1.06	.25	
1.52	1.46	1.42	1.38	1.34	1.31	1.28	1.24	1.22	1.20	1.17	1.14	.10	200
1.72	1.62	1.57	1.52	1.46	1.41	1.39	1.32	1.29	1.26	1.22	1.19	.05	
2.13	1.97	1.89	1.79	1.69	1.63	1.58	1.48	1.44	1.39	1.33	1.28	.01	
1.22	1.19	1.18	1.16	1.14	1.13	1.12	1.09	1.08	1.07	1.04	1.00	.25	
1.49	1.42	1.38	1.34	1.30	1.26	1.24	1.18	1.17	1.13	1.08	1.00	.10	∞
1.67	1.57	1.52	1.46	1.39	1.35	1.32	1.24	1.22	1.17	1.11	1.00	.05	
2.04	1.88	1.79	1.70	1.59	1.52	1.47	1.36	1.32	1.25	1.15	1.00	.01	

Source: Taken from Appendix B.4 of D.E. Hinkle, W. Wiersma, and S. G. Jurs: *Applied Statistics for the Behavioral Sciences,* (Boston: Houghton Mifflin, 1979). Abridged from Table 18 of E. S. Pearson and H. O. Hartley, eds: *Biometrika Tables for Statisticians,* Vol. 1, 3rd ed., (New York: Cambridge, 1966). Used with permission of the Biometrika Trustees and Houghton Mifflin Company.

APPENDIX D

Probability Values for the Distribution of Chi-square

df	.99	.98	.95	.90	.80	.70	.50
1	.0³157	.0³628	.00393	.0158	.0642	.148	.455
2	.0201	.0404	.103	.211	.446	.713	1.386
3	.115	.185	.352	.584	1.005	1.424	2.366
4	.297	.429	.711	1.064	1.649	2.195	3.357
5	.554	.752	1.145	1.610	2.343	3.000	4.351
6	.872	1.134	1.635	2.204	3.070	3.828	5.348
7	1.239	1.564	2.167	2.833	3.822	4.671	6.346
8	1.646	2.032	2.733	3.490	4.594	5.527	7.344
9	2.088	2.532	3.325	4.168	5.380	6.393	8.343
10	2.558	3.059	3.940	4.865	6.179	7.267	9.342
11	3.053	3.609	4.575	5.578	6.989	8.148	10.341
12	3.571	4.178	5.226	6.304	7.807	9.034	11.340
13	4.107	4.765	5.892	7.042	8.634	9.926	12.340
14	4.660	5.368	6.571	7.790	9.467	10.821	13.339
15	5.229	5.985	7.261	8.547	10.307	11.721	14.339
16	5.812	6.614	7.962	9.312	11.152	12.624	15.338
17	6.408	7.255	8.672	10.085	12.002	13.531	16.338
18	7.015	7.906	9.390	10.865	12.857	14.440	17.338
19	7.633	8.567	10.117	11.651	13.716	15.352	18.338
20	8.260	9.237	10.851	12.443	14.578	16.266	19.337
21	8.897	9.915	11.591	13.240	15.445	17.182	20.337
22	9.542	10.600	12.338	14.041	16.314	18.101	21.337
23	10.196	11.293	13.091	14.848	17.187	19.021	22.337
24	10.856	11.992	13.848	15.659	18.062	19.943	23.337
25	11.524	12.697	14.611	16.473	18.940	20.867	24.337
26	12.198	13.409	15.379	17.292	19.820	21.792	25.336
27	12.879	14.125	16.151	18.114	20.703	22.719	26.336
28	13.565	14.847	16.928	18.939	21.588	23.647	27.336
29	14.256	15.574	17.708	19.768	22.475	24.577	28.336
30	14.953	16.306	18.493	20.599	23.364	25.508	29.336

df	.30	.20	.10	.05	.02	.01	.001
1	1.074	1.642	2.706	3.841	5.412	6.635	10.827
2	2.408	3.219	4.605	5.991	7.824	9.210	13.815
3	3.665	4.642	6.251	7.815	9.837	11.345	16.266
4	4.878	5.989	7.779	9.488	11.668	13.277	18.467
5	6.064	7.289	9.236	11.070	13.388	15.086	20.515
6	7.231	8.558	10.645	12.592	15.033	16.812	22.457
7	8.383	9.803	12.017	14.067	16.622	18.475	24.322
8	9.524	11.030	13.362	15.507	18.168	20.090	26.125
9	10.656	12.242	14.684	16.919	19.679	21.666	27.877
10	11.781	13.442	15.987	18.307	21.161	23.209	29.588
11	12.899	14.631	17.275	19.675	22.618	24.725	31.264
12	14.011	15.812	18.549	21.026	24.054	26.217	32.909
13	15.119	16.985	19.812	22.362	25.472	27.688	34.528
14	16.222	18.151	21.064	23.685	26.873	29.141	36.123
15	17.322	19.311	22.307	24.996	28.259	30.578	37.697
16	18.418	20.465	23.542	26.296	29.633	32.000	39.252
17	19.511	21.615	24.769	27.587	30.995	33.409	40.790
18	20.601	22.760	25.989	28.869	32.346	34.805	42.312
19	21.689	23.900	27.204	30.144	33.687	36.191	43.820
20	22.775	25.038	28.412	31.410	35.020	37.566	45.315
21	23.858	26.171	29.615	32.671	36.343	38.932	46.797
22	24.939	27.301	30.813	33.924	37.659	40.289	48.268
23	26.018	28.429	32.007	35.172	38.968	41.638	49.728
24	27.096	29.553	33.196	36.415	40.270	42.980	51.179
25	28.172	30.675	34.382	37.652	41.566	44.314	52.620
26	29.246	31.795	35.563	38.885	42.856	45.642	54.052
27	30.319	32.912	36.741	40.113	44.140	46.963	55.476
28	31.391	34.027	37.916	41.337	45.419	48.278	56.893
29	32.461	35.139	39.087	42.557	46.693	49.588	58.302
30	33.530	36.250	40.256	43.773	47.962	50.892	59.703

Source: Taken from Table IV, p. 47 of Ronald A. Fisher and Frank Yates: *Statistical Tables for Biological, Agricultural and Medical Research,* published by Longman Group Ltd., London, (previously published by Oliver and Boyd, Edinburgh), and by permission of the authors and publishers.

For df > 30, the expression $\sqrt{2\chi^2} - \sqrt{2\,df - 1}$ may be used as a normal deviate with unit variance.

Index

A

Ability and Ability Plus, 160–161
ABstat, 171
ADDaSTAT, 171–172
Analysis of covariance, 111–112
   Minitab, 362–363
   SAS, 218–219
   SPSS-X, 311–312
   SYSTAT, 272–274
Analysis of variance, 42–43, 84–114
   factorial, 85–94
      Minitab, 349–356
      SAS, 208–211
      SPSS-X, 308–311
      SYSTAT, 259–263
   nested/hierarchical, 104–108
      Minitab, 359–360
      SAS, 217–218
      SYSTAT, 269–271
   repeated measures/split plot, 94–103
      Minitab, 356–359
      SAS, 211–217
      SYSTAT, 263–269
   See also Randomized-Blocks Designs and Latin Square Designs
ANOVA (*see* Analysis of variance)
AppleWorks, 156–157
Assigning subjects to groups, 132–133

B

Bartlett test of homogeneity of variance
   SYSTAT, 254
Biserial correlation, 53–54
   SAS, 196
   SPSS-X, 293
BMDP PC, 164–165
Boeing Calc, 146

C

Central tendency, measures of, 44–45
Checklist for research, 135–137
Chi-square
   Friedman two-way analysis of variance by ranks, 83–84
      Minitab, 347–348
      SAS, 207
      SPSS-X, 307
      SYSTAT, 258–259
   McNemar's Test, 79–80
      SPSS-X, 306
   Median test, 75–76
      SAS, 206
      SPSS-X, 305–306
   Sign test for K-independent samples, 82
      SAS, 207
      SPSS-X, 307
   test for goodness of fit, 68–69
      Minitab, 339
      SAS, 202
      SPSS-X, 300, 301
      SYSTAT, 250
   test of independence (homogeneity), 78
      Minitab, 330
      SAS, 206–207
      SPSS-X, 291, 306
      SYSTAT, 258
   variance (sample) compared to some value, 66–67
      Minitab, 339
      SAS, 200–201
      SPSS-X, 300
      SYSTAT, 250
ClearCut, 152
Cluster sampling, 131–132
Coefficient of variation
   SAS, 192
   SYSTAT, 234

Comparisons
    Multiple-comparison procedures, 43–44, 114–120
    One sample, 41–42, 66–69
    Two samples, nonparametric, 42, 75–81
    Two samples, parametric, 42, 70–75
    Two samples or more, analysis of variance, 42–43, 84–114
    Two samples or more, nonparametric, 42, 81–84
    see also Analysis of variance
Contingency coefficient, 55–56
    Minitab, 332
    SAS, 196
    SPSS-X, 291, 294
    SYSTAT, 239, 242
Correlation (see also individual correlation procedures and Measures of Relationships)
    correlation (sample) compared to 0, 67–68
        Minitab, 339, 341, 342
        SAS, 201
        SPSS-X, 293, 300, 301
        SYSTAT, 250
    correlation (sample) compared to some value, 68
        Minitab, 339, 341, 342
        SAS, 202
        SPSS-X, 300, 301
        SYSTAT, 250
Correlation coefficients
    significant difference between dependent samples, 74–75
        Minitab, 344, 345
        SAS, 205
        SPSS-X, 302, 305
        SYSTAT, 255
    significant difference between independent samples, 74
        Minitab, 344
        SAS, 204–205
        SPSS-X, 302, 304
        SYSTAT, 255
Covariance, 50–51
    Minitab, 332
    SYSTAT, 238
Cramer V correlation
    SYSTAT, 239
Critical features for statistical programs, 176–177
CRUNCH, 172
CSS—Complete Statistical System, 172
Curvilinear regression, 64–65
    Minitab, 336, 337
    SAS, 199
    SPSS-X, 298
    SYSTAT, 248

D

Database manager profiles, 148–155
DataEase, 148–149
Data Set 1, 188
Data Set 2, 195
Data Set 3, 208
Data Set 4, 212
Data Set 4B, 213
Data Set 5, 215
Data Set 6, 217
Data Set 7, 218
dBase II, 153
dBase III Plus, 149
Descriptive statistics, 40–41, 44–50
Designs (*see* Research Designs)
Disarray (Nonparametric), 47–48
Dispersion (*see* Variability)
Distribution shapes, 48–50
    Minitab, 328, 329
    SAS, 193
    SPSS-X, 289–290
    SYSTAT, 236
Duncan multiple range test, 119–120
    Minitab, 363
    SAS, 220
    SPSS-X, 312
    SYSTAT, 274, 275
Dunnett's test for control groups
    Minitab, 363
Durbin-Watson statistic, 333
Dyna-Stat, 165–166

E

Enable, 157
Estimators, 126–127
Eta-squared, 59
    SPSS-X, 291

F

F-statistic
    analysis of covariance, 111–112
        Minitab, 362–363
        SAS, 218–219
        SPSS-X, 311–312
        SYSTAT, 272–274
    factorial analysis of variance, 85–94
        Minitab, 349–356
        SAS, 208–211
        SPSS-X, 308–311
        SYSTAT, 259–263
    independent sample variances, 71–72
        SAS, 203
        SYSTAT, 254

Latin square designs, 109–111
  Minitab, 361
  SAS, 218
  SYSTAT, 272
 nested/hierarchical ANOVA, 104–108
  Minitab, 359–361
  SAS, 217–218
  SYSTAT, 269–271
 orthogonal contrasts, 114–115
  SAS, 220
  SPSS-X, 312
 orthogonal polynomial contrasts, 115–116
  SAS, 220
  SPSS-X, 312
 randomized blocks designs, 108–109
  Minitab, 361
  SAS, 218
  SYSTAT, 272
 repeated-measures/split-plot ANOVA, 94–103
  Minitab, 356–359
  SAS, 211, 212–217
  SYSTAT, 263–269
 Scheffé contrasts, 116–117
  Minitab, 363
  SAS, 220
  SPSS-X, 312
Factor analysis, 59–60
 SAS, 196
 SPSS-X, 295
Factorial analysis of variance, 85–94
 one-way ANOVA, 85–86
  Minitab, 350
  SAS, 208
  SPSS-X, 309–310
  SYSTAT, 260
 two-way ANOVA—fixed effects model, 86–87
  Minitab, 352–356
  SAS, 209
  SPSS-X, 310–311
  SYSTAT, 260–262
 two-way ANOVA—random effects model, 87–88
  Minitab, 352–356
  SYSTAT, 260–262
 two-way ANOVA—mixed effects model, 88–89
  Minitab, 352–356
  SAS, 210
  SYSTAT, 260–262
 three-way ANOVA—fixed effects model, 89–90
  Minitab, 356
  SAS, 211
  SPSS-X, 311
  SYSTAT, 262
 three-way ANOVA—mixed effects model with one factor fixed and two factors random, 90–91
  Minitab, 356
  SAS, 211
  SYSTAT, 262
 three-way ANOVA—mixed effects model with two factors fixed and one factor random, 92–93
  Minitab, 356
  SYSTAT, 262
 three-way ANOVA—random effects model, 93–94
  Minitab, 356
  SYSTAT, 262
File, 153
Fisher's z-transformation
 Minitab, 341, 342
 SAS, 202
 SPSS-X, 301
 SYSTAT, 252–253
Framework II, 158
Frequency distribution, 48–49
 Minitab, 329
 SAS, 193
 SPSS-X, 289–290
 SYSTAT, 235, 236
Friedman two-way analysis of variance by ranks, 83–84
 Minitab, 347–348
 SAS, 207
 SPSS-X, 307
 SYSTAT, 258–259

G

Goodman-Kruskal Gamma correlation
 SYSTAT, 239

H

Helix, 153
Homogeneous sampling, 130
Hsu's multiple comparisons
 Minitab, 363

I

Infoscope, 153–154
Integrated Software Package Profiles, 156–162
Intra-class correlation, 57

J

Jazz, 158–159

384 INDEX

K

Kendall's coefficient of concordance, 56
Kendall's tau correlation, 52
   SAS, 196
   SPSS-X, 291, 294
   SYSTAT, 238, 239, 241
KnowledgeMan, 154
Kolmogorov-Smirnov one-sample test, 69
   Minitab, 339
   SAS, 202
   SPSS-X, 300, 301, 302
   SYSTAT, 250
Kolmogorov-Smirnov two-sample test, 78–79
   SYSTAT, 256–257
   SPSS-X, 305–306
Kruskal-Wallis rank test for K-independent samples, 82–83
   Minitab, 347, 348
   SAS, 207
   SPSS-X, 307
   SYSTAT, 256, 258
Kurtosis, 50
   Minitab, 328, 329
   SAS, 192
   SPSS-X, 288
   SYSTAT, 234

L

Latin square designs, 109–111
   Minitab, 361
   SAS, 218
   SYSTAT, 272
Linking Research Questions to Designs, Statistical Procedures, 6–14
LOTUS 1-2-3, 142–143

M

Mac Base, 154
Mann-Whitney U-test, 77
   Minitab, 346, 347
   SAS, 206
   SPSS-X, 305–306
   SYSTAT, 256
Math Plan, 146
McMax, 154
McNemar's test for significance of change, 79–80
   SPSS-X, 306
Mean, computation of, 44
   Minitab, 328
   SAS, 192
   SPSS-X, 288, 289
   SYSTAT, 234

Mean deviation, 46
Mean (sample) compared to some value, 66
   Minitab, 339, 341
   SAS, 200–201
   SPSS-X, 300
   SYSTAT, 250
Means
   significant difference between two dependent samples, 71
      Minitab, 343
      SAS, 203–204
      SPSS-X, 302, 303
      SYSTAT, 253
   significant difference between two independent samples, 70–71
      Minitab, 342–343
      SAS, 203
      SPSS-X, 302
      SYSTAT, 253–254
Measures of Relationships (*see* Relationships, Measures of)
Median, 45
   Minitab, 328
   SAS, 192, 193
   SPSS-X, 306
   SYSTAT, 235–236
Median test, 75–76
   SAS, 206
   SPSS-X, 305–306
Microsoft Works, 161
Microstat, 173
Minitab, 166, 313–364
   ABSOLUTE operation, 327
   ALL subcommand, 329
   ALTERNATE subcommand, 343
   analysis of covariance, 362–363
   analysis of variance (ANOVA), 349–361
   ANCOVA command, 362
   ANOVA command, 350, 353–354
   AOVONEWAY command, 350
   arithmetic operations, 324, 327
   Boolean logic arguments, 327
   BOXPLOTS command, 331
   BREG command, 337
   BRIEF command, 334
   cases, number of, 328
   CDF function, 345
   chi-square
      Friedman two-way analysis of variance by ranks, 347–348
      test for goodness of fit, 339
      test of independence (homogeneity), 330
      variance (sample) compared to some value, 339
   CHISQUARE command, 330–331
   CHISQUARE subcommand, 330
   CODE command, 324
   column operations, 327

command files, 338
comparison operations, 324, 327
CONCATENATE command, 315
contingency coefficient, 332
contingency table, creating a, 331
CONVERT command, 324, 325, 349
COPY command, 324
CORRELATE command, 332
correlation (sample) compared to 0, 339, 341, 342
correlation (sample) compared to some value, 339, 341, 342
correlation coefficients, significant difference between dependent samples, 344, 345
correlation coefficients, significant difference between independent samples, 344
COUNT operations, 327
COUNTS subcommand, 329
covariance correlation, 332
COVARIANCE FOR C_I C_{I+1} command, 332
CPLOT command, 331
CTABLE command, 331
CUMCOUNTS subcommand, 329
CUMPERCENTS subcommand, 329
curvilinear regression, 336, 337
data manipulation, 322–326
data storage format, 314
DBMS\COPY, 320
DELETE command, 324
DESCRIBE command, 327, 328, 340
descriptive statistics, general procedure for obtaining, 327–331
distribution shapes, 328, 329
distributions, functions for, 339, 345
DOTPLOT command, 329
Duncan multiple range test, 363
Dunnett's test for control groups, 363
Durbin-Watson statistic, 333
ECHO command, 340
EMS subcommand, 354
END command, 318–319, 338
ERASE command, 324
F-statistic
  analysis of covariance, 362–363
  factorial analysis of variance, 349–356
  Latin square designs, 361
  nested/hierarchical ANOVA, 359–361
  randomized blocks designs, 361
  repeated measures/split-plot ANOVA, 356–359
  Scheffé contrasts, 363
factorial analysis of variance, 349–356
  one-way ANOVA, 350
  two-way ANOVA—fixed effects model, 352–356
  two-way ANOVA—mixed effects model, 352–356
  two-way ANOVA—random effects model, 352–356
  three-way ANOVA—fixed effects model, 356
  three-way ANOVA—mixed effects model with one factor fixed, two factors random, 356
  three-way ANOVA—mixed effects model with two factors fixed, one factor random, 356
  three-way ANOVA—random effects model, 356
Fisher's LSD, 363
Fisher's z-transformation, 341, 342
FITS subcommand, 353
FORMAT subcommand, 315–317, 321
frequency distribution, 329
frequency sets compared to some expected frequency values, 339
Friedman two-way analysis of variance by ranks, 347–348
GLM command, 350, 351–352
Hsu's multiple comparisons, 363
IF, THEN, ELSE substitutions, 325
INFORMATION command, 319
inputting data and file creation, 315–320
INSERT command, 319, 324
INSET command, 315
INVCDF function, 345
Kolmogorov-Smirnov one-sample test, 339
Kruskal-Wallis rank test for K-independent samples, 347, 348
kurtosis, 328, 329
Latin square designs, 361
LET command, 322–323, 325, 327
LOGE operation, 327
LOGTEN operation, 327
Mann-Whitney U-test, 346, 347
MAX (maximum) operation, 327
mean
  computation of, 328
  MEAN operation, 327
  sample mean compared to some value, 339, 341
  standard error, 328, 340
  trimmed mean, 328
means, significant difference between two dependent sample, 343
means, significant difference between two independent sample, 342–343
median
  computation of, 328
  MEDIAN operation, 327
MIN (minimum) operation, 327
minimum and maximum values, 328
MINITAB entry command, 314

Minitab (continued)
  mode, 329
  multiple comparison methods, undocumented, 363
  multiple regression, 336
  N operation, 327
  NAME command, 314
  nested/hierarchical ANOVA, 359–361
  Newman-Keuls test, 363
  NMISS operations, 327
  NOBS=x subcommand, 319
  NOECHO command, 340
  NOOUTFILE command, 338
  NOPAPER command, 338
  NSCORES operation, 327
  OMIT subcommand, 324
  ONEWAY command, 350, 351
  orthogonal contrasts, 364
  orthogonal polynomial contrasts, 364
  OUTFILE command, 338
  outputting and saving data, 320–322
  PAPER command, 338
  part correlation, 335
  partial correlation, 335
  Pearson product-moment correlation coefficient, 332
  PERCENTS subcommand, 329
  Phi-coefficient, 332
  PLOT command, 331
  point biserial correlation, 332
  POOLED subcommand, 343
  PORTABLE subcommand, 321
  PRINT COMMAND, 321, 334
  programming (customized) within *Minitab*, 338, 354
  proportion (sample) compared to some value, 339
  proportions, comparing dependent sample, 344
  proportions, comparing independent sample, 344
  quartiles, values of first and third, 328
  R-squared, 334
  RANDOM subcommand, 354, 355
  randomized blocks designs, 361
  range, 328
  RANK command, 332
  RANK operation, 327
  READ command, 315
  REGRESS command, 333
    subcommands, 333
  regression coefficient, 334
  repeated-measures/split-plot ANOVA, 356–359
    lack of special command/subcommand for repeated measures, 356
    one-way repeated measures, subjects by occasions, 356–358
    two-way split-plot, 358–359
    three-way split-plot, one between- and two within-subjects factors, 359
    three-way split-plot, two between- and one within-subjects factors, 359
  RETRIEVE command, 315, 319, 328
  ROUND operation, 327
  row operations, 327
  SAVE command, 318, 320–322
  saving and outputting data, 320–322
  scatter plots, 331
  Scheffé contrasts, 363
  semi-colon usage, 315
  SET command, 315, 318, 319, 357
  significance levels for z, Chi-square, t, and F statistics, 345–346
  Sign test for dependent samples, 346–347
  simple linear regression, 333–334
  skewness, 328, 329
  SORT operation, 327
  Spearman rank order correlation (Rho), 332
  SQRT operation, 327
  STACK command, 323, 357
  standard deviation, 328
  standard error of estimate, 334
  STEM-AND-LEAF command, 331
  STEPWISE command, 337
  stepwise multiple regression, 337
  STORE command, 338
  straight line (|), 353
  SUM operation, 327
  t-test
    sample correlation compared to 0, 339, 341, 342
    sample dependent correlations, 344, 345
    sample dependent means, 343
    sample dependent variances, 344
    sample independent means, 342–343
    sample mean compared to some value, 339
  TABLE/TABLES commands, 329, 330
  Tables, two-way (coded), 331
  TALLY command, 329
  TEST command, 355
  test statistics, command files for, 339–340
  TTEST command, 343
  Tukey's honestly significant difference, 363, 364
  TWOSAMPLE command, 342–343
  TWOT command, 342–343
  TWOWAY command, 350
  USE subcommand, 324
  variance, computation of, 328
  variance (sample) compared to some value, 339
  variances, significant difference between two dependent sample, 344

variances, significant difference between two independent sample, 344
WRITE command, 321
Y-intercept, 334
z-statistic
   sample correlation compared to some value, 339, 341, 342
   sample dependent proportions, 344
   sample independent correlations, 344
   sample independent proportions, 344
   sample mean compared to some value, 339, 341
   sample proportion compared to some value, 339
   ZTEST command, 341
Mode, 45
   Minitab, 329
   SAS, 192, 193
   SPSS-X, 290
   SYSTAT, 235
Model sampling (nonrandom sampling), 125–126
Multiplan, 143–144
Multiple comparison procedures, 43–44, 114–120
   A Priori, planned comparisons
      SAS, 220
      SPSS-X, 312
   Post Hoc comparisons
      SAS, 220
      SPSS-X, 312
Multiple correlation, 58
   SAS, 199
   SPSS-X, 296
   SYSTAT, 239
Multiple matrix sampling, 128
Multiple regression, 62–63
   Minitab, 336
   SAS, 199
   SPSS-X, 298–300
   SYSTAT, 248
Multistage sampling, 128

N

Nested/hierarchical ANOVA, 104–108
   Minitab, 359–361
   SAS, 217–218
   SYSTAT, 269–271
   two-factor design, 104–105
   three-factor designs (C nested in B, B nested in A), 105–107
   three-factor designs (B nested in A, C crossed with B nested in A), 107–108
Newman-Keuls test, 118–119
   Minitab, 363
   SAS, 220
   SPSS-X, 312
   SYSTAT, 274, 275
NWA Statpak, 173

O

Omnis 3 Plus, 154–155
1,2,3 Forecast!, 171
Open Access, 161
Orthogonal contrasts, 114–115
   Minitab, 364
   SAS, 220
   SPSS-X, 312
Orthogonal polynomial contrasts, 115–116
   Minitab, 364
   SAS, 220
   SPSS-X, 312

P

Paradox, 149
Part correlation, 57–58
   Minitab, 335
   SAS, 197–198
   SPSS-X, 295
   SYSTAT, 239, 246
Partial correlation, 58
   Minitab, 335
   SAS, 197–198
   SPSS-X, 294–295
   SYSTAT, 239, 246
PC-File/R, 150
PC Statistician, 173
Pearson product-moment correlation coefficient, 51
   Minitab, 332
   SAS, 195
   SPSS-X, 291, 292–293
   SYSTAT, 238, 239
PFS First Choice, 161–162
PFS Professional Plan, 146–147
PFS: File, 150
Phi-coefficient, 54–55
   Minitab, 332
   SAS, 196
   SPSS-X, 291, 294
   SYSTAT, 239, 242
Point biserial correlation, 52–53
   Minitab, 332
   SAS, 196
   SPSS-X, 293
   SYSTAT, 238, 241
Population (sampling), 124
Practibase, 155
Probability sampling (random sampling), 126
Probase, 155

Proportion (sample) compared to some value, 67
   Minitab, 339
   SAS, 200–201
   SPSS-X, 300
   SYSTAT, 250
Proportions
   comparing dependent sample proportions, 73–74
      Minitab, 344
      SAS, 204
      SPSS-X, 302, 304
      SYSTAT, 255
   comparing independent sample proportions, 72–73
      Minitab, 344
      SAS, 204
      SPSS-X, 302, 304
      SYSTAT, 255

Q

Quartiles, values of first and third
   Minitab, 328

R

R-Base for DOS, 151
R-Base System V, 151–152
R-Base 5000, 150–151
R-squared, 63–64
   Minitab, 334
   SAS, 199
   SPSS-X, 296
   SYSTAT, 246
Randomized blocks designs, 108–109
   Minitab, 361
   SAS, 218
   SYSTAT, 272
Range, 46
   Minitab, 328
   SAS, 192
   SPSS-X, 288
   SYSTAT, 234, 235
Reflex, 152
Regression analysis, 41, 60–65
Regression coefficient, 61
   Minitab, 334
   SAS, 197
   SPSS-X, 296
   SYSTAT, 246
Relationships, Measures of, 41, 50–60
Repeated-measures ANOVA (one-way), subjects by occasions, 94–95
   Minitab, 356–358
   SAS, 211
   SYSTAT, 263–264
Research designs
   analyses, 16–17
   design perspective, 17–18
   statistical perspective, 18
   symbols explained, 17
   validity, external and internal, 15–16
Research design profiles, 19–38
   analysis of covariance (ANCOVA), 37–38
   factorial ANOVA, 28–30
   Latin square, 37
   multiple groups, 25–28
   nested/hierarchical ANOVA, 34–36
   one group, 19–21
   randomized blocks, 36–37
   repeated-measures /split-plot, 30–33
   Solomon Four-Group, 27–28
   two groups, 21–25
Research questions linked to their proper designs and statistical procedures (summary table), 14

S

S statistic, 47–48
Sampling
   assigning subjects to groups, 132–133
   cluster sampling, 131–132
   estimators, 126–127
   homogeneous sampling, 130
   model sampling, 125–126
   multistage sampling, 128
   multiple matrix sampling, 128
   population, 124
   probability sampling, 126
   sampling frame, 124–125
   sampling unit, 125
   sampling without replacement, 127
   simple random sampling, 128–129
   stratified sampling, 129–130
   systematic sampling, 130–131
Sampling frame, 124–125
Sampling unit, 125
Sampling without replacement, 127
SAS *(Statistical Analysis System)*, 166–168, 185–222
   analysis of covariance (ANCOVA), 218–219
   analysis of variance (ANOVA), 207–219
   arithmetic operation symbols, 191
   bar charts, 193
   biserial correlation, 196
   BY statements, 187
   category control for charts, 194
   CHART, 193–194
   chi-square
      Friedman two-way analysis of variance by ranks, 207

Median test, 206
Sign test for K-independent samples, 207
test for goodness of fit, 202
test of independence (homogeneity), 206–207
variance (sample) compared to some value, 200–201
CHISQ=, 201
coefficient of variation, 192
comment statements, 187
contingency coefficient, 196
contingency table, 193
CONTRAST, 220
correlation (sample) compared to 0, 201
correlation (sample) compared to some value, 202
correlation coefficients, significant difference between dependent samples, 205
correlation coefficients, significant difference between independent samples, 204–205
curvilinear regression, 199
data manipulation, 190–192
data set 1, 188
data set 2, 194–195
data set 3, 208
data set 4, 211–212
data set 4B, 213
data set 5, 215
data set 6, 216–217
data set 7, 217–218
descriptive data plots, 192
DO statements, 187
Duncan multiple range test, 220
exponent symbol, 191
F-statistic
  analysis of covariance, 218–219
  factorial analysis of variance, 208–211
  independent sample variances, 203
  Latin square designs, 218
  nested/hierarchical ANOVA, 217–218
  orthogonal contrasts, 220
  orthogonal polynomial contrasts, 220
  randomized blocks designs, 218
  repeated-measures/split-plot ANOVA, 211–217
  Scheffé contrasts, 220
factor analysis, 196
factorial analysis of variance
  one-way ANOVA, 208
  two-way ANOVA—fixed effects model, 209
  two-way ANOVA—mixed effects model, 210
  three-way ANOVA—fixed effects model, 211
  three-way ANOVA—mixed effects model, 211
Fisher's z-transformation, 202
FREQ, 193, 196
frequency distribution, 193
frequency sets compared to some expected frequency values, 200–201
Friedman two-way ANOVA by ranks, 207
graphic representations, 192
groups of variables, 187
HBAR TRSP, 193
IF statements, 191–192
INPUT formats, 187–190
JCL, 186
Kendall's tau correlation, 196
Kolmogorov-Smirnov one-sample test, 202
Kruskal-Wallis rank test for K-independent samples, 207
kurtosis, 192
Latin square designs, 218
LEVELS, 194
logical operators with IF, 191–192
Mann-Whitney U-test, 206
mean, computation of, 192
mean (sample) compared to some value, 200–201
means, significant difference between two dependent sample, 203–204
means, significant difference between two independent sample, 203
median, 192, 193
Median test, 206
minimum and maximum values, 192
mode, 192, 193
multiple correlation, 199
multiple regression, 199
naming rules, 186
nested/hierarchical ANOVA, 217–218
Newman-Keuls test, 220
orthogonal contrasts, 220
orthogonal polynomial contrasts, 220
OUTPUT Option, 187
part correlation, 197–198
partial correlation, 197–198
Pearson product-moment correlation coefficient, 195
percentiles, 193
Phi-coefficient, 196
point biserial correlation, 196
PROC ANOVA, 207, 208
PROC CHART, 192
PROC CORR, 194–196, 201
PROC FACTOR, 196
PROC FREQ, 193, 196
PROC general format, 186
PROC GLM, 196–197, 207
PROC MEANS, 192
PROC NESTED, 217–218

SAS (Statistical Analysis System) (continued)
  PROC NPAR1WAY, 206, 207
  PROC PRINT, 187
  PROC PLOT, 192, 194
  PROC REG, 196–197
  PROC RSQUARE, 199
  PROC SORT, 205
  PROC STEPWISE, 199–200
    backward, 200
    forward, 200
    maximum R-squared improvement, 200
    minimum R-squared improvement, 200
    stepwise, 200
  PROC TTEST, 203
  PROC UNIVARIATE, 192–193
  proportion (sample) compared to some value, 200–201
  proportions, comparing dependent sample, 204
  proportions, comparing independent sample, 204
  randomized blocks designs, 218
  range, 192
  regression coefficient, 197
  REPEATED, 214–216, 221
  repeated-measures/split-plot ANOVA, 211–217
    one-way repeated measures, subjects by occasions, 211
    two-way split-plot, 212–215
    three-way split-plot, one between and two within-subjects factors, 215–216
    three-way split-plot, two between and one within-subjects factors, 216–217
  R-squared, 199
  Scheffé contrasts, 220
  Sign test for K-independent samples, 207
  significance levels for z, Chi-square, t, and F statistics, 205–206
  simple linear regression, 196
  skewness, 192
  Spearman rank order correlation (Rho), 196
  standard deviation, 192
  standard error of estimate, 197
  standard error of the mean, 192
  stepwise multiple regression, 199–200
  T=, 201
  t-test
    sample correlation compared to 0, 201–202
    sample dependent correlations, 205
    sample dependent means, 203–204
    sample dependent variances, 204
    sample independent means, 203
    sample mean compared to some value, 200–201
  TITLE cards, 187
  Tukey's honestly significant difference, 220
  VAR, 187
  variable statement, 187
  variables, relationship between pairs of, 192
  variables, sum of values, 191, 192
  variance, computation of, 192
  variance (sample) compared to some value, 200–201
  variances, significant difference between two dependent sample, 204
  variances, significant difference between two independent sample, 203
  VBAR TRSP, 193
  Wilcoxon matched-pairs signed-rank test, 207
  Wilcoxon rank sum test, 206
  WITH, 194
  Y-intercept, 197
  z-statistic
    sample correlation compared to some value, 202
    sample dependent proportions, 204
    sample independent correlations, 204–205
    sample independent proportions, 204
    sample mean compared to some value, 200–201
    sample proportion compared to some value, 200–201
Satterthwaite's correction formula, 254, 344
Scheffé contrasts, 116–117
  Minitab, 363
  SAS, 220
  SPSS-X, 312
The Sensible Solution, 155
Sign test for dependent samples, 80
  Minitab, 346–347
  SPSS-X, 306–307
  SYSTAT, 256, 257
Sign test for K-independent samples, 82
  SAS, 207
  SPSS-X, 307
Significant differences (*see* Comparisons and individual tests of significance)
Simple linear regression, 60
  Minitab, 333–334
  SAS, 196
  SPSS-X, 295–296
  SYSTAT, 244
Simple random sampling, 128–129
Skewness, 49
  Minitab, 328, 329
  SAS, 192
  SPSS-X, 288
  SYSTAT, 234
Smart Software System, 159

Spearman's rank order correlation—Rho,
    51–52
  Minitab, 332
  SAS, 196
  SPSS-X, 294
  SYSTAT, 238, 240
Split-plot ANOVA, 96–103
  two-way with one between-subjects factor
      and one within-subjects factor, 96–97
    Minitab, 358–359
    SAS, 212–215
    SYSTAT, 264–267
  three-way with one between-subjects factor
      and two within-subjects factors, 97–99
    Minitab, 359
    SAS, 215–216
    SYSTAT, 267–268
  three-way with two between-subjects factors
      and one within-subjects factor, 99–101
    Minitab, 359
    SAS, 216–217
    SYSTAT, 268–269
  four-way with two between-subjects factors
      and two within-subjects factors,
      101–103
Spreadsheet software profiles, 142–147
SPSS/PC+, 168–169
SPSS-X, 277–312
  analysis of covariance (ANCOVA),
      311–312
  analysis of residuals, 298
  analysis of variance, 308–311
  ANOVA command, 308, 310
  arithmetic operation symbols, 287
  BACKWARD, 297
  /BARCHART subcommand, 290
  BEGIN DATA, 278
  biserial correlation, 293
  BY, 294
  /CASEWISE subcommand, 298
  cell frequency
    expected frequency, 290–291
    residual between observed and expected,
        290–291
  /CELL subcommand, 289, 290
  chi-square
    Friedman two-way analysis of variance by
        ranks, 307–308
    McNemar's Test, 306
    Median test, 305–306
    Sign test for K-independent samples, 307
    test for goodness of fit, 300, 301
    test of independence (homogeneity), 291,
        306
    variance (sample) compared to some
        value, 300
  COLLECT, 297
  COMMAND, 278–279

COMMENT lines, 279
COMPUTE CHISQ=, 300
COMPUTE command, 286–287, 300
COMPUTE T=, 300
contingency coefficient, 291, 294
contingency table, 290–291
/CONTRAST subcommand, 312
correlation (sample) compared to 0, 293,
    300, 301
correlation (sample) compared to some
    value, 300, 301
correlation coefficients, significant
    difference between dependent samples,
    302, 305
correlation coefficients, significant
    difference between independent
    samples, 302, 304
CORRELATIONS command, 278,
    292–293, 300, 301, 305
/CRITERIA subcommand, 297–298
  criteria for entering/removing variables
      in regression analysis, 297–298
/CROSSBREAK subcommand, 289
CROSSTABS command, 290–291, 294,
    306
curvilinear regression, 298
DATA LIST, 278
data manipulation, 285–288
/DEPENDENT subcommand, 297
DESCRIPTIVES command, 278, 288–289
DO commands, 280–281
Duncan multiple range test, 312
ELSE, 286
END DATA, 278
END LOOP command, 281
ENTER, 297
Eta-squared, 291
EXCLUDE, 281
exponent symbol, 287
F-statistic
  analysis of covariance, 311–312
  factorial analysis of variance, 308–311
  independent sample variances, 303
  orthogonal contrasts, 312
  orthogonal polynomial contrasts, 312
  Scheffé contrasts, 312
factor analysis, 295
factorial analysis of variance, 308–311
  one-way ANOVA, 309–310
  two-way ANOVA—fixed effects model,
      310–311
  three-way ANOVAs, 311
Fisher's z-transformation, 301
FORWARD, 297
FREQUENCIES command, 289–290, 300
frequency distribution, 289–290
frequency sets compared to some expected
    frequency values, 300

SPSS-X (continued)
  Friedman two-way analysis of variance by ranks, 307
  groups (strings) of variables, 280
  HI, 286
  /HISTOGRAM subcommand, 290
  HOMOGENEITY, 309
  IF statements, 287–288
  INCLUDE, 281
  input formats, 281–285
  INTO, 286
  JCL, 277
  /K-S (Kolmogorov-Smirnov two-sample test) subcommand, 306
  Kendall's tau correlation, 291, 294
  Kolmogorov-Smirnov one-sample test, 300, 301, 302
  Kolmogorov-Smirnov two-sample test, 305–306
  Kruskal-Wallis rank test for K-independent samples, 307
  kurtosis, 288
  LIST command, 279
  LISTWISE, 281
  LO, 286
  logical operators with IF, 287–288
  LOOP command, 281
  /M-W (Mann-Whitney) subcommand, 306
  Mann-Whitney U-test, 305–306
  /MCNEMAR (McNemar's Test) subcommand, 306
  McNemar's Test, 306
  mean, computation of, 288, 289
  mean (sample) compared to some value, 300
  MEANS command, 289, 308, 310
  means, significant difference between two dependent sample, 302, 303
  means, significant difference between two independent sample, 302
  /MEDIAN (Median Test) subcommand, 306
  median, 306
  Median test, 305–306
  /METHOD subcommand
    BACKWARD option, 297
    ENTER option, 297
    FORWARD option, 297
    REMOVE option, 297
    STEPWISE option, 297
    TEST option, 297
  minimum and maximum values, 288
  /MISSING subcommand, 281
  missing data
    EXCLUDE option, 281
    INCLUDE option, 281
    LISTWISE option, 281
    /MISSING subcommand, 281
    MISSING VALUES command, 281
    PAIRWISE option, 281
  MISSING VALUES command, 281
  mode, 290
  multiple correlation, 296
  multiple regression, 298–300
  Newman-Keuls Test, 312
  /NTILES subcommand, 290
  NONPAR CORR command, 294
  NPAR TESTS command, 301–302, 305, 306, 307
  ONEWAY command, 308, 309, 312
  orthogonal contrasts, 312
  orthogonal polynomial contrasts, 312
  /PAIRS subcommand, 303
  PAIRWISE, 281
  part correlation, 295
  PARTIAL CORR command, 294–295
  partial correlation, 294–295
  Pearson product-moment correlation coefficient, 291, 292–293
  percentages—cell, row, column, 290–291
  /PERCENTILES subcommand, 290
  Phi-coefficient, 291, 294
  PLOT command, 291
  point biserial correlation, 293
  /POLYNOMIAL subcommand, 312
  /PRINT=BOTH subcommand, 294
  /PRINT=TWOTAIL subcommand, 294
  PRINT command, 279
  proportion (sample) compared to some value, 300
  proportions, comparing dependent sample, 302, 304
  proportions, comparing independent sample, 302, 304
  R-squared, 296
  range, 288
  /RANGES subcommand, 312
  RECODE command, 285–286
  regression coefficient, 296
  REGRESSION command, 295, 298–299
  REMOVE, 297
  /RESIDUALS subcommand, 298
  /SAVE, 289
  /SCATTERPLOT subcommand, 298
  Scheffé contrasts, 312
  /SIGN (Sign Test) subcommand, 306
  Sign test for dependent samples, 306–307
  Sign test for K-independent samples, 307
  simple linear regression, 295–296
  skewness, 288
  SORT command, 280
  SORT CASES BY command, 304–305
  Spearman rank order correlation (Rho), 294
  SPLIT FILE BY command, 280, 304–305
  SPLIT FILE OFF command, 280

standard deviation, 288, 289
standard error of estimate, 296
standard error of the kurtosis, 290
standard error of the mean, 288
standard error of the skewness, 290
/STATISTICS=ALL, 290
/STATISTICS=CORR DESCRIPTIVES
   subcommand, 294–295
/STATISTICS subcommand, 278, 288,
   289, 290, 298, 309
STEPWISE, 297
stepwise multiple regression, 299
/SUBCOMMAND, 279
SUM, 287
summing variables, 287
t-test
   sample correlation compared to 0, 300,
      301
   sample dependent correlations, 302, 305
   sample dependent means, 302, 303
   sample dependent variances, 302, 304
   sample independent means, 302
   sample mean compared to some value,
      300
T-TEST command, 302–303, 304
TEST, 297
THRU, 286
TITLE, 280
TO, 289
Tukey's honestly significant difference, 312
/TWOTAIL subcommand, 293
/VARIABLES subcommand, 296–297
/VARIABLES=ALL subcommand, 296
VALUE LABELS, 278
VARIABLE LABELS, 277–278
VARIABLES=, 280
variance, computation of, 288
variance (sample) compared to some value,
   300
variances, significant difference between
   two dependent sample, 302, 304
variances, significant difference between
   two independent sample, 302, 303, 304
/WILCOXON (Wilcoxon matched-pairs)
   subcommand, 306
Wilcoxon matched-pairs signed-rank test,
   307
Wilcoxon rank sum test, 306
WITH, 292, 294
Y-intercept, 296
z-statistic
   sample correlation compared to some
      value, 300, 301
   sample dependent proportions, 302, 304
   sample independent correlations, 302,
      304
   sample independent proportions, 302,
      304
   sample mean compared to some value,
      300
   sample proportion compared to some
      value, 300
   ZPP, 295, 298
Standard deviation, 47
   Minitab, 328
   SAS, 192
   SPSS-X, 288, 289
   SYSTAT, 234, 235
Standard error of estimate, 62
   Minitab, 334
   SAS, 197
   SPSS-X, 296
   SYSTAT, 246
Standard error of the kurtosis
   SPSS-X, 290
Standard error of the mean
   Minitab, 328, 340
   SAS, 192
   SPSS-X, 288
Standard error of the skewness
   SPSS-X, 290
Stat I—A Statistical Toolbox, 173–174
STATA, 174
Statgraphics, 169
Statistical Software Package Profiles,
   164–175
StatPac Gold, 174
StatPlan III, 175
Statpro, 175
Stepwise multiple regression, 65
   Minitab, 337
   SAS, 199–200
   SPSS-X, 299
   SYSTAT, 249
Stratified sampling, 129–130
Sun of D-squared, 47
SuperCalc 3, 144
SuperCalc 5, 144–145
Symphony, 160
SYSTAT, 169–171, 223–276
   ALL, 234
   analysis of covariance (ANCOVA),
      272–274
   analysis of variance (ANOVA), 259–272
   ANOVA command, 260
   BAR command, 235
   Bartlett test of homogeneity of variance,
      254
   bivariate distribution, picture of, 238
   BY GROUP command, 254
   chi-square,
      Friedman two-way analysis of variance by
         ranks, 258–259
      Kruskal-Wallis rank test, 258
      test for goodness of fit, 250
      test of independence (homogeneity), 258

SYSTAT (continued)
   chi-square *(continued)*
      variance (sample) compared to some value, 250
   Cluster module, 223
   CMATRIX command, 276
   coefficient of variation, 234
   command files, 249
   contingency coefficient, 239, 242
   CONTRAST command, 275–276
   corr module, 223, 238, 239, 255
   correlation (sample) compared to 0, 250
   correlation (sample) compared to some value, 250
   correlation coefficients, significant difference between dependent samples, 255
   correlation coefficients, significant difference between independent samples, 255
   covariance correlation, 238
   COVARIANT command, 274
   Cramer V correlation, 239
   curvilinear regression, 248
   data manipulation, 226–227, 233–234
   data module, 223, 227–229, 232, 233
   DBMS/COPY, 232
   descriptive statistics, general procedure for obtaining, 234–238
   distribution shapes, 236
   distributions, functions for, 250
   Duncan multiple range test, 274, 275
   edit module, 223, 224–227
   EFFECT command, 262
   ERROR command, 262
   ESTIMATE command, 244, 262
   F-statistic
      analysis of covariance, 272–274
      factorial analysis of variance, 259–263
      independent sample variances, 254
      Latin square designs, 272
      nested/hierarchical ANOVA, 269–271
      randomized blocks designs, 272
      repeated-measures/split-plot ANOVA, 263–269
   factor module, 223
   factorial analysis of variance, 259–263
      one-way ANOVA, 260
      two-way ANOVA—fixed effects model, 260–262
      two-way ANOVA—mixed effects model, 260–262
      two-way ANOVA—random effects model, 260–262
      three-way ANOVA—fixed effects model, 262
      three-way ANOVA—mixed effects model with one factor fixed, two factors random, 262
      three-way ANOVA—mixed effects model with two factors fixed, one factor random, 262
      three-way ANOVA—random effects model, 262
   FEDIT (editor), 249
   filename profile, 226
   Fisher's z-transformation, 252–253
   fixed format data sets, 227, 231
   free format data sets, 227, 230, 231
   frequency distribution, 235, 236
   frequency sets compared to some expected frequency values, 250
   Friedman two-way ANOVA by ranks, 258–259
   full-screen editor, 224–227
   GET command, 228
   Goodman-Kruskal Gamma correlation, 239
   graph module, 223, 235, 236
   histogram, 236
   HYPOTHESIS command, 262
   IF, 226–227
   IMPORT command, 227, 232
   INPUT command, 228–229, 231
   input symbols, 225, 229
   inputting data and data set creation, 224–232, 233–234
   Kendall's tau correlation, 238, 239, 241
   Kolmogorov-Smirnov one-sample test, 250
   Kolmogorov-Smirnov two-sample test, 256–257
   Kruskal-Wallis rank test for K-independent samples, 256, 258
   kurtosis, 234
   Latin square designs, 272
   logical operant symbols, 227
   macro module, 223, 250
   Mann-Whitney U-test, 256
   MDS module, 223
   mean, computation of, 234
   mean (sample) compared to some value, 250
   means, significant difference between two dependent sample, 253
   means, significant difference between two independent sample, 253–254
   median, 235–236
   merging data sets, 233
   MGLH module, 223, 239, 244, 260, 274
   minimum and maximum values, 234, 235
   mode, 235
   MODEL command, 244
   modules, listing of, 223–224
   multiple R correlation, 239
   multiple regression, 248
   nested/hierarchical ANOVA, 269–271
   Newman-Keuls test, 274, 275

NONLIN module, 223
NPAR module, 224, 256
number of cases, 234, 235
OUTPUT @ command, 269
OUTPUT [filename] command, 269
OUTPUT * command, 269
part correlation, 239, 246
partial correlation, 239, 246
Pearson product-moment correlation coefficient, 238, 239
Phi-coefficient, 239, 242
PLOT command, 238
point biserial correlation, 238, 241
PRINT = LONG command, 242, 254
/ PROB, 239
programming (customized) within *SYSTAT*, 249–250
proportion (sample) compared to some value, 250
proportions, comparing dependent sample, 255
proportions, comparing independent sample, 255
QUIT command, 226
R-squared, 246
randomized blocks designs, 272
range, 234, 235
regression coefficient, 246
REPEAT, 264
repeated-measures/split-plot ANOVA, 263–269
 one-way repeated measures, subjects by occasions, 263–264
 two-way split-plot, 264–267
 three-way split-plot, one between- and two within-subjects factors, 267–268
 three-way split-plot, two between- and one within-subjects factors, 268–269
RUN command, 229
SAVE command, 244
scatterplot, 238
Series module, 224
Sign test for dependent samples, 256, 257
significance levels for z, Chi-square, t, and F statistics, 256
simple linear regression, 244
skewness, 234
SORT, 233, 236
Spearman rank order correlation (Rho), 238, 240
SSORT module, 233
standard deviation, 234, 235
standard error of estimate, 246
standard error of measure, 234
STATISTICS command, 234
Statistics module, 224, 234, 274
Stem-and-leaf plot, 235
STEMLEAF command, 235

STEP command, 249
stepwise multiple regression, 249
SUBMIT command, 263
sum, 234
SWITCHTO command, 227
sygraph module, 224
SYMBOL='*', 238
Sytran, 232
t-test
 sample correlation compared to 0, 250
 sample dependent correlations, 255
 sample dependent means, 253
 sample dependent variances, 255
 sample independent means, 253–254
 sample mean compared to some value, 250
tables module, 224, 236, 242
TABULATE command, 236, 242
TEST command, 262
test statistics, command files for, 251
Tukey's honestly significant difference, 274
TTEST command, 253
USE command, 234
variance, Bartlett test of homogeneity, 254
variance, computation of, 234
variance (sample) compared to some value, 250
variances, significant difference between two dependent sample, 255
variances, significant difference between two independent sample, 254
Wilcoxon matched-pairs signed-rank test, 256, 257–258
Y-intercept, 246
z-statistic
 sample correlation compared to some value, 250
 sample dependent proportions, 255
 sample independent correlations, 255
 sample independent proportions, 255
 sample mean compared to some value, 250
 sample proportion compared to some value, 250
Systematic sampling, 130–131

T

T-statistic, 81
t-test
 sample correlation compared to 0, 67–68
  Minitab, 339, 341, 342
  SAS, 201–202
  SPSS-X, 293, 300, 301
  SYSTAT, 250
 sample dependent correlations, 74–75
  Minitab, 344, 345

t-test *(continued)*
  sample dependent correlations *(continued)*
    SAS, 205
    SPSS-X, 302, 305
    SYSTAT, 255
  sample dependent means, 71
    Minitab, 343
    SAS, 203–204
    SPSS-X, 302, 303
    SYSTAT, 253
  sample dependent variances, 72
    Minitab, 344
    SAS, 204
    SPSS-X, 302, 304
    SYSTAT, 255
  sample independent means, 70–71
    Minitab, 342–343
    SAS, 203
    SPSS-X, 302
    SYSTAT, 253–254
  sample mean compared to some value, 66
    Minitab, 339
    SAS, 200–201
    SPSS-X, 300
    SYSTAT, 250
Tests of differences (*see* Comparisons or the individual tests of significance)
Tetrachoric correlation, 55
Time series analysis, 113–114
Tukey's honestly significant difference, 117–118
  Minitab, 363–364
  SAS, 220
  SPSS-X, 312
  SYSTAT, 274

V

Validity of research designs, 15–16
Variability, 46–47
Variance, computation of, 46–47
  Minitab, 328
  SAS, 192
  SPSS-X, 288
  SYSTAT, 234
Variance (sample) compared to some value, 66–67
  Minitab, 339
  SAS, 200–201
  SPSS-X, 300
  SYSTAT, 250
Variances
  significant difference between two dependent samples, 72
    Minitab, 344
    SAS, 204
    SPSS-X, 302, 304

    SYSTAT, 255
  significant difference between two independent samples, 71–72
    Minitab, 344
    SAS, 203
    SPSS-X, 302, 303, 304
    SYSTAT, 254
VP-Planner, 145

W

Wilcoxon matched-pairs signed-rank test, 81
  SAS, 207
  SPSS-X, 307
  SYSTAT, 256, 257–258
Wilcoxon rank sum test, 76–77
  SAS, 206
  SPSS-X, 306

Y

Y-intercept, 61–62
  Minitab, 334
  SAS, 197
  SPSS-X, 296
  SYSTAT, 246

Z

z-statistic
  sample correlation compared to some value, 68
    Minitab, 339, 341, 342
    SAS, 202
    SPSS-X, 300, 301
    SYSTAT, 250
  sample dependent proportions, 73–74
    Minitab, 344
    SAS, 204
    SPSS-X, 302, 304
    SYSTAT, 255
  sample independent correlations, 74
    Minitab, 344
    SAS, 204–205
    SPSS-X, 302, 304
    SYSTAT, 255
  sample independent proportions, 72–73
    Minitab, 344
    SAS, 204
    SPSS-X, 302, 304
    SYSTAT, 255
  sample mean compared to some value, 66
    Minitab, 339, 341
    SAS, 200–201
    SPSS-X, 300

SYSTAT, 250
sample proportion compared to some value, 67
  Minitab, 339

SAS, 200–201
SPSS-X, 300
SYSTAT, 250

Other Books of Interest by Macmillan and ACE/Macmillan

Touchton: *Fact Book on Women in Higher Education*
1990-91 Fact Book on Higher Education
The Community College Fact Book
1990-91 Accredited Institutions of Postsecondary Education
Freed et al: *Educator's Desk Reference*
Linn: *Educational Measurement*
Astin: *Assessment for Excellence*
Bauer: *"How To" Grants Manual*, 2/e
Brievik/Gee: *Information Literacy*
Johnson/Foa: *Instructional Design*
Wittrock: *Handbook of Research on Teaching*
Houston: *Handbook of Research on Teacher Education*
Shaver: *Handbook of Research on Social Studies Teaching and Learning*
Flood et al.: *Handbook of Research on Teaching the English Language Arts*